Intelligent Designs, Innovations and Sustainability in Agriculture 4.0

Editors

Dr. Anand B. Deshpande
Principal & Director
SAEF's Angadi Institute of Technology and Management
(Affiliated to Visvesvaraya Technological University)
Belagavi, Karnataka, India

Dr. Vania V. Estrela
Department of Telecommunications
Federal Fluminense University, Rio de Janeiro, Brazil

Dr. Asiya Khan
School of Engineering, Computing and Mathematics
University of Plymouth, Plymouth, UK

CRC Press
Taylor & Francis Group
Boca Raton London New York

CRC Press is an imprint of the
Taylor & Francis Group, an **informa** business

First edition published 2025
by CRC Press
2385 NW Executive Center Drive, Suite 320, Boca Raton FL 33431

and by CRC Press
4 Park Square, Milton Park, Abingdon, Oxon, OX14 4RN

CRC Press is an imprint of Taylor & Francis Group, LLC

Library of Congress Cataloging-in-Publication Data (applied for)

ISBN: 978-1-032-46259-2 (hbk)
ISBN: 978-1-032-46260-8 (pbk)
ISBN: 978-1-003-38080-1 (ebk)

DOI: 10.1201/9781003380801

Typeset in Times New Roman
by Prime Publishing Services

Acknowledgement

Embarking on the journey of creating this book, "*Intelligent Designs, Innovations, and Sustainability in Agriculture 4.0*", has been a collaborative effort that spanned across continents and brought together the collective wisdom of diverse institutions. It is with gratitude and appreciation that we extend our acknowledgments to the following institutions that played an integral role in the realization of this project.

Our sincere appreciation goes to Fluminense Federal University, Brazil, for their steadfast commitment to academic excellence and their invaluable contributions to the field of agriculture. The rich academic environment provided by UFF has undoubtedly enriched the content of this book.

We extend our gratitude to the Management and staff members of Suresh Angadi Education Foundation's Angadi Institute of Technology and Management, Belagavi, Karnataka, India, affiliated to Visvesvaraya Technological University (VTU), Belagavi, for their unwavering support and dedication to fostering innovation in technology and management. Their commitment to education and research has greatly enhanced the depth and breadth of perspectives covered in this book.

The University of Plymouth has been a beacon of academic rigor and global collaboration. We are thankful for the intellectual vibrancy and diverse perspectives contributed by the academic community at the University of Plymouth, which have significantly enriched the content of this book.

The collaborative spirit demonstrated by these institutions reflects a shared commitment to advancing knowledge and addressing the challenges facing agriculture in the 21st century. This book stands as a testament to the power of international collaboration and the potential that emerges when scholars and institutions from different corners of the world come together in pursuit of a common goal.

We also express our gratitude to all the contributors, whose expertise and dedication have shaped this book into a comprehensive exploration of the intelligent designs, innovations, and sustainability initiatives driving Agriculture 4.0. Together, we have laid the foundation for a discourse that transcends borders and seeks to chart a sustainable and innovative course for the future of agriculture.

Preface

In the intricate tapestry of human progress, few domains are as integral to our survival and well-being as agriculture. As we stand on the cusp of a new era, the fusion of intelligence, innovation, and sustainability in agriculture beckons us toward a future that promises not just abundance, but also responsible stewardship of our planet.

This book, *Agriculture 4.0 and Smart Development: Intelligent Designs, Innovation, and Sustainability*, serves as a compass, guiding readers through the unfolding landscape of the fourth industrial revolution in agriculture. The pages within encapsulate a mosaic of ideas, technologies, and paradigms that are propelling farming into an era marked by unprecedented connectivity, precision, and ecological consciousness. The term "Agriculture 4.0" embodies a paradigm shift – a seismic transformation where the digital realm converges with the green fields. This book is a testament to the tireless efforts of researchers, innovators, and thought leaders who are pioneering this revolution. It is an exploration into the realms of intelligent designs that harness the power of emerging technologies like artificial intelligence, the Internet of Things, and sustainable practices that harmonize with the delicate balance of our environment.

As you embark on this intellectual journey, you will traverse the realms of precision farming, smart agriculture, autonomous machinery, data-driven decision-making, and the ethical considerations that accompany the infusion of technology into the soil of our planet. The chapters unfold like seeds, each one holding the promise of new growth and insight, as contributors from diverse backgrounds converge to share their expertise and visions for a sustainable agricultural future. Sustainability is not just a buzzword within these pages; it is a guiding principle. We explorc how technology can be wielded not only to boost yields but to do so responsibly, mitigating the environmental footprint of agriculture. From smart irrigation systems to predictive analytics, the tools at our disposal are as varied as the challenges we face. This book endeavors to illuminate the path toward a more sustainable, resilient, and equitable agricultural landscape.

We extend our gratitude to the contributors for their invaluable insights and expertise. May this book inspire and inform, fostering a deeper understanding of the transformative potential embedded in *Agriculture 4.0 and Smart Development: Intelligent Designs, Innovation, and Sustainability*.

May this book serve as a source of inspiration, a reference for those navigating the ever-evolving field of Agriculture 4.0, and a catalyst for a global conversation on how we can feed the world's growing population while safeguarding the planet we call home.

Dr. Anand B. Deshpande
Dr. Vania V. Estrela
Dr. Asiya Khan

Contents

About the Editors

Dr. Anand B. Deshpande is currently serving as the Principal and Director of the Suresh Angadi Education Foundation's, Angadi Institute of Technology and Management (AITM), Belagavi, Karnataka, India, affiliated to Visvesvaraya Technological University, Belagavi, India. He received his Ph.D. in Electrical and Electronics Engineering Sciences in 2019 and a Master of Technology degree in Digital Communication and Networking from Visvesvaraya Technological University, Belagavi, India. His research work has been published in international journals, international conferences, and books, and he has filed patents in several areas. He is a guest Editor for Elsevier Neuroscience Informatics special issue *Computational Intelligence in Neuroinformatics, Technologies and Data Analytics (CINTDA)*. He is Editor of Springer's *Computational Intelligence Methods for Super-Resolution in Image Processing Applications*. Dr. Deshpande is a reviewer for a number of journals published by the IEEE Access, IET, Springer-International Journal of Fuzzy Systems, Elsevier, etc. His research interests include Artificial Intelligence, Image and Video Analytics, Data Analytics, and Machine Vision.

Dr. Vania V. Estrela has ample experience teaching postgraduate and undergraduate courses. She has a B.Sc. degree from the Federal University of Rio de Janeiro (UFRJ) in Electrical and Computer Engineering (ECE); M.Sc. from the Technological Institute of Aeronautics (ITA), Brazil; M.Sc. in ECE from the Northwestern University, USA; and Ph.D. in ECE from the Illinois Institute of Technology (IIT), Chicago, IL, USA. She has taught at the DePaul University, USA, and Universidade Estadual do Norte Fluminense (UENF), Brazil. She was a visiting professor at the Polytechnic Institute of Rio de Janeiro (IPRJ)/State University of Rio de Janeiro (UERJ) in Brazil. She works at Universidade Federal Fluminense's (UFF) Department of Telecommunications. She has proposed and participated in various pedagogical projects for the specialties of "Computer Engineering" at UENF; "Computer Technology" and "Material Science and Engineering with emphasis on Polymers" at UEZO/UERJ. Research interests include Cyber-Physical Systems, Signal/Image/Video Processing, Multimedia,

Biomedical Engineering, Neuroscience, Electronic Instrumentation, Computer Architecture, Unmanned Aerial Systems, Modelling/Simulation, Sustainable Projects, Smart Designs, Inverse Problems, Communications, Motion Estimation and Understanding, Artificial Intelligence, and Geoprocessing. She edits and reviews for several prestigious publishers. She is engaged in Humanitarian Engineering, Technology Transfer, STEAM Education, Environmental Issues, Digital Inclusion, and all UN Sustainable Development Goals (SDGs). She has served as editor of more than 15 books and special issues and on a plethora of Technical and Organizational Committees. She is a member of IEEE.

Dr. Asiya Khan is an associate professor of Multimedia Communication and Intelligent Control at the University of Plymouth. She received her B.Eng. (Hons) in Electrical & Electronic Engineering from the University of Glasgow, M.Sc. in Communications, Control & Digital Signal Processing from Strathclyde University, and Ph.D. from the University of Plymouth. She is a Chartered Engineer and a Senior Fellow of the higher education Academy. She is currently the associate Head for teaching and learning in her school. Dr. Asiya Khan is an editor of *Neuroscience Informatics*.

Her current research interests include prediction and control of video quality using artificial intelligence, machine learning, cloud computing, fuzzy logic, applying computer vision techniques and deep learning in pedestrian recognition, disease identification in cotton crops and damage recognition in wind turbine blades. She is a reviewer for IEEE, Elsevier, IET, and Springer journals.

CHAPTER

1

How Industry 4 Can Reshape Agribusiness: The Meaning of Agriculture 4.0

Vania V. Estrela[1*[0000-0002-4465-7691]], **Anand Deshpande**[2[0000-0002-1500-0981]], **Aline Chaves Intorne**[3,4[0000-0001-8015-6926]], **Asiya Khan**[5[0000-0003-3620-3048]], **Dhanashree Kulkarni**[6[0000-0003-3906-3298]], **Shoulin Yin**[7[0000-0002-5367-1372]] and **Nikolaos Andreopoulos**[8[0000-0001-9975-7043]]

[1] Telecommunications Department, Fluminense Federal University (UFF), RJ, Brazil
[2] Department of Artificial Intelligence and Data Science, Angadi Institute of Technology and Management, Savagaon Road, Belagavi, Karnataka, India
[3] Laboratory of Physiology and Biochemistry of Microorganisms, Universidade Estadual do Norte Fluminense Darcy Ribeiro, Campos dos Goytacazes, RJ, 28013-602, Brazil
[4] Laboratory of Chemistry and Biology, Instituto Federal de Educação, Ciência e Tecnologia do Rio de Janeiro, Volta Redonda, RJ, 27213-100, Brazil
[5] University of Plymouth, Plymouth, United Kingdom
[6] Department of Computer Science & Engineering, Angadi Institute of Technology and Management, Belagavi, Karnataka, India
[7] Shenyang Normal University, Shenyang, Liaoning Province, 110034, China
[8] Computer Science Department, Technological Institute of Iceland, Sæmundargata 2, 102, Reykjavik, Ssn. 600169-2039, Iceland

This chapter describes Agriculture 4.0 (Agri4) and examines how "Industry 4.0" (I4) must fit agriculture, accounting for the convergence of "operational technologies" (OT) and "information and communications technology". Agri4 fuses ICT and OT to address pitfalls such as supply chain supervision, productivity regulation, quality improvement, secure data storage/management, data-driven breakthroughs for ingenious agribusiness, optimized capital handling, growing income, process/asset/labor/human life traceability, and smart farming insurance. Likewise, the merits, restrictions, and potential disruptive factors bring some multidisciplinary flavors to people working with/for rural communities.

* Corresponding author: vania.estrela.phd@ieee.org

1. Introduction

When it comes to agribusiness, some scholars, entrepreneurs, and administrators ponder the effects of new "artificial intelligence" (AI) interventions and further avant-garde equipment to engender a 4th agricultural revolution, aka Agriculture 4.0 or Agri4.

Given the Agri4 potentially transformational impacts on imminent farming, society must reflect on its consequences (i.e., advantages, caveats, and disruption). It must serve everybody: farmers (aside from their size or enterprise or money), property owners, farmhands, rustic communities, and the all-inclusive public. Yet, when it comes to developing countries, policymakers, NGOs, educational institutions, cooperatives/collectives, and media tend to either paint Agri4 as colossally progressive or pay no attention to it. These countries still treat workforces as replaceable and discardable elements, without focusing on potential outcomes.

The agri-revolution commenced when humans realized they could cultivate and farm to satisfy their needs in grim times (circa 12 millennia ago). The second restructured the countryside when European feudalism terminated in the 17th century. The 3rd wave ushered in chemical pesticides, fertilizers, and high-yield selections of human-altered plants/animals alongside far-reaching machinery usage.

Industry 4.0 (I4) technologies may draw colossal capital and speculation to heighten food availability while minimizing associated ecological degradation. Past farming technologies have also encountered excessively optimistic praise, already followed by dreadful consequences regarding technical advances such as UVs and nanotechnology.

Agri4 anticipates changes to agrarian practices, the environment, and all types of life after the Industry 4.0 (I4) novel technologies usage. These shifts have elected AI as a panacea for smarter decision-making and all genera of energy harvesting/dispensing for the most part. For example, intelligence-driven equipment could grow and harvest crops, disperse weeds, allot agrochemicals, and benefit from livestock utilizing unmanned vehicles (UVs), which come/go throughout the land, water, air, and space. Additional farming technologies incorporate genetic modifications (GMs) for higher-producing, disease-resilient harvests. More radical disruptions include vertical farms, in addition to artificially-developed meat.

Adopting new farming technology requires reflections on human well-being and removing interdisciplinary hurdles. According to precedent, advantages are unlikely to be disseminated evenhandedly throughout humanity, and some individuals will lose out. All stakeholders must identify who loses the quality of life and whatever people can do about upcoming hitches and ponder whether and how I4 will deliver on its promises.

Robotics is a touchy subject since it allows rural workers to reclaim work-life balance and handicapped farmworkers to continue farming. Nonetheless, these developments stress animals and humans owing to excessive information, a sense of not fitting within society, and a loss of privacy caused by farmers' constant surveillance.

Another issue is the integration of elderly farmers. By luring a younger, more technologically savvy labor force, older hands jeopardize their existence in favor of equipment. Additionally, a paradigm move will unavoidably harm someone who lacks expertise in "information and communication technologies" (ICT) and/or "operational technology" (OT). People's cognitive abilities may become less required, resulting in emotional/psychological consequences and less flexibility/creativity as an equipment-dependence outcome than their ancestors' and individual expertise.

The "cyber-physical system" (CPS) concept behind the Agri4 structure entails Hardware (HW)/Software (SW) platforms and applications, everything as service (XaaS) paradigm [4–6], secure cryptographic data interactions, rich ecosystems backing up innovation, project, deployment, and leadership systematized in a four-level fashion as seen in Fig. 1 and discussed below.

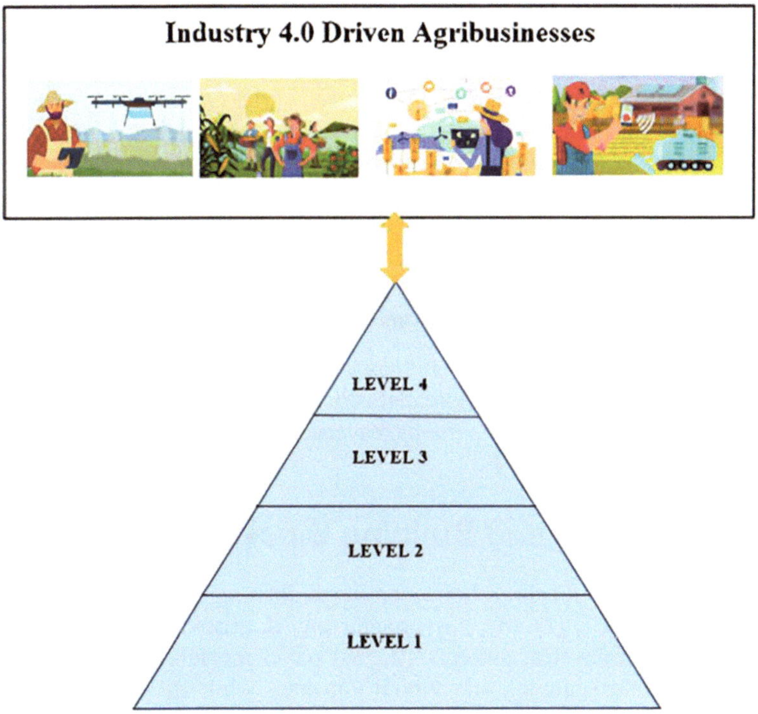

Fig. 1. Four-level Agriculture 4.0 pyramid.

Level 1: Computing capacity, storage, communication at the edge of networks (lowermost HW/SW junctures), a plethora of sensors/actuators, and control modules are some facets of Edge Computing (EC). It ties digital and physical assets nearby the laypeople, permitting information to be broadcast through connected and wireless modes. Addressable, identifiable services exist for sensing, processing data, and sending the necessary commands locally or to an upper layer in the event of failures. EC stands adjacent to the "human-machine interface" (HMI), not utilizing the "supervisory control and data acquisition" (SCADA) building [6, 7]. Stage 1 holds "enterprise resource planning" (ERP), linking blocks without a glitch transducers/sensors/actuators/microcontrollers/ automata like programmable logic controllers (PLC), trailblazing data storage/ interchange strategies, manufacturing execution systems (MES), besides manifold micro-electromechanical systems (MEMS).

Level 2: It takes more asset monitoring, multifaceted HW/SW frameworks, service assistance, energy management, in addition to handling more advanced control insights than Level 1. Value chain organization utilizes management, multi-scale surveillance, and data-driven services.

Level 3: Connectivity is the foundation of Level 3, encompassing the conservation, asset tracing/tracking, productivity issues, and enlarged complexity compared with the preceding planes with advanced services and applications.

Level 4: This portion routinely gains excessive varieties of e-governmental, technical, and corporative ecosystems whose intricacy combines issues, viz. monetization, payments, revenues, innovation, collaborations, business models, and interaction among producers, customers, and business partnerships, among others. Novel strategies can establish value appraisal, conversion, and equivalency among entities/commodities. Some old and recent ways to deal with value are plain money, blockchain-reliant e-governance, carbon footprints measurement/ equivalence/ appraisal, crypto coins, and quality of life measures to restrain and mitigate maladies, besides expertise, amid other concerns.

The previous 4-level description has a close relationship with the network layer model. Section 2 lists the essential concepts and building blocks related to Agri4. Section 3 covers Agri4's merits, caveats, yet-to-come work, and possible disruptive traits.

2. I4 Essentials and Building Blocks

I4 has a more significant robotization level than in previous eras via the Internet of Things (IoT) and has brought an AI misuse tendency. IoT plus the canny "cyber-physical systems" (CPSs) paradigm amalgamates and helps handle the physical, digital, and virtual universes while decentralizing efforts smartly over closed-loop models (namely, via feedback) and personalization/ customization of products [7–10]. These characteristics enable: (i) autonomous management, (ii) real-time (RT) monitoring of resources and processes, and (iii) RT stakeholder participation in value creation networks via vertical and horizontal arrangements. I4 remains a dynamic methodology with sets of

CPSs' architectures, standardizations, designations, and support tools. The CPS logic implies that partial/complete autonomy will shift to leverage information utilization, better transfers, and autonomous answers among CPSs constituents.

2.1 Reference Architectural Model Industry 4.0 (RAMI 4.0)

I4 architectures, global nature, and standardization stemmed first from German initiatives, followed by other countries with subcontinental and transnational adherence. The main idea was to enforce and promote the driving concepts behind the Industrial Internet (II) "Consortium and Platform Industry 4.0". By intercommunicating voluminous efforts, documents, various data, and opinions, companies/organizations/nations evolved, to a greater extent, members of both previously-mentioned platforms. Such a line of attack would lead to groups' ongoing similarities' observations and foster supplements/collaborations. The "Reference Architecture Model for Industry 4.0" (RAMI 4.0) has been created to keep I4 concepts, entities, and means under the same umbrella [1-3]. Contemporary I4 covers a snowballing amount of 'industries' in a broader sense other than manufacturing. CPSs tackle this propensity and move towards extending the systemic scope, which stimulates and encircles, under this deep jargon ocean, states' monitoring and remote control/sensing/actuation prospects. As an outcome, the expression 'smart' became ubiquitous.

2.2 Real-time Capability

Streaming or Real-Time (RT) architectures crunch inward data and progressively deliver outputs, differing from batch architectures, which superfluously pause for all input data processing to end to attain by-products [2, 33]. The RT notion strongly leans on the analysis background after processing periods stretching from milliseconds to a few minutes. A streaming structure can (i) use micro-batch in which tiny data chunks are processed within temporal windows with treated outputs, in the end, the corresponding computing tasks, or (ii) with an RT policy handling each new data chunk straightaway and quickly to convey processed information but at the expense of being restricted to data flow processing architectures.

2.3 Cyber-physical Systems (CPSs)

CPSs form the I4 foundation through modern control frameworks, hardware (HW) with embedded software (SW) modules, and an Internet address (IP) related to the IoT [8, 9]. In layperson's terms, it combines the real world with digital/virtual structures. Products and production entities can network, exchange data, support novel production ways, participate in value creation, and intensify RT optimization. CPSs craft smart IoT agrarian frameworks with features like access to remote sensing information, tracking and tracing entities, performing cloud computing, and query content information retrieval DBs.

2.4 Internet of Things (IoT) and Communications

IoT integrates physical objects like sensors, SW, and other technologies via the Internet to interconnect and interchange information items with people, animals, other gadgets, and systems. IoT constituents can be inputs viz. pesticides, fertilizers, complete agricultural products, crops, orders, deliveries, packing, and other materials/assets for processing. These entities move between transformational facilities and stakeholders, propitiating all kinds of local and global smart devices-to-objects connections and interactions, often wirelessly. Therefore, precision agriculture (PA) arises as a "big data" (BD) by-product and imperative driver. Drastic farm administration reforms can result from clear information access and decision-making. So, agribusinesses have advanced, and each agrarian activity will engender and convey data relying mainly on wireless transfers.

The high freedom of movement and heterogeneous nature of agrarian production facilities complicates planning and control even more as surrounding conditions and circumstances are mostly unclear, bringing in inconsistencies and glitches to communication every so often from wireless connections' shortage, low quality, and ill-respected bandwidths [11–15]. Wireless technologies can bring more suitable alternative ways to circumvent misery and lack of education, as with low-power consumption equipment and multifaceted communication range in agriculture.

Narrowband-IoT (NB-IoT) systems have built on Long-Term Evolution (LTE) functionalities, especially those permeating 5G. The NBIoT's interconnection and communication features tie together vast amounts of ordinary and more sophisticated objects/entities with manifold applicability goals to agrarian domains, such as people's well-being, access to climate as well as remote sensing facts generated by personal gadgets, provincial knowledge valuation, supply chain management, easy transportation with the cheapest least quantity of intermediaries, high-tech logistics, and ecologically-focused demands/requirements/applications. NB-IoT's design goals comprise greater coverage paths/areas/volumes, prolonged battery lives, wireless energy gathering, astonishing growth in network sizes/varieties, and cheap, readily available devices/assets. NB-IoT frameworks will ease rural applications' expansion thanks to low-down power designs and rural information broadcasting over long distances.

2.5 Artificial Intelligence (AI)

AI delivers synthetic cognitive characteristics to inanimate agrisystems agents/parts with conditioning learning abilities akin to humans. AI in agriculture can (supposedly) ameliorate eclectic purposes and fallouts, noting that crop production, monitoring, processing, and commercialization can enrich RT AI employment. There are several prospective benefits of state-of-the-art agrisystems, like those conceived to spot diseases, decide on quality yield, and foster robot-like irrigation/spraying, consequently downsizing the workload of the new farmers/rural population. Miscellaneous sensors, actuators, and robots can work with unmanned vehicles (UV) to map processes, some equipped with other inherited/sophisticated transducers/actuators that operationalize functions such as irrigation actions [16–20].

2.6 Blockchain

Agri4 integrates clever know-how, viz. blockchain technology (BT), in agriculture to cope with issues akin to supply chain, quality improvement, regulation of productivity, and product traceability as the core [21]. BT can secure data storing and managing, thus assisting in developing and using data-driven modernizations for smart agribusiness, better capital management, revenue generation, traceability of processes/ assets/labor/ human life, and clever farming insurance. Since the deployment of a huge amount of IoT-based devices is in an open arena, that becomes even more dangerous when one looks at many possible new Agri4 threats related to security. BT can help ensure agrisystems' safety before adversaries can initiate many cyberattacks, such as "distributed denial-of-service" (DDoS) outbreaks that can make services inaccessible by injecting false information concerning agricultural equipment health or task completion, accompanied by data theft [22–24].

2.7 Big Data Space and Ecosystems

The BD entails data creation from a greater variety of sources in snowballing volumes, extremely fast, containing more intricate data sets, endowed with meaning (i.e., semantics and semiotics become an issue), particularly from newfangled information sources. Old-style data processing SW has plenty of glitches when treating this heterogeneous knowledge [25, 26].

IoT interconnects many intelligent or not entities, as said before. In consequence, PA has become an important BD development driver. Far-reaching agrarian changes demand access to clear knowledge, records, and superior decision-making means. So, rural groups can broadcast their data through innumerable wireless transfer technology modalities [27–29].

Today's agrarian equipment has followed industries with cumulative information usage, availability, and knowledge-derived by-products for farming practices. HW/SW robots and the declining costs of sensors, actuators, and controllers have driven the automation of farm duties such as RT monitoring of soil proprieties and animal handling at detailed spatial and chronological scales. Accessible and affordable personal devices have created new decision-support alternatives for better management and improved rural practices. BD analytical platforms can rely on cloud/fog computing and AI algorithms. These innovations bring relevant growth concerning agribusiness data volume, velocity, diversity, and legitimacy. Thus, Agri4 can revolutionize outputs, the food chain, and eco-friendly management.

2.8 Decentralization, Decision-making, and Autonomy

Decentralization connotes a broad-serviceable practice that reassigns authority and responsibility for public/corporational functions from a prevailing governmental/managerial to other governance/corporational levels [30–32]. Hence, it transfers responsibilities (e.g., resource acquisition, design, and management). Additionally, apportionment may swing from the central administration to:

- Domain divisions of central governments or transnational/international agencies;
- Provincial or *de facto* authorities;
- Organizations contemplating the private and voluntary sectors;
- Subsidiary units or governmental levels related to a nation; or
- Semi-independent public reps or corporations.

The processes' decentralization empirical nature demands systematic analyses for decision-making. Strategic characteristics help examine policy implantation and reform agendas comprehensively, cutting across such initiatives as rural revenue generation, poverty decline, foodstuff security, and nutrition improvements while managing and protecting natural and eco-friendly assets.

Decentralization applicability spreads further training contexts, counting those on rural growth schemes, partaking development approaches, sustainable livelihoods, global/local/microregional development, and ecological planning with analysis. As perceived by ICT, decentralization can boost and intensify even more I4 impacts on the Agri4 realm while facilitating, bringing revenue, and giving voice to rural communities. Blockchain is the I4 feature that may speed up transactions, and transportation, safeguarding laypeople from corruption. Blockchain can legally legitimize and digitally protect/register all stakeholders' contacts in one cryptographed cyber-protected record.

2.9 Virtualization with Information Transparency

Virtualization creates an immaterial layer over HW through programming and exploiting single processors, memory varieties, and data storage set up into Virtual Machines (VMs). Virtualization issues may appear and be addressed manifold in a non-mutually exclusive fashion like these possibilities [2, 34–36]:

Transparent virtualization: Operating systems (OSs) or their specific components may run via VMs without amendment.

Paravirtualization: The OS necessitates alterations before a VM executes contingent on the OS part undergoing alterations. OS manufacturers may forestall and maintain these revisions or not (conditional on the OS kernel).

Hardware virtualization: Although somewhat rare currently, HW virtualization can affect overtly better designs for the reason that its purpose has gone sky-high while leaning towards a standard layer within data centers, HW retailers, as well as other players, promising more resilience as part of digital twins.

These advances combined, i.e., applied to the elementary hypervisor features (CPU, memory administration, and I/O), will upsurge performance, dependability, and compatibility humongously to several intermediary stakeholders until reaching out to the end-user. Hence, virtualization lessens obstacles from a limited reality:

(i) Geographical location as virtualization dismisses mandatory on-site attendance to surveil, handle, control, and actuate upon agrarian elements; and
(ii) Time-given virtual depictions of physical entities tolerate keeping a historical data record along with some simulations of prospective evolution scenarios besides extreme, threatening outer stimuli to test physically (e.g., forms to deal with the scarcity of materials triggered by adverse climatic events).

"Virtual reality" (VR) fosters cybernetic settings with virtual groups (as per virtual workplaces where rural and urban actors collaborate). Laymen perceive VR as tangible through interfaces that emulate human tactile, visual, and acoustic competencies. In this way, virtualization arises as lively organizational configurations that briefly tie different organizations' assets to better react to agribusiness prospects. However, virtual equivalents (twins) of products and assets contain detail galore (locally or globally accessible). They allow connecting all present-day and historically relevant evidence about objects' (or system's) properties, origin, sensory features, etc.

2.10 I4 Security and Relationships to Life, Environment, and Society

I4 can transcend technologies by looking at their impacts on people, society, and the environment, assuming collaboration between beings and gadgets (e.g., collaborative robots aka cobots), new essential skillsets for hands amidst modifications (i.e., disruption), and, inescapably, the loss of occupations due to automation and how to remediate this major side-effect [27–29].

Likewise, I4 needs robust cybersecurity with cryptography based on blockchain for general information and communication networks, besides collective/individual data protection that supports unprecedented and upcoming rules for green facilities, energy, and ecology, while shielding the dignity/welfare of human beings and other full of living creatures, properties, critical infrastructure health, and studies to lessen its effects on people's healthiness and emotions. This latter trait is the Industry 5.0 heart [37, 38].

The UN "sustainable development goals" (SDGs) working with I4 call for irrefutable engineering practices with the support of and assistance from other expertise subjects to devise critical power designs with multiple energy sources. Several energy grids must work disruptively as plausible and be steady, trusty, operational, and fair.

Cybersecurity entails redundancy, blockchain-reliant smart contracts, disaster relief, business continuity, and services via virtualization/digital twins/additional replication practices. Other cyber resilience expedients are indispensable to sustain novel and legacy structures functioning under constant local/global transforming conditions. This physical and cybersecurity challenge ranks tall among the I4 challenges and threats.

CPSs may combine welfare, cloud/fog computing, EC, 3D printing, digital twins, virtual/augmented reality, innovative robotics and cobots, support for green capitalism/circular economy praxes, autonomous nondiscriminatory production, and prompt prototyping/development. These innovations must be consistent and auditable practices across an all-inclusive value chain with VI/HI systems. I4 entities examples are IoT, BD gathering and provision, cybersecurity, surveillance of critical facilities/organisms employing many electromagnetic bands, AI, on-body or in-body (e.g., implantables, injectables, ingestibles) wearables for monitoring life forms. A vast range of interoperable, compatible, legacy-friendly standards/protocols should keep existing protocols while making room for upcoming designs supporting and carrying out justice through technologies like blockchain.

It is paramount to look at the behavioral, social, educational/semantic, and semiotic problems/demands related to the quick technological/scientific terminologies insertion into the mainstream without disturbing heritage and family structures. Alternative monetization arrangements and commerce must come with judicial practices and systems to handle data while boosting, protecting, exchanging products, dealing with richness, and blending in businesses well.

2.11 Value Chain Data and Optimization

The in-depth I4 value chain takes account of suppliers/origins of miscellaneous provisions and elements for innumerable smart manufacturing forms, from side-to-side of the digital supply chain, to reach manufacturing/production through all in-between segments and actors towards the end clientele. Such rationale enables more direct models for service provision, personalized production, and stakeholders RT interaction (with data acquisition from physical product on-spot usage) while lessening irrelevance, inefficiencies, and expenses of intermediary elements belonging to a digital supply chain archetypal. Agribusinesses must adapt to some I4 goals in a stakeholder-centric sense whenever possible, becoming compatible with other demands, e.g., valuing speed, cost-effectiveness, and ground-breaking value-added services [39, 40].

Ultimately, it remains an industry with the agribusiness models' innovation and transformation twists along with agrarian processes. Changes embrace increasing profit, diminishing costs, augmenting end-user experience satisfaction, service quality amelioration, optimizing stakeholders' lifespan value and, where viable, client fidelity, selling more, and updating CPS elements to nurture and keep relevancy.

2.12 Strategy and Deployment

I4 aims to be the subsequent organization and control phase concerning the all-embracing of the product value chain lifecycle [41]. CPSs afford enhanced competencies like (a) design, prototyping, and development of products, (b) remote control, (c) services and assessments, (d) conditions' inspecting, (e) proactive and predictive maintenance, (f) localization of entities, (g) structural health monitoring/maintenance, (h) planning, (i) innovation know-how, (j) agility, (k) RT applications, (l) the well-being of organic creatures, and so on. The prior items also lead to brand new use cases and deployments, not counting the benefits of I4 compliance such as personalization abilities, RT alarms/interventions, the possibility of mitigating emergencies by communicating with superior network layers, ultramodern services, dynamic product updating, augmented productivity, greater up-time and, in the long run, fresh agribusiness models. I4 capabilities have combined the expression 'smart' with anything like grids, energy harvesting equipment, logistics, facilities, edifications, sets, workplaces, manufacturing, agrofactories, farms, cities, etc.

2.13 Horizontal and Vertical Integration

Agri4 principles, rationale, and elements go further than trade by handling the inexorable transformations affecting other industries. Despite HI and VI dissimilarities, they share a goal: ecosystem-wide stored and flowing data available across many systems and all processes, employing information transfer standards and crafting the ultra-modern supply and value chain groundwork [2, 42].

A. **Horizontal integration**

"Horizontal integration" (HI) mixes ICT structures for innumerable production/business preparation practices through several processes, materials, energy, and data flows. Additionally, they impact internal and external associates, suppliers, and clientele, besides other ecosystem elements, from logistics to novelty flows and participants. HI deals with the full value and supply chain digitization with information exchanges via ICT systems as centerpieces, which is not trivial. First, there exist loose ICT systems challenging all types of organizations. Complexity escalates when looking at unified integration and information exchanges involving suppliers, consumers, and outer stakeholders.

HI pertains to the way from suppliers to clients, end-to-end ICT integration with information handled by IoT, sophisticated analytics, and BD knowledge. RAMI 4.0 life cycle and value stream dimensions influence the first parts of data gathering and provisioning. Considerable work still exists regarding product data or evidence related to assorted horizontal processes right through the value chain. Still, it is critical for I4 and for agribusiness as a whole. Gains and drivers for horizontally coupled ICT structures entice those handling knowledge despite the shortcomings of non-integrated systems. Benefits include stakeholders' service quality within supply chains, formulation, performance, stakeholders' approval, and speediness, among other matters. Agri4 means more partakers, highly inter-reliant processes and people, and far more data to allude to some HI aspects smoothing horizontal harmonization, value formation, teamwork, budget savings, the efficient and speedy realization of smooth services and operations for markets, workers, and the prospects to construct horizontal value ecosystems hinging on ICT.

B. Vertical integration

Suppose HI is about ICT structures, supply/value chain flows, and countless processes. Then, "vertical integration" (VI) blends ICT systems/subsystems/components at several hierarchical production and manufacturing levels/layers.

These hierarchical levels relate to ICT guts, i.e., interfacing with the production process via sensors/actuators, HMI, controls/regulations of HW/SW, the *de facto* deployment/operation of the monitored/controlled process line or actual production process levels. The operational echelon entails production orderliness, quality control, and enterprise planning.

Archetypal VI solutions and technologies embrace PLCs controlling manufacturing and settle on the control level, e.g., SCADA is genuinely employed in control systems and can assist a myriad of production procedure levels and managing tasks. MES assists in the managerial and sharp ERP levels on the VI top, where the MES plays a substantial role in the initial I4 transformation stages and works as the digital information hub and connectivity facilitator.

2.14 Business Models

The I4 opportunities go beyond objectives like automation and higher performance to the true prospects regarding identifying and deploying new business models. Even though many agrarian cooperatives, collectives, and companies struggle to think of these steps and goals, such behavior is an almost unmanageable task. Undoubtedly, they will mainly concentrate on a step-by-step attitude. New business models using data, innovative ecosystems, and original methods will help rural stakeholders meet hassles uniquely, in addition to devising new income sources.

These Agri4 transformational intentions mainly revolve around the service dimension or automation pyramid as perceived by VI (Fig. 2). Now, one has the IoT growing prominence, a lot about ICT influencing OT and convergence.

As in HI and VI, data from interconnected assets are indispensable, going back and forth between several HW/SW elements to more refined processing services. These top units identify assets and establish correspondences between sensors, actuators, specialists, laypeople, systems, and action plans. Services' results from knowledgeable analyses go crosswise networks to other platforms/layers where more services can be built upon. The I4 automation pyramid enactment embraces five areas resembling the layers of the OSI network model as below.

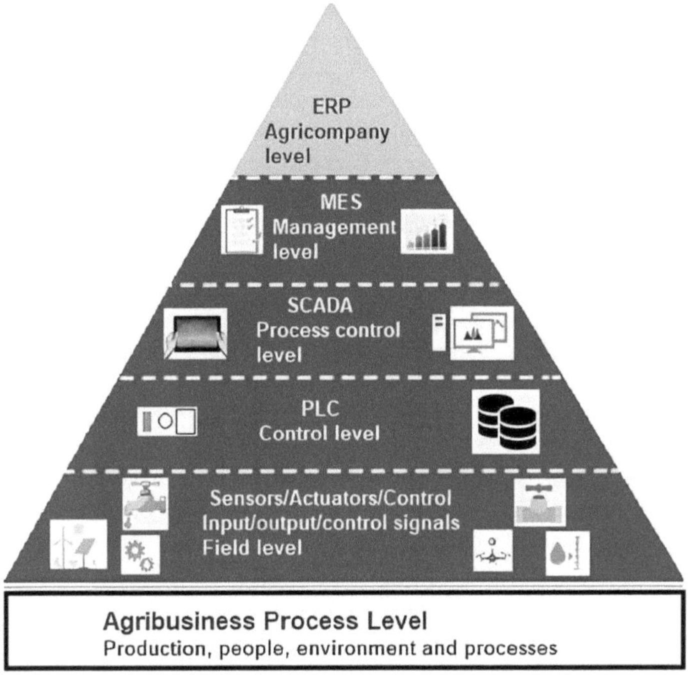

Fig. 2. Agribusiness 4.0 automation pyramid.

(i) **ERP** has the top means to monetize IoT while congregating data and, for instance, blockchain-empowered platforms. Paper and legacy structures may challenge interconnected resources while leveraging performance and keeping smooth relationships with previous steps.

(ii) **Systems/Control Management Services:** This stage monitors and manages data between the basic level and (if crucial) services and systems from a superior layer. This structure enables new ways to organize and manage the value chain by handling applications such as energy harvesting/monitoring, observation, and management of systems/conditions on machinery, facilities, plants, buildings, and infrastructure. Monitoring and management aimed at enhancing, understanding, and building new experiences. This part is the first to allow monetization established upon performance, albeit without much sophistication.

(iii) **SCADA Connectivity/Control:** It connects new applications and competencies. This additional layer emphasizes connectivity besides just interconnected assets. IoT-reliant, IP-dotted data and monitoring systems and service models facilitate cannier applications and aptitudes, e.g., preventive/predictive maintenance, module tracking, etc. These modifications have already changed many organizations' businesses. In the beginning, maintenance, search, and supplementary applications concerted on internal operations, but some can engender additional revenue if deployed and presented within a customer ecosystem setting, as is the case with offering maintenance contracts or making them available as a service while decreasing service and support costs for stakeholders. New services/ecosystems toil, transmute, and leverage means, services, plus data/intelligence in countless amenities dedicated and oriented agribusiness models to agrarian contexts. These landscapes can range from sanctioning stakeholders to refashioning merchandise they order and trade to cutting-edge services comprising revenue opportunities. Indeed, when people develop services within data ecosystems and prospective partners, they can create new services that bring significant shifts in agribusiness by aggregating/treating data intelligently and carrying out tasks that balance/complete systems while fostering entirely innovative segments.

(iv) **Internal Control Services and Systems:** This stage contains programmable logic controllers (PLCs), microcontroller-based boards, along with assorted microcomputers. These HW units monitor and manage data between the basic level and (if indispensable) the more top-quality services and systems layers. This structure enables new ways to organize and manage the value chain by handling applications viz. energy harvesting/checking, surveillance, and management of machinery/subsystems conditions, amenities, plants, buildings, and remaining infrastructure categories. Monitoring and management aimed at enhancing, understanding, and building new experiences. This part allows applications to monetize yields and labor hinging on performance, notwithstanding without much cleverness.

(v) **Sensors/Actuators/Control:** According to the IoT paradigm, this bottommost stage joins sensors, actuators, as well as control units. Hence, it is all about a layer with things in addition to wired or wireless entities encircling products and manufacturing tools/resources carrying information. These IP-addressable articles partake geographical localization traits that are reachable, owning to sensors besides the IoT foundation weaving whatever craves sensing. The data collected by transducers can be processed to perform necessary actions via actuators, bridging digital and physical realms.

2.15 Design Principles

I4 comprises six design principles (DPs) that help the Agri4 vision and elucidate guidelines for agrarian businesses to realize, identify, and implement I4 ventures. Plenty of academic works targeted these DPs using other terms. Some research mentions four instead of six DPs to explain what I4 means in other situations [43].

Utilizing the six-piece I4 DPs entails: (1) Interoperability (aka interconnection or just connectivity), (2) data transparency (alternatively, virtualization), (3) decentralization (autonomous verdicts or autonomy), (4) RT functions, (5) specialized support, service orientation (SO), and perhaps together with HMI, and (6) modularity.

Interoperability, interconnection, and connectivity mean moving to intelligent connected manufacturing, cooperatives, collectives, factories, or industries conciliate things like real physical entities, people, standards, work processes (especially those related to man-machine interactions), and more. This interconnection boils down to thinking about the application of data and networks so that these cyber-physical objects all interoperate and interconnect.

Specialists must tie IT and OT to assets' understanding, establish HW/SW connections and convey this info thanks to transducers, actuators, and other equipment. Multi-, inter-, and transdisciplinary efforts must merge people, data, and machines, among other pieces. From a broader perspective, IoT has spawned neologisms exemplified by the "Internet of Services", "Internet of Medical Devices", the "Internet of People/Services/Things", and the "Internet of Everything".

Interoperability is also about collaborations involving too many standards that talk to each other, leveraging data from various sources. IoT gateways/platforms can work concurrently with putting together ICT and OT. Interoperability implies connected devices, correlated communication technologies, linked people, associated data, people collaborating with devices, and machines-to-machines cooperation operating under an interoperable, unified, holistic information pool/flow within a protected data layer. Interoperation and interconnection relate to the HI and VI concepts.

3. Discussion, Future Trends, and Disruption

The I4 has to achieve the desired features, principles, technological maturity levels, benefits, and realizations. Answers to manifold questions are similar to those in digital changeovers across any business and digital transformational strategy challenge. The forthcoming Agri4 production comprehends:

1. An extraordinary individualization (personalization) of products degree with a vastly malleable production;
2. The prompt stakeholders' and business partners' inclusion/involvement in the design and value-making processes; and
3. Establishing links between production and high-grade services may engender hybrid merchandise.

Governments expressed their intention to back up the business and science sectors to develop and grasp I4 with an extra focus on security and its effects on labor quality and employment rates. Other digital urgencies include smart services, data, cloud/fog computing, heterogeneous/multilayer networking, science, educational practices, and living environments, whose majority is closely tied to I4 and require the examination along with high-tech schemes' prioritization to devise innovation recommendations.

HI makes a more grounded, interconnected, efficient value organization, yet it is far from clarity. Ask anyone from any industry niche. Other than data, its related information, necessary wisdom, and doings will, in due course, become noteworthy.

3.1 Benefits

In more than one respect, I4, as deemed the digital transformation of agribusiness, is still mainly focused on the first changes and maturity aspects, bringing benefits and requiring a broader potential perspective. Among the previously cited issues, one can mention enhancing productivity, automation, improving operational processes' optimization, agribusiness legacy practices, and adopting IoT in agribusiness operations keeping in mind predictive and intelligent maintenance services with (all forms of) life preservation.

The essential Agri4 foundation blocks facilitate, operationalize and democratize agrarian production ways and correlated goings-on, like logistics along with revenue increase/creation, faster, more efficient, and more client-centric while incorporating legacy and novel automation with several types of optimization and becoming aware of new agribusiness prospects and models.

Most Agri4 boons mirror their I4 counterparts, with IoT-intensive usage, operational agribusiness process optimization, overall ICT-driven ecosystems, and many other matters. I4 is the current automation drift and data trade in agribusiness technologies, including CPSs, IoT, and cloud/fog computing. However, a lot has to take place in developing countries to have smart farms.

3.2 Disruption and Challenges

I4 is the first stage towards a broader ecosystem of possibilities as groups migrate from more or fewer patent objectives to spot-on novelty and even disruption. One or two I4 insulated traits are implemented *de facto*, e.g., big data and some robotics.

Developing countries lack education, wisdom, and systems that normalize innovation besides augmenting productivity. Doubts arise concerning higher alertness, immediate prospects, availability of training and veritable innovation, recognizing green information-driven together with service-oriented revenue sources, and countless other goals.

The answers are challenges fostering the original abilities expansion, finding intelligent deployments, labor relationships, local/global traditions' balance, and understandings for publics, processes, innovation, creating competitive paybacks and services whereas impacting agroindustry paradigms and even the rural landscape as a whole, necessitating additional projects. Still, most agribusinesses are in premature CPS-convergence stages. Critical I4 aspects that puzzle many researchers follow:

- Definition of Agri4 strategies with regional DNAs.
- Agri-processes organization and rethinking to make the most out of products.
- Better agribusiness understanding.

- Performing fruitful pilot studies along with local representatives.
- Conscientization about the need to act towards overcoming poverty.
- Transform and adapt management.
- Understand cultural, historical, and emotional differences between urban, semi-urban, and rural settings.
- How to interconnect rural activities and people without causing distress.
- Education, capacitation, sharing/transferring local knowledge, and establishing data organization at the necessary pace.

Information management quality entails usable knowledge, interconnected information, agribusiness processes' merit/significance, modernization, and well-timed availability/satisfaction of all stakeholders' goals.

Cybersecurity plus cyberprivacy have become indispensable and vital given IoT attacks and abuses exploiting the often conflicting needs of ICT and OT convergence. Certain IoT every day, technological, as well as ecosystem-related distresses are:

- The granted-for ICT and OT integration assumption.
- Data compliance complications.
- Handling and embracing risks while decreasing budgets during crises.
- The interconnected supply chain complexity.
- Ameliorated discernment and expertise on ICT running with OT and how to leverage both worlds' perceptions.
- Shifting stakeholders' demands.
- Competition faced by Agri4 from I4 already established practices.
- Preservation of life forms.

4. Conclusion

This chapter has delineated Agriculture 4.0 (Agri4) while talking over Industry 4.0 (I4) adaptation to agribusiness, bearing in mind the ICT and Operational Technologies (OT) convergence. Agri4 mingles and bridges ICT and OT to stitch together supply chain, excellence improvement, productivity regulation, secure data storing/managing, data-driven innovations for smart agribusiness, better capital management, revenue generation, traceability of processes/assets/labor/human life, and smart farming insurance. Benefits, challenges, and possible disruptive aspects are also analyzed to establish Agri4 in developing nations with the help of smart farming frameworks and applications that will effectively and efficiently cover societal needs.

References

1. Bücker, I., Hermann, M., Pentek, T. and Otto, B. (2016). Towards a methodology for Industrie 4.0 transformation. *BIS*.

2. Vogel-Heuser, B., Bauernhansl, T. and Hompel, M.T. (2017). Handbuch Industrie 4.0 Bd. 4, Allgemeine Grundlagen. *Handbuch Industrie 4.0.*
3. https://www.i-scoop.eu/industry-4-0/ Accessed on December 26, 2021.
4. Monteiro, A.C., França, R.P., Estrela, V.V., Fernandes, S.R., Khelassi, A. et al. (2020). UAV-CPSs as a test bed for new technologies and a primer to Industry 5.0. *In*: Estrela, V.V. et al. (Eds.), *Imaging and Sensing for Unmanned Aircraft Systems: Volume 2: Deployment and Applications.* IET.
5. Klerkx, L., Jakku, E. and Labarthe, P. (2019). A review of social science on digital agriculture, smart farming, and agriculture 4.0: New contributions and a future research agenda. *Njas-wageningen J. Life Sciences*, 100315.
6. López-Leyva, J.A., Talamantes-Álvarez, A., Ponce-Camacho, M.A., Meza-Arballo, O., Valadez-Rivera, B. et al. (2019). Industrial IoT Projects based on automation pyramid: Constraints and minimum requirements. *In*: The Internet of Things in the Industrial Sector. Springer International Publishing.
7. Fan, C., Chan, C., Yu, H. and Yih, S. (2018). A simulation platform for human-machine interaction safety analysis of cyber-physical systems. *International Journal of Industrial Ergonomics,* 68, 89–100.
8. Estrela, V.V., Saotome, O., Loschi, H.J., Hemanth, J., Farfan, W.S. et al. (2018). Emergency response cyber-physical framework for landslide avoidance with sustainable electronics. *MDPI, Technologies*, 6(2), 42.
9. Dumitrache, I., Caramihai, S.I., Sacala, I.S. and Moisescu, M.A. (2017). A cyber-physical systems approach for agricultural enterprise and sustainable agriculture. *2017 21st International Conference on Control Systems and Computer Science* (*CSCS*), 477–484.
10. Szász, C. (2020). Cyber-physical platform development and implementation for industry 4.0. *International Review of Applied Sciences and Engineering*, 11, 66–72.
11. Zambon, I., Cecchini, M., Egidi, G., Saporito, M.G. and Colantoni, A. (2019). Revolution 4.0: Industry vs. agriculture in a future development for SMEs. *MDPI Processes 2019*, 7, 36.
12. Balmos, A., Layton, A.W., Ault, A.C., Krogmeier, J.V. and Buckmaster, D. (2016). Investigation of Bluetooth communications for low-power embedded sensor networks in agriculture. *Transactions of the ASABE*, 59, 1029.
13. Solodovnik, A.I., Savkin, V. and Amelina, A.V. (2021). The role of the Internet of Things as direction for the development of Agriculture 4.0 for rural areas. *IOP Conference Series: Earth and Environmental Science*, 839.
14. Loschi, H.J., Estrela, V.V., Hemanth, D.J., Fernandes, S.R., Iano, Y., Laghari, A. et al. (2020). Communications requirements, video streaming, communications links, and networked UAVs. *In*: Estrela, V.V. et al. (Eds.), *Imaging and Sensing for Unmanned Aircraft Systems: Volume 2: Deployment and Applications*, IET.
15. Aroma, J., Raimond, K., Razmjooy, N., Estrela, V.V. and Hemanth, J.D. (2020). Multispectral vs. hyperspectral imaging for unmanned aerial vehicles: Current and prospective state of affairs. *In*: Estrela, V.V. et al. (Eds.), *Imaging and Sensing for Unmanned Aircraft Systems: Volume 2: Deployment and Application*s. IET.
16. Spanaki, K., Karafili, E. and Despoudi, S. (2021). AI applications of data sharing in agriculture 4.0: A framework for role-based data access control. *Int. J. Inf. Manag.*, 59, 102350.
17. Megeto, G.A., Silva, A.G., Bulgarelli, R.F., Bublitz, C.F., Valente, A.C. and Costa, D.R. (2020). Artificial intelligence applications in the agriculture 4.0. *Revista Ciencia Agronomica*, 51, 1–8.

18. Monteleone, S., Moraes, E.A. and Maia, R.F. (2019). Analysis of the variables that affect the intention to adopt precision agriculture for smart water management in Agriculture 4.0 context. *2019 Global IoT Summit* (*GIoTS*), 1–6.
19. Gupta, N., Khosravy, M., Gupta, S., Dey, N. and Crespo, R.G. (2020). Lightweight artificial intelligence technology for health diagnosis of agriculture vehicles: Parallel evolving artificial neural networks by genetic algorithm. *International Journal of Parallel Programming*, 50, 1–26. Springer Link.
20. Ruiz-Real, J.L., Uribe-Toril, J., Torres Arriaza, J.A. and de Pablo Valenciano, J. (2020). A look at the past, present, and future research trends of artificial intelligence in agriculture. *Agronomy*, 10, 1839.
21. Xu, D., Wang, W., Zhu, L. and Li, R. (2020). Research status and prospect of blockchain technology in agriculture field. *In*: Lu, W. et al. (Eds.) *Cyber Security. CNCERT 2020. Communications in Computer and Information Science*, 1299. Singapore: Springer.
22. Ferrag, M.A., Shu, L., Djallel, H. and Choo, K. (2021). Deep learning-based intrusion detection for distributed denial of service attack in Agriculture 4.0. *Electronics*, 10, 1257.
23. Khan, A.A., Laghari, A., Shaikh, A.A., Dootio, M.A., Estrela, V.V. and Lopes, R.T. (2021). *A Blockchain Security Module for Brain-Computer Interface (BCI) with Multimedia Life Cycle Framework* (*MLCF*). Elsevier, Neuroscience Informatics.
24. Jumani, A.K., Laghari, A. and Khan, A.A. (2021). Blockchain and big data: Supportive aid for daily life. *In*: David, S., Anand, R.S., Jeyakrishnan, V., Niranjanamurthy, M. (Eds.), *Security Issues and Privacy Concerns in Industry 4.0 Applications*. John Wiley & Sons. https://doi.org/10.1002/9781119776529.ch8.
25. Kamilaris, A., Kartakoullis, A. and Prenafeta-Boldú, F.X. (2017). A review on the practice of big data analysis in agriculture. *Comput. Electron. Agric.*, 143, 23–37.
26. Tseng, F.H., Cho, H.H. and Wu, H.T. (2019). Applying big data for intelligent agriculture-based crop selection analysis. *IEEE Access*, 7, 116965–116974.
27. Weigel, R., Koellner, T., Poppenborg, P. and Bogner, C. (2018). Crop diversity and stability of revenue on farms in Central Europe: An analysis of Big Data from a comprehensive agricultural census in Bavaria. *PLoS ONE*. https://doi.org/10.1371/journal.pone.0207454.
28. Shuo, Q. (2017). Construction of the industry chain of ecological agriculture combined with Big Data. *Agro Food Industry Hi-Tech.*, 28.
29. Ruan, J.H., Hua, J., Xiaoyu, L., Yan, S., Felix, C. and Weizhen, R. (2019). A Granular GA-SVM predictor for big data in agricultural cyber-physical systems. *IEEE Trans. Industrial Informatics*, 15(12), 6510–6521.
30. Lai, K.C. and Cistulli, V. (2005). Decentralized development in agriculture: An overview. FAO, UN, FAO Economic, and Social Development Paper. Rome, Italy.
31. Retallack, M. (2020). Paradigmatic policy change or unintended subordination of rural autonomy: The case of source water protection in Ontario, Canada. *Policy Sciences*, 53, 85–100.
32. Torky, M. and Hassanein, A.E. (2020). Integrating blockchain and the Internet of Things in precision agriculture: Analysis, opportunities, and challenges. *Comput. Electron. Agric.*, 178, 105476.
33. Debauche, O., Mahmoudi, S., Manneback, P. and Lebeau, F. (2021). Cloud and distributed architectures for data management in agriculture 4.0: Review and future trends. *J. King Saud University – Computer and Information Sciences*, 34(9), 7494–7514.

34. Corallo, A., Latino, M.E. and Menegoli, M. (2020). Agriculture 4.0: How to use traceability data to tell food product to the consumers. *2020 9th International Conference on Industrial Technology and Management (ICITM)*, 197–201.
35. Kaiwartya, O., Abdullah, A.H., Cao, Y., Lloret, J., Kumar, S., Shah, R.R. et al. (2018). Virtualization in wireless sensor networks: Fault tolerant embedding for internet of things. *IEEE Internet of Things Journal*, 5, 571–580.
36. Taskin, D., Taşkın, C. and Yazar, S. (2021). Container-based virtualization for Bluetooth low energy sensor devices in Internet of Things Applications. *Tehnicki Vjesnik – Technical Gazette*, 28(1), 13–19.
37. Leakey, R.R. (2020). A re-boot of tropical agriculture benefits food production, rural economies, health, social justice, and the environment. *Nat. Food*, 1, 260–265.
38. An, C., Cai, M. and Guy, C. (2020). Rural Sustainable Environmental Management. *Sustainability*, 12, 6688.
39. Anitha, K., Palaksha Reddy, K., Krishnamoorthy, N. and Jaiswal, S. (2021). IoT's in enabling the supply chain visibility and connectivity and optimization of performance. *Materials Today: Proceedings*. doi: 10.1016/j.matpr.2020.12.343.
40. Wu, J., Shi, H., Wang, L., Feng, Y. and Du, Y. (2019). How to help raise the status of agricultural operators in the agricultural value chain: Design of a one-stop agricultural service platform. *ISJ Theoretical & Applied Science*, 10(78), 1–10.
41. Zhai, Z., Martínez, J., Beltran, V. and Martínez, N.L. (2020). Decision support systems for agriculture 4.0: Survey and challenges. *Comput. Electron. Agric.*, 170, 105256.
42. Despoudi, S., Spanaki, K., Rodríguez-Espíndola, O. and Zamani, E.D. (2021). From Industry 4.0 to Agriculture 4.0, *In*: Agricultural Supply Chains, and Industry 4.0.: Technological Advance for Sustainability, 101 pp. Springer Nature.
43. Dikhanbayeva, D., Shaikholla, S., Suleiman, Z. and Turkyilmaz, A. (2020). Assessment of Industry 4.0 maturity models by design principles. *Sustainability*, 12, 9927.

CHAPTER

2

Internet of Things-based Devices in Agriculture 4.0 – Wireless Sensor and Actuators Networks

Gulbir Singh[1*][0000-0002-3113-2474] **and Kuldeep Kumar Yogi**[2][0000-0001-9509-0114]

[1] Department of Computer Science and Engineering, Graphic Era Hill University, Haldwani Campus, Uttarakhand, India

[2] Department of Computer Science, Banasthali Vidyapith, Rajasthan, India

Agriculture 4.0 emphasizes precision agriculture (PA) in several ways: cultivation practices' refinement, better crop varieties, less risk/instability, hydric control, optimized pesticide treatment, and land/crop surveillance with negligible ecological influence. The Internet of Things (IoT)-built devices and technology (IoTDT) afford the most viable ways to attain PA. The IoTDT brisk developments have impacted every industry, comprising 'agriculture'. Trailblazing PA methodologies' reforms provoke new hopes and limitations. This chapter explores the IoTDT, communication techniques, and "wireless sensors and actuators networks" (WSANs) in agronomic affairs. Specific sensors/actuators exist for particular agrarian scenarios like soil handling, crop status, pest and insect discovery, fertilizers' spraying, and irrigation. The IoTDT use helps farmers over the yield junctures, from seeding to harvesting. This chapter concludes by listing the prospects as well as hindrances faced while applying IoTDT in PA.

1. Introduction

The global economy majorly hinges on agriculture for survival, which is the primary way to fulfill human needs, playing noteworthy roles in the development of nations as listed below:

* Corresponding author: gulbir.rkgit@gmail.com

- Influence on nations' income,
- As a paramount food supply source,
- As materials' vital font for different industries,
- Reducing inequality,
- It is a foreign exchange source for the nations, and
- It fosters capital formation.

The agrarian industry is experiencing a revolution since "cyber-physical systems" (CPSs) are revolutionizing "precision agriculture" (PA) by combining hardware (HW), software (SW), sensing/actuating technology that facilitates equipment to interact with their physical (i.e., real-world (RW)) environment. CPSs make PA more resourceful and practical, bringing in the prospect of gigantic economic and eco-friendly benefits [19, 20].

PA manages harvests on an exact scale, involving sensors, actuators, controllers, drones, and other technologies to observe crop health and earth conditions, permitting farmers to engender targeted and timely intercessions to maximize their yields. Agrarian CPSs give agribusinesses a much more directed and data-driven tactic to cope with crops. Conventional farming practices blended with the newest advances in "information and communication technology" (ICT) allow agriculturalists to make more accurate measurements and surveillance asset conditions. This data can update managers to improve decision-making on time and correct methods to fertilize, water, and harvest.

CPSs help agriculturalists to adapt to mutable environmental circumstances. Real-time data concerning weather, earth conditions, and yield health facilitate farmers' work since this material can prompt more knowledgeable decisions about possible farming adjustments. Real-time facts also stimulate quick responses/actions to unexpected environmental changes like unforeseen weather patterns or malady outbreaks. CPSs also help lessen energy, water, and fertilizer handling, cutting expenses, and fostering sustainability [21–23].

CPSs can record and survey the whereabouts of an animal. By employing a system to track the current location of animals, agriculturalists may effectively maintain control over their livestock, preventing them from wandering outside the boundaries of their property or becoming lost. There are various advantages associated with the utilization of CPS for livestock monitoring. By utilizing real-time data, farmers can promptly respond to any signs of suffering in their animals or any indications of more harsh environmental conditions. This intervention can potentially enhance the health and productivity of farmers' livestock while concurrently mitigating their financial expenditures. In general, CPS serves as an indispensable instrument for cattle farmers. By utilizing CPS, farmers can continuously monitor their animals' health and well-being in real-time, ensuring the maintenance of a healthy and productive herd.

PA intends to achieve smart farming (SF), also called smart agriculture (SA). It refers to farm management utilizing cutting-edge technologies to boost agrarian products' excellence and amount. This technique implies the usage of the "Internet of Things" (IoT)-based devices and technologies (IoTDT), data administration, soil/harvest groundwork, crop sowing, pest/yield reconnaissance, irrigation, and lastly, cultivation.

Over the past years, SF has become more useful to farmers as it grants them access to modern technologies and devices, augmenting the agricultural product's quantity and quality to cut farming costs. SF benefits include: (i) a high crop production rate, (ii) decreasing fertilizers, pesticides, and water usage, (iii) lessening environmental pollution, and (iv) boosting the safety of farms and farmers.

Recent research suggested a variety of smart devices to fetch and monitor crop-related information and field statuses. Many manufacturers now provide various sensors and actuators, e.g., "unmanned aerial vehicles" (UAVs), communicating devices, agricultural robots (agribots), and other heavy PA machinery.

The IoT helps agriculture retrieve information about the crop field through various sensors/actuators for observing, trespassing, and controlling the spaces [1]. IoTDT enables sensing objects in a controlled environment via "remote sensing" (RS) devices. It helps reduce the gap between the computer-based cyber-physical and real worlds. The main advantage of IoTDT in agriculture is maximizing efficiency and accuracy with low to moderate anthropological intervention.

As per Business Insider Intelligence, IoTDT devices' shipment for SA is rising rapidly. Figure 1 displays a rapid IoTDT growth of 150% from 2015–2020 [2].

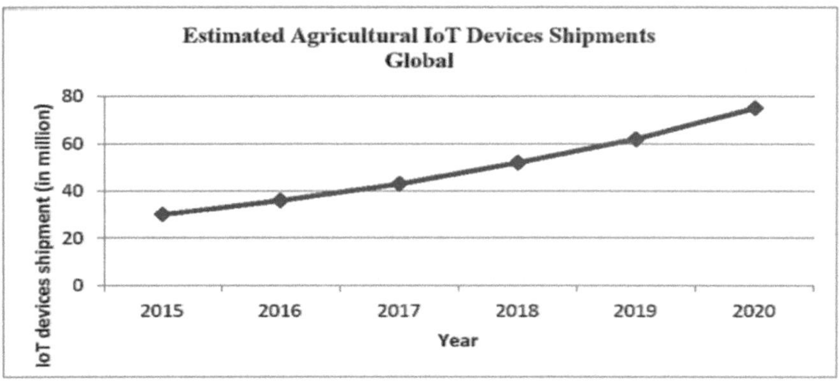

Fig. 1. Growth of shipment of IoT-based devices from 2015–2020.

This chapter can assist engineers and researchers in implementing IoT-based smart agriculture by first overviewing the primary IoT application in agriculture. Different IoT-based devices/sensors/actuators are also settled, followed by a conclusion and challenges in implementing IoT in agriculture.

2. Literature Review

Research centers and SW/HW industries focus on the impact and usage of IoT technologies and suggest significant issues in this field. This section focuses on the past work done by researchers that can use IoTDT in agriculture (Fig. 2 and Table 1) and shows the work done by researchers using IoT-based technologies.

Table 1. Past research using IoT-based technologies in agriculture

Authors and Year	Subdomain	Observations
Cheng-Jun, Z. (2014) in [3]	Monitoring of field	An online crop monitoring system captures multiple variables like soil moisture, water pH levels, humidity, temperature, etc. It consists of a three-level architecture. The first perception layer has WSN support, followed by the network layer for data communication, and concludes with an application-layer webserver for data analyses.
Mafuta et al. (2012) in [4]	Monitoring of Soil	It utilizes IoT with WSN to monitor soil moisture and heat via communication technology like ZigBee, GPRS, and the Internet.
Singh, G. et al. (2019) in [5]	Air pollution monitoring	It has a system with the help of Arduino and several gas sensors and a web application to store and process the processed data.
Singh, G. and Yogi, K.K. (2020) in [6]	Crop disease monitoring	A critical review on crop/plant disease detection employing IoT devices that focus on machine learning (ML) and artificial intelligence (AI) for plant disease detection.
Jain et al. (2008) in [7]	Animal monitoring	It monitors animals with the IoT system for watching swamp deer, simultaneously collecting animal behavior and climate data.
Shuwen & Changli, (2015) in [8]	Irrigation control and monitoring	It has a solar power-based irrigation monitoring system that uses the ZigBee protocol for data communication.
Pahuja et al. (2013) in [9]	Pesticides and fertilizers monitoring	It has a microclimate monitoring system for the greenhouse via a WSN to collect and process data.
Cozzolino, D. et al. (2015) in [10]	An infrared (IR) sensor in the field	It is a critical review of IR sensors in agronomy to monitor cereal crop harvesting.
Bhatnagar, V. et al. 2020 in [11]	IoT devices in agriculture	It analyzes IoT-based devices' usage in agriculture and the challenges due to the use of these devices.

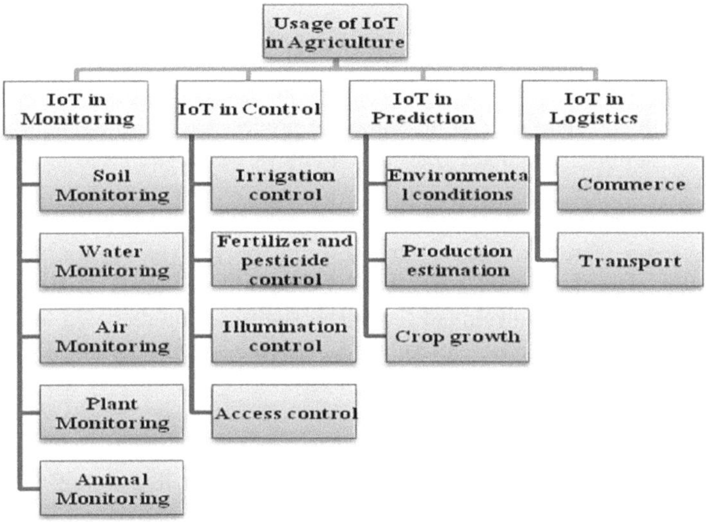

Fig. 2. Usage of IoT-based devices in agriculture.

3. Applications of IoT-based Devices in Agriculture

Traditional agronomy can be transformed into modern/intelligent/precise agribusiness by implementing IoT devices' latest sensors/actuators, i.e., smart farming (Fig. 3 lists all the major applications of IoTDT).

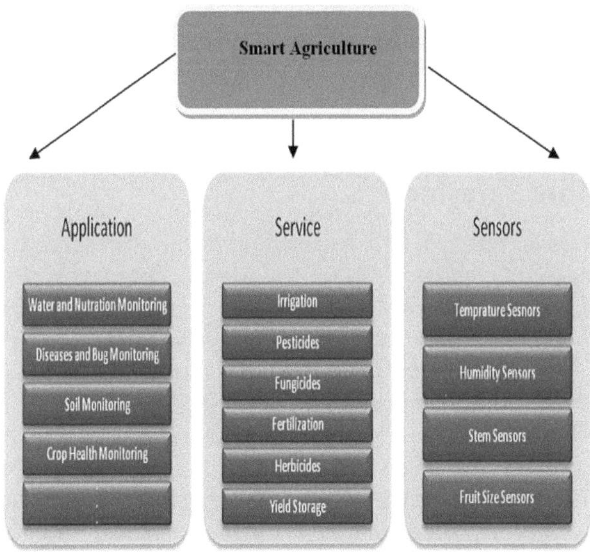

Fig. 3. Major SF applications of IoT-based technologies.

3.1 Soil Monitoring

Soil is the major building block for plants as it delivers all the nutrients and water supply necessary for plant growth. The first step of soil monitoring will be done through soil sampling, as the sampling process examines the field-specific details for farming. The main goal of soil analysis is to get information on the nutrient status so that measures can be taken when the crop faces issues related to nutrient deficiency.

The main factors to analyze the soil are its type, fertilizers'/pesticides' application, cropping history, topography, irrigation level, etc. This factor details the soil's physical, chemical, and biological statuses. Soil mapping helps solve many issues of sowing different crop varieties on a specific farmland.

The IoTDT, like sensors and vision-based units, can identify the depth and distance for sowing the seeds. The sowing activity can be performed via agribots implemented at every farm location that can benefit from them and incorporate knowledge from global and local maps generated by GPSs. The vision-based system is connected to a computer.

3.2 Farm Irrigation

Irrigation can happen by sprinkling and dripping, among other methods, and it can solve water wastage issues in farmlands. While in traditional farming, irrigation methods such as furrow irrigation and flood irrigation were used that can badly affect the quality and quantity of the crop. It is an arduous task to determine the water requirement of crops where other factors are involved, like soil moisture, irrigation methods, crop type, etc.

Better irrigation and rapid growth are expected in crop efficiency employing IoTDT, e.g., the green city-wide inclusive sanitation strategy (CWIS) irrigation management systems. The CWIS is a wireless sensor-based system for monitoring irrigation where all field sensors are connected to collect the field data from the farm.

3.3 Fertilizers Spraying

Fertilizers arrange for nutrients for plant growth. Plants need three major micronutrients (NPK): nitrogen (N) for leaf growth, phosphorus (P) for flowers, fruits, and root development, and potassium (K) for stem development and water movement [12]. The nutrient shortage and excessive applications may wrongly harm plants. The excessive usage of fertilizers may cause financial losses and damage the crop, soil, and environment.

The fertilization is done with IoTDT in SA to help estimate the required nutrients, minimizing the adverse effects on the crop and the environment. The IoT-based fertilization practice helps identify the nutrient requirements by crop with high accuracy and minimum labor requirements. IoT-based technologies like geo-mapping, GPS accuracy, VRT technology, and autonomous vehicles are employed for SF fertilization.

3.4 Pest and Crop Disease Management

A report by the "Food and Agriculture Organization" (FAO) estimated that 20 to 40% of global crops are lost annually due to disease and pests. Agrochemicals and pesticides were considered to diminish the loss in past years.

IoTDT, like wireless sensors, drones, and agribots, helps farmers reduce the usage of pesticides in fields or only apply the precise amount when required. These pesticides are very harmful to humans and the environment. They have even been sprayed in massive amounts also detrimental to crops. The advanced IoTDT can help real-time pest/disease management, provide live monitoring, and forecast pests and diseases, and hence are proven more efficient [13]. Advanced detection techniques for pests and diseases depend upon three levels: sensing, evaluation, and treatment. Real-time farm imageries can be captured to feed more advanced tools and techniques with pest and disease forecasting. Techniques like VRT chemigation [14] and precise vehicle spray are applied in SF for disease and pest treatment.

3.5 Harvesting, Forecasting, and Yield Monitoring

The yield monitoring mechanism can scrutinize the different aspects of agriculture, like harvested grain quality, moisture content, and grain mass flow. Crop forecasting predicts yield and production before crop harvesting. This forecasting helps farmers plan for the near future and make decisions. The yield surveillance contains multiple development stages and utilizes product conditions, such as size, color, etc.

4. SF Equipment and Technologies

The primary modern agriculture tasks can be done by large-scale heavy and urban tools like harvesters, robots (agribots), tractors, etc., fully or partially supporting remote sensing and other ICTs. In PA, the task is mainly performed by vehicles equipped with GIS and GPS technologies to accomplish precisely, independently, and accurately. Figure 4 shows the significant applications of IoTDT in SF.

4.1 Wireless Sensors

Wireless sensors play a noteworthy role in collecting information about crop conditions and their related data. The wireless sensors can work stand-alone when required and be integrated into every advanced agriculture tool. The following discusses the significant sensor types per their operational purpose and procedure.

4.1.1 Optoelectric Sensors

Optoelectric sensors are generally used to differentiate plants and can help detect herbicides, weeds, and other unwanted plants in crops [15]. When united with location information, one can map weeds and resolution. Optoelectric sensors use reflection spectra to differentiate vegetation and soil.

4.1.2 Airflow Sensors

The airflow sensors can gauge the soil air permeability and moisture percentage and detect the soil structure to identify the different soil types. The measurement can be done at singular or dynamic locations while in motion or at a fixed location or mobile node. It pushes the desired air the ground requires at a predefined soil depth. It can identify soil properties like compaction, moisture level, soil structure, etc.

4.1.3 Electromagnetic Sensors

These sensors identify the conductivity, responses to stimuli (transducers), and electromagnetic reactions in the actual situation. Electromagnetic sensors (EMSs) exploit electric circuits to evaluate the capacity of soil particles to accumulate or conduct electric charge via contact or non-contact techniques. EMSs can also measure soil nitrates and organic matter [16].

4.1.4 Optical Sensors

Optical sensors (OSs) utilize the light reflection phenomena to measure the soil's organic substances, color, moisture, and minerals. OSs reflect light on different terrain portions for earth testing. These sensors can assess plants, especially to observe fruit maturation. OSs together with microwave scattering, can help distinguish grove canopies viz. olives and similar crops [17].

4.1.5 Acoustic Sensors

Acoustic sensors are used for various applications in the agriculture farm, like weeding, fruit harvesting, soil cultivation, etc. Their main advantage is the low-cost and quick response when considering mobile equipment. They work by measuring changes in the noise as the tool is connected with other materials, e.g., soil particles [18].

4.1.6 Electrochemical Sensors

These sensors are paramount in agriculture as they measure soil features (e.g., pH) and identify soil nutrient levels. The standard soil chemical examination is costly and time-consuming and can be exchanged with electrochemical sensors.

4.1.7 Mechanical Sensors

Mechanical sensors measure the soil compaction (resistance) and can be put into the ground to record the force assessed by a strain gauge. The pressure units identify the earth's mechanical compaction level.

Fig. 4. Major SF technologies.

Fig. 5. Harvesting robots.

4.1.8 Mass Flow Sensors

The mass flow sensors are used to monitor yield as they provide yield information by measuring the quantity of grain flow. The yield monitoring system contains server modules such as grain moisture sensors, data storage HW, and dedicated SW to analyze data collected from the agriculture field.

4.2 IoT-based Tractors

Due to the growth in the agriculture industry, rural labor resources have come under heavy stress and pressure, resulting in tractors and other heavy machinery entering agriculture to provide efficient and progressive efforts. The major agriculture-based industries have begun to offer better solutions to farmers' requirements to fulfill the rising demand of farmers. The self-driving tractors are now available in the market and can lessen the revisiting of the same row/area of field by reducing the overlap. These IoT-built tractors provide better precision and decreased error, mainly when spraying pesticides and other related tasks unavoidable when a human operates a machine.

Most farmers cannot afford this advanced machinery due to their elevated cost. To overcome this issue, "Hello Tractors" has designed a cost-efficient monitoring device that can be mounted to any ordinary tractor, and that device provides SW and analytical tools. Figure 5 illustrates an IoT-reliant tractor in agribusiness.

4.3 Harvesting Robots

Harvesting robots provide a major role in the harvesting of crops. Some yields can be gathered a single time, or cropping is performed several times in some crops. Harvesting early or late may affect crop production, so harvesting the produce at the right time is critical. Agribots' role was to achieve precision harvesting to automate the harvesting process. Several researchers have increased the capability of fruit detection by its size, shape, color, and localization [19, 20]. The automated harvesting of fruits requires dedicated sensors that are efficient for unambiguous and precise details of that particular fruit. This process requires sophisticated and specialized tools to identify the fruit's conditions. Considering this issue, many robots have been developed for specific crops. Octinion, SW 6010, and FF Robot exemplify a few of the leading agribots used for crop harvesting, as Fig. 6 shows different agrarian harvesting robots [21].

4. Unmanned Aerial Vehicles (UAVs) in Agriculture

The IoT has developed to higher levels, affecting several industries like farming, fishing, poultry, etc. Blending agricultural automation with human activity has some restrictions because the amount of restricted information varies (and so do its measures). Adding the newest ICT is exceptionally arduous to implement in rural areas. During this state of affairs, the remote-controlled "unmanned aerial

Fig. 6. IoT-based tractor in the agriculture field.

vehicles" (UAVs or drones) area unit the choice resolution [22, 23, 48, 49]. Drones will communicate across the whole field through the wireless sensors/ actuators touching the sector and collecting the field information for more processing. They must have high-resolution cameras and other sensors/actuators flying through thousands of hectares. However, cheap equipment like shown in Fig. 7 illustrates a typical UAV that can be utilized in agrarian settings and benefit from advanced super-resolution equipment [48, 49].

Fig. 7. A typical over-the-counter drone can help smart farming via cloud and fog computing.

5. IoT Protocols for Agriculture 4.0

Agrarian IoT applications have been at the front position for several years. In this context, simple, small-scale, complete prototypes are implemented in medium- or large-scale usages. Evaluating IoT application layer protocols in SF implementations follows based on the key parameters.

5.1 Review of SF IoT Implementations

5.1.1 Simple, Small-Scale Applications

Simple, experimental applications exploiting sensors and actuators fall mainly in this category. An application of soil moisture measurement using MQTT is presented in work [24], while in [25], a relay is used to control a water pump, also with the MQTT protocol. Somewhat more complex frameworks employing more sensors (e.g., measuring luminance, temperature, atmospheric pressure, humidity) and actuators for watering control are presented in [26–29], using the MQTT protocol in all cases. A simple gateway system for collecting measurements, an edge computing equipment, appears in [30], where sensor communication uses the MQTT protocol. In contrast, REST HTTP transfers the measurements from the gateway to the cloud. Measurement acquisition with LoRaWAN and "The Things Network" (TTN) platform [31] uses the REST API for data transfer from the TTN to the data management application. WebSocket transmits data to the web. Finally, simple measurement sensors' collection and actuators management applications built on REST HTTP and WebSocket emerged [32]. An imaginative, small-scale hydroponic configuration relying on MQTT can transfer temperature, humidity, pH, and electrical conductivity measurements to control actuators [33, 34].

5.1.2 Medium and Large-Scale Full Prototypes

Integrated prototypes developed for medium-scale up to combined large-scale intelligent agro-industrial systems can monitor small crops by MQTT [33]. The MQTT protocol also sustained the designs from [35-37]. The latest is a plant-monitoring system using various sensors, including a camera. An agribusiness temperature and humidity measurement module handled desiccation with the MQTT protocol [38]. A mid-scale application submitted an intelligent watering system using multiple sensors and machine learning [39], where the REST protocol collected metrics while the HTTP handled actuators.

Moving to somewhat more complex applications and prototypes, [40] presented an integrated MQTT-based application relying on numerous sensors to observe vines. An original PA structure with many sensors and actuators appears in [41], where the MQTT and CoAP protocols were the primary choices.

An integrated prototype for precision farming that manages various sensors and actuators was also presented in [42, 43]; it applied MQTT and REST HTTP in parallel. An application via MQTT and REST HTTP/NGSI was submitted in [44]. Complete prototypes for watching farmstead facilities tasks and other environmental parameters, employing REST HTTP appear in [45, 46]. In [47], a case study of an innovative irrigation system developed using the MQTT protocol was related to the measured data (temperature and soil moisture) collected and managed by the AWS IoT cloud.

A large-scale integrated SF application with many sensors and actuators exists [47–52]. The CoAP protocol gathers sensor readings into a REST HTTP gateway to converse with the controlling platform.

The MQTT protocol is the most employed and favored for simple and complex IoT deployments. Following MQTT, the next most favored choice is the REST HTTP, while the CoAP protocol handles a few cases. It is worth perceiving that MQTT and CoAP are combined with REST HTTP in some larger-scale implementations.

5.2 SF and IoT Application Layer Protocols' Evaluation

Some essential aspects are identified by summarizing and elaborating on the preceding descriptions and comparisons of the IoT application layer protocols, besides discussing SF issues and challenges. The CoAP protocol (i.e., based mainly on UDP) has been shown to achieve the lowest latency while simultaneously raising bandwidth and energy prerequisites in almost all research works under consideration. MQTT comes next, followed by AMQP, XMPP, and REST HTTP. These last three with considerably lower performance. However, the situation is different regarding reliability: the capability to deliver the information effectively. Effective throughput and reliability are paramount for SF since unreliable data can instigate erroneous decisions and inappropriate automated procedures, resulting in high or even intolerable costs. This circumstance complicates further owning to a vast data variety, going from low-periodicity simple quantities to real-time multimedia, presenting huge variability in velocity and need appropriately supported in PA.

Further, the diverse and possibly unpredictable (beforehand) propagation conditions, ranging from almost free space to severe attenuation and fading (e.g., passing through a heavy harvest canopy), are critical factors affecting data veracity and accuracy. It might be inadequate even if the underlying physical/MAC/network layers' technology is high (leading to augmented cost). Thus, MQTT (which uses TCP, in contrast to CoAP UDP) seems to be the best option, outperforming the rest of the protocols in almost all cases surveyed.

Contrariwise, challenges are still there when it comes to real-life implementations. Scalability is critical to add newer devices to existing infrastructure without disturbing an established framework's performance, functionality, and Quality of Service (QoS) [50]. For data collection and reliable information transfer, many sensor and router nodes must be diffused in an IoT agrarian system. Moreover, implementing an IoT husbandry system network should not be physically delimited; in most cases, farming applications refer to broad geographical spaces, meaning that the architecture should be widely deployable and scalar. A large-scale deployment or the addition/duplication of several smaller networks could minimize the operational hazards over a vast pastoral area. The scalability of the abovementioned IoT application protocols affects bandwidth utilization, throughput, reliability, and QoS. So, the IoT protocols in existing or future agrarian IoT systems should be verified. While some recent investigation works attempted to emulate such stratagems, the performance of a downright settled IoT network with many nodes under different environmental conditions creates a large, diverse data volume not yet tested and evaluated. Hence, this state of affairs remains a challenging task to achieve.

Another critical aspect when choosing the application layer protocol when developing an IoT-driven SF framework is the best choice for farmers, manufacturers, and developers. The MQTT and CoAP protocols are elementary and 'light' in computational requirements to be applied sensibly and effortlessly to nodes with solid constraints on computational load and power autonomy. This chapter shows that the MQTT and HTTP protocols are at the researchers' and manufacturers' choice forefront, experimentally and commercially, for all SF applications. This fact is chiefly due to the protocols' maturity and their support by large corporations and communities. Still, the designers' choice relates directly to the price of implementing or integrating protocols for superior bandwidth or throughput rates between IoT devices. SF products' total HW and SW costs must be low to be globally available to affluent and underprivileged country markets. It is a significant challenge to cut software and hardware costs further. Still, the developers' choice can be biased by the IoT application protocol availability, meaning the development easiness or the existence of suitable HW and SW that can deliver the sought-after IoT services. Many versions of MQTT and HTTP application protocols are available that may be easily integrated into off-the-shelf HW and middleware.

SW modules for protocols like CoAP do not yet have adequate plurality and may need more sophisticated programming and hardware settings. Hence, the easy association of application protocol SW modules with IoT apparatuses or, more preferably, the amalgamation and multi-functionality of different protocol versions in IoT HW can ameliorate the effective IoT performance in SF. Finally, it is imperative to keep their services modernized and broadly accepted by as many IoT devices as possible to be adequate for the numerous varying SF applications. This statement means that each protocol's performance is directly associated with the underlying software design and its integration into the IoT hardware equipment. An extra factor in the same framework is interoperability. Many heterogeneous devices are applied in the farming sector to collect, broadcast, and process data. Data traffic management among these HW pieces is thought-provoking. The lack of standards (e.g., in the supply chain, weather data, semantics, data modeling, agri-machinery, e-commerce retail stores, etc.), technologies, or protocols do not hinder functionality but their proper number and selection among them.

Data security is very challenging, and it can employ various communication devices and protocols, making data vulnerable in terms of safety, invasion, threats, and breaches. Hence, maximum emphasis should go to design or apply IoT protocols. Security SF implications in modern highly networked food systems depend on information exchange networks, some vulnerable to cyber-attacks and hybrid warfare [51–53].

Concerning security, MQTT and WebSocket may use the TLS safeguarding connectivity protocol. In contrast, the CoAP may depend on the DTLS protocol at transport protocol, albeit none has the matching security protocol embedded. XMPP contains a built-in TLS to afford secrecy and data integrity reliability, exploiting a unique SASL profile to identify entities. AMQP blends communications security and identification by SASL and TLS, and DDS delivers the TLS or DTLS security options, depending on whether TCP or UDP is used. However, the security issue remains open for DDS. The recently protected MQTT extended version, Secure MQTT (SMQTT), deals with security concerns and seems promising for SF applications. Even though IoT application layer protocols embrace security solutions (either embedded or not), these must be further enhanced to ensure maximum security in all cases. At the same time, their overall performance concerning bandwidth, latency, and throughput must be assessed.

5.3 Discussion

The safest option now seems to be the MQTT protocol for an end-to-end network framework or when utilizing a gateway-server architecture to perform measurements. This statement is strengthened by comparing the IoT application protocols analyzed to the most critical research works. Table 1 ranks the IoT protocols from the utmost to the least optimistic. This table hinges on the prior survey and essential indicators of performance analysis: latency, necessary bandwidth, reliability, energy intake, throughput, and the developers' choice that uncovers the maturity and adoption easiness for off-the-shelf protocol solutions. Researchers' preferences for agricultural applications and development also appear. From the table and the analysis above, it can be derived that, in most cases, MQTT outperforms other IoT protocols. MQTT attains the first or (less often) second-best ranking for most of the examined performance indicators. CoAP protocol use can be considered the next best option since it attains auspicious performance in most network performance indicators. Idyllically, an identical protocol could be in all communication architecture parts, i.e., from

1. IoT devices to the gateway,
2. Gateway to the cloud (via IoT platform),
3. Cloud to the end-user equipment, and
4. Reverse directions when considering actuators' management (see also [3, 51] for IoT device-to-fog-to-cloud architecture considerations).

Hence, no protocol-to-protocol translator is required. Yet, this framework is not compulsory so other protocols may suit different communication network architecture parts. This choice is valid contingent on the additional IoT platform requirements for use/development, the HW and SW requirements of the gateways (if any), etc. In any case, utilizing MQTT for IoT device-to-gateway communications would benefit agrarian applications already deployed using HTTP.

Overall, MQTT looks like the best choice now. Nevertheless, as summarized above, many issues and challenges must be considered to choose safely since no solution suits all designs. Newer and evolving protocols (as per Section 2) must be further analyzed and verified in natural settings. The development of efficient agricultural applications with long-lasting choices entails the following:

1. Specific analysis on the implementation scale (either geographical or concerning the number of devices),
2. The physical environment, and
3. The breakdown of existing HW, SW, underlying network technology, and specific requirements.

Next, some IoT caveats in smart farming are listed [51–56].

- PA requires new ICT implementations for crop production. Setting up IoT architecture and sensor networks for an average farmer in this field could be tricky. So, the lack of knowledge in implementing these technologies can be dangerous [53–56].
- The instability of wireless networks can affect SF in many remote or rural areas worldwide.
- Lack of configuration and scalability problems.
- Energy depletion risk.
- Technical and corollary losses.
- Loss of human employment.
- IoT-based devices' security and trustworthiness in natural disasters like heavy floods, storms, cyclones, and fires.
- Optimal resource utilization is a big SF challenge.
- Mobility and cost analysis is a critical SF challenge.

6. Conclusion

This chapter has discoursed the work of researchers/engineers and the major application areas of IoT-based devices in smart agriculture. This chapter also debated the various equipment and technologies used in smart agriculture. Finally, this research highlights multiple challenges that arise while implementing.

This comprehensive examination of Internet of Things (IoT) agribusiness-enabled devices holds significant importance and utility for researchers, agriculturists, and industry experts engaged in contemporary agricultural practices. This study offers a comprehensive and current examination of IoT message (application) protocols considered significant possibilities for IoT applications, including MQTT, CoAP, DDS, XMPP, AMQP, REST-HTTP, and WebSocket. The offered entities were analyzed and compared based on performance, as determined by pertinent indicators like latency, energy and bandwidth prerequisites, throughput, reliability, and security. The discussion highlighted significant concerns raised by present-day and prospective smart agricultural applications. These issues aim to establish a strong foundation for practical decision-making and to guide future research endeavors that will tackle the mentioned obstacles to advance smart husbandry systems and applications that adequately meet societal requirements.

Acknowledgment

The authors thank the Department of Computer Science & Engineering, Banasthali Vidyapith, Rajasthan, India, for their motivation and support in completing this research.

References

1. Sales, N., Remédios, O. and Arsenio, A. (2015). Wireless sensor and actuator system for smart irrigation on the cloud. *Proc. 2nd IEEE World Forum on Internet of Things (WF-IoT)*, 693–698. Milan, Italy. IEEE.
2. Putjaika, N., Phusae, S., Chen-Im, A., Phunchongharn, P. and Akkarajitsakul, K. (2016). A control system in an intelligent farming by using Arduino technology. *Proc. 5th IEEE International Student Project Conference (ICT-ISPC)*, 53–56. Nakhonpathom, Thailand. IEEE
3. Cheng-Jun, Z. (2014). Research and implementation of agricultural environment monitoring based on internet of things. *Proc. Fifth International Conference on Intelligent Systems Design and Engineering Applications*, 748–752. Hunan, China. IEEE.
4. Mafuta, M., Zennaro, M., Bagula, A., Ault, G., Gombachika, H. and Chadza, T. (2012). Successful deployment of a Wireless Sensor Network for precision agriculture in Malawi. *Proc. IEEE 3rd International Conference on Networked Embedded Systems for Every Application (NESEA)*, 1–7. Liverpool, UK. IEEE.
5. Singh, G., Kumar, G., Bhatnagar, V., Srivastava, A. and Jyoti, K. (2019). Pollution management through internet of things: A substantial solution for society. *Humanities & Social Sciences Reviews*, 7(5), 1231–1237.
6. Singh, G. and Yogi, K.K. (2020). A review on recognition of plant disease using intelligent image retrieval techniques. *Asian J. Biol. Life Sci.*, 9(3), 274–285.
7. Jain, V.R., Bagree, R., Kumar, A. and Ranjan, P. (2008). wildCENSE: GPS based animal tracking system. *Proc. 2008 International Conference on Intelligent Sensors, Sensor Networks and Information Processing*, 617–622. Sydney, NSW, Australia. IEEE.
8. Shuwen, W. and Changli, Z. (2015). Study on farmland irrigation remote monitoring system based on ZigBee. *Proc. 2015 International Conference on Computer and Computational Sciences* (*ICCCS*), 193–197. Greater Noida, India. IEEE.
9. Pahuja, R., Verma, H.K. and Uddin, M. (2013). A wireless sensor network for greenhouse climate control. *IEEE Pervasive Computing,* 12(2), 49–58.
10. Cozzolino, D., Porker, K. and Laws, M. (2015). An overview on the use of infrared sensors for in field, proximal, and at harvest monitoring of cereal crops. *Agriculture-Basel*, 5, 713–722.
11. Bhatnagar, V., Singh, G., Kumar, G. and Gupta, R. (2010). Internet of Things in smart agriculture: Applications and open challenges. *Students' Research in Technology & Management,* 8(1), 11–17.
12. Kiiski, H., Dittmar, H., Drach, M., Vosskamp, R., Trenkel, M.E., Gutser, R. et al. (2009). Fertilizers, 2. types. *Ullmann's Encyclopedia of Industrial Chemistry.*

13. Kim, S., Lee, M. and Shin, C. (2018). IoT-based strawberry disease prediction system for smart farming. *Sensors*, 18, 4051.
14. Oberti, R., Marchi, M., Tirelli, P., Calcante, A., Iriti, M., Tona, E. et al. (2016). Selective spraying of grapevines for disease control using a modular agricultural robot. *Biosystems Engineering*, 146, 203–221.
15. Andújar, D., Ribeiro, Á., Fernández-Quintanilla, C. and Dorado, J. (2011). Accuracy and feasibility of optoelectronic sensors for weed mapping in wide row crops. *Sensors*, 11, 2304–2318.
16. Yunus, M.A.M. and Mukhopadhyay, S.C. (2011). Novel planar electromagnetic sensors for detection of nitrates and contamination in natural water sources. *IEEE Sensors Journal*, 11(6), 1440–1447.
17. Molina, I., Morillo, C., García-Meléndez, E., Guadalupe, R. and Roman, M.I. (2011). Characterizing olive grove canopies by means of ground-based hemispherical photography and spaceborne RADAR data. *Sensors* (*Basel*), 11(8), 7476–7501.
18. Kong, Q., Chen, H., Mo, Y.L. and Song, G. (2017). Real-time monitoring of water content in sandy soil using shear mode piezoceramic transducers and active sensing—A feasibility study. *Sensors*, 17, 2395.
19. Estrela, V.V., Saotome, O., Loschi, H.J., Hemanth, D.J., Farfan, W.S., Aroma, R.J. et al. (2018). Emergency response cyber-physical framework for landslide avoidance with sustainable electronics. *Technologies*, 6, 42. doi:10.3390/technologies6020042.
20. Estrela, V.V. (Ed.). (2023). *Intelligent Healthcare Systems* (1st Edn.). CRC Press. https://doi.org/10.1201/9781003196822.
21. Razmjooy, N. and Estrela, V.V. (2019). *Applications of Image Processing and Soft Computing Systems in Agriculture*, 1–300. Hershey, PA, USA: IGI Global.
22. Abreu, L.M., Hora, H.R.M., Rangel, J.J.A., Erthal, Jr. M., Razmjooy, N., Estrela V.V. et al. (2020). A multi-criteria modelling for ranking CO_2 emitting G20 countries from the Kaya Identity and their impacts on elderly health. *In*: Iano Y. et al. (Eds.), *Proc. of the 5th Brazilian Technology Symposium (BTSym'19). BTSym 2019. Smart Innovation, Systems, and Technologies.* Cham.: Springer. doi: 10.1007/978-3-030-16053-1_62.
23. Frąckiewiczin, M. (2023). How cyber-physical systems are revolutionizing precision agriculture. Artificial intelligence, Drones, TS2 Spaceon, April 2023. https://ts2.space/en/the-benefits-of-cyber-physical-systems-for-agriculture/#:~:text=CPS%20for%20agriculture%20is%20a,in%20a%20number%20of%20ways.
24. Kodali, R.K. and Sahu, A. (2016). An IoT based soil moisture monitoring on losant platform. *Proc. 2nd International Conference on Contemporary Computing and Informatics* (*IC3I*), 764–768. Noida, India, Dec. 14–17.
25. Kodali, R.K. and Sarjerao, B.S. (2016). A low cost smart irrigation system using MQTT protocol. *Proc. IEEE Region 10 Symposium* (*TENSYMP*), July 14–16.
26. Pooja, S. Uday, D.V., Nagesh, U.B. and Talekar, S.G. (2017). Application of MQTT protocol for real time weather monitoring and precision farming. *Proc. International Conference on Electrical, Electronics, Communication, Computer, and Optimization Techniques* (*ICEECCOT*), 814–819. Mysuru, India, Dec. 15–16.
27. Cao-Hoang, T. and Duy, C.N. (2017). Environment monitoring system for agricultural application based on wireless sensor network. *Proc. 7th International Conf. Information Science and Technology*, 99–102. Da Nang, Vietnam, Apr. 16–19.
28. Raikar, M.M., Desai, P., Kanthi, N. and Bawoor, S. (2018). Blend of cloud and Internet of Things (IoT) in agriculture sector using lightweight protocol. *Proc. International*

Conference on Advances in Computing, Communications and Informatics (*ICACCI*), 185–190. Bangalore, India, Sep. 19–22.

29. Syafarinda, Y., Akhadin, F., Fitri, Z.E., Yogiswara, B. and Widiawan, E. Rosdiana. (2016). The precision agriculture based on wireless sensor network with MQTT protocol. *IOP Conf. Ser.: Earth Environ. Sci.*, 207, 012059.
30. Suma, M.R. and Madhumathy, P. (2019). Acquisition and mining of agricultural data using ubiquitous sensors with Internet of Things. *In:* Smys, S., Bestak, R., Chen, J.Z., Kotuliak, I. (Eds.), *International Conference on Computer Networks and Communication Technologies*. Lecture Notes on Data Engineering and Communications Technologies, vol 15. Springer, Singapore.
31. Davcev, D., Mitreski, K., Trajkovic, S., Nikolovski, V. and Koteli, N. (2018). IoT agriculture system based on LoRaWAN. *Proc. 14th IEEE Int'l Workshop on Factory Communication Systems* (*WFCS*), Imperia, Italy, June 13–15.
32. Marcelino, R., Casagrande, L.C., Cunha, R., Crotti, Y. and Gruber, V. (2017). Internet of Things applied to precision agriculture. *Lecture Notes in Networks and Systems*, 499–509.
33. Gomez, J., Fernandez, A. and Sánchez, M.Z. (2018). Monitoring of small crops for the measurement of environmental factors through the Internet of Things (IoT). *Proc. International Conference on Technology Trends* (*CITT*), 16–28.
34. Ruengittinun, S., Phongsamsuan, S. and Sureeratanakorn, P. (2017) Applied Internet of Things for smart hydroponic farming ecosystem (HFE). *Proc. 10th Int'l Conf. Ubi-media Computing and Work* (Ubi-Media), Pattaya, Thailand, Aug. 1–4.
35. Bauer, J. and Aschenbruck, N. (2017). Measuring and Adapting MQTT in cellular networks for collaborative smart farming. *Proc. IEEE 42nd Conference on Local Computer Networks* (*LCN*), 294–302. Singapore, Oct. 9–12.
36. Bauer, J. and Aschenbruck, N. (2018). Design and implementation of an agricultural monitoring system for smart farming. *IoT, Vertical and Topical Summit on Agriculture – Tuscany* (*IoT Tuscany*), Tuscany, Italy, May 8–9.
37. Tran, H.A.M., Ngo, H.Q.T., Nguyen, T.P. and Nguyen, H. (2018). Design of green agriculture system using Internet of Things and image processing techniques. *Proc. 4th International Conf. Green Technology and Sustainable Development* (*GTSD*), 28–32. Ho Chi Minh City, Vietnam, Nov. 23–24.
38. Grgić, K., Špeh, I. and Ivan Heđi. (2016). A web-based IoT solution for monitoring data using MQTT protocol. *Proc. International Conf. Smart Systems and Technologies* (*SST*), Osijek, Croatia, 249–253. Oct. 12–14.
39. Goap, A., Sharma, D., Shuklab, A.K. and Krishna, C.R. (2018). An IoT based smart irrigation management system using machine learning and open source technologies. *Computers and Electronics in Agriculture*, 155, 41–49.
40. Trilles, S., González-Pérez, A. and Huerta, J. (2018). A comprehensive IoT node proposal using open hardware. A smart farming use case to monitor vineyards. *Electronics*, 7, 419.
41. Ahmed, N., De, D. and Md. I. Hussain. (2018). Internet of Things (IoT) for smart precision agriculture and farming in rural areas. *IEEE Internet of Things Journal*, Nov.
42. Ferrández-Pastor, F.J., García-Chamizo, J.M., Nieto-Hidalgo, M., Mora-Pascual, J. and Mora-Martínez, J. (2016). Developing ubiquitous sensor network platform using Internet of Things: Application in precision agriculture. *Sensors*, 16, 1141.

43. Ferrández-Pastor, F.J., García-Chamizo, J.M., Nieto-Hidalgo, M. and Mora-Martínez, J. (2018). Precision agriculture design method using a distributed computing architecture on Internet of Things context. *Sensors*, 18, 1731.
44. Zamora-Izquierdo, M.A., Santa, J. Martinez, J.A., Martinez, V. and Skarmeta, A.F. (2019). Smart farming IoT platform based on edge and cloud computing. *Elsevier Biosystems Engineering, Special Issue Intelligent Systems for Environmental Applications*, 177, 4–17.
45. Bajceta, M., Sekulic, P., Krstajic, B., Djukanovic, S. and Popovic, T. (2016). A private IoT cloud platform for precision agriculture and ecological monitoring. *Proc. Int'l Conference on Electrical, Electronic and Computing Engineering.* June, 2016, Zlatibor, Serbia.
46. Popovic, T., Latinovic, N., Pešicc, A., Zarko Zecevic, Krstajic, B. and Djukanovic, S. (2017). Architecting an IoT-enabled platform for precision agriculture and ecological monitoring: A case study. *Elsevier Computers and Electronics in Agriculture*, 140, 255–265.
47. Raikar, M.M., Desai, P., Kanthi, N. and Bawoor, S. (2018). Blend of cloud and Internet of Things (IoT) in agriculture sector using lightweight protocol. *2018 International Conference on Advances in Computing, Communications and Informatics* (*ICACCI*). Bangalore, India, pp. 185-190.
48. Deshpande, A., Estrela, V.V., Patavardhan, P. and Kallimani G. (2023). Super-resolution in a world of scarce resources for medical imaging applications. *In*: Vyas, S. et al. (Eds.), *Edge-AI in Healthcare: Trends and Future Perspectives.* Chapter 13. London, UK: CRC Press.
49. Deshpande, A., Razmjooy, N. and Estrela, V.V. (2021). Introduction to computational intelligence and super-resolution. *In*: Deshpande, A. et al. (Eds.), *Computational Intelligence Methods for Super-Resolution in Image Processing Applications.* Cham, Switzerland: Springer.
50. Laghari, A.A., Zhang, X., Shaikh, Z.A., Khan, A., Estrela, V.V. and Izadi, S. (2023). A review on quality of experience (QoE) in cloud computing. *J. Reliable Intell. Environ.* https://doi.org/10.1007/s40860-023-00210-y.
51. Glaroudis, D., Iossifides, A. and Chatzimisios, P. (2020). Survey, comparison, and research challenges of IoT application protocols for smart farming. *Comput. Networks*, 168.
52. Lee, M., Kim, H. and Yoe, H. (2019). ICBM-based smart farm environment management system. *In:* Lee, R. (Ed.), *Software Engineering, Artificial Intelligence, Networking and Parallel/Distributed Computing.* SNPD 2018. Studies in Computational Intelligence, vol 790. Springer, Cham.
53. Jahn, M.M., Oemichen, W.L., Treverton, G.F. et al. (2019). *Cyber Risk and Security Implications in Smart Agriculture and Food Systems.* White Paper, Jahn Research Group, University of Wisconsin–Madison, College of Agriculture and Life Sciences.
54. Zujevs, A., Osadcuks, V. and Ahrendt, P. (2015). Trends in robotic sensor technologies for fruit harvesting: 2010–2015. *Procedia Computer Science*, 77, 227–233.
55. Bargoti, S. and Underwood, J. (2017). Deep fruit detection in orchards. *Proc. 2017 IEEE International Conference on Robotics and Automation* (*ICRA*), 3626–3633. IEEE, Singapore.
56. Kondratieva, O.V., Fedorov, A.D., Slinko, O.V., Voytyuk, V. and Alekseeva, S.A. (2022). New solutions in the horticultural industry. *IOP Conference Series: Earth and Environmental Science*, 1010.

CHAPTER
3

Precision-Agri: IoT-Smart Contract Enables Novel and Secure Supply Chain and Forecasting Architecture for Smart Business Agriculture

Abdullah A. Khan[1,2*][0000-0003-2838-7641], **Aftab Ahmed Shaikh**[1][0000-0002-1533-6311], **Asif Ali Laghari**[1][0000-0001-5831- 5943], **Mazhar Ali Dootio**[2][0000-0002-6846-0439], **Mudassir Hussain**[3][0000-0001-5898-4709], **Awais Khan Jumani**[4][0000-0001-9468-0446], **Rukhsar Ayub**[5], **Vania V. Estrela**[6][0000-0002-4465-7691] and **Andrey Terziev**[7][0000-0002-7069-367X]

[1] Faculty of Computer Science, Sindh Madressatul Islam University, Karachi (74000), Sindh, Pakistan

[2] Research Lab of Artificial Intelligence and Information Security, Faculty of Computing Science and Information Technology, Benazir Bhutto Shaheed University, Lyari, Karachi (75660), Sindh, Pakistan

[3] Department of Commerce, Benazir Bhutto Shaheed University, Lyari, Karachi (75660), Sindh, Pakistan

[4] Department of Computer Science, Ilma University, Karachi, Sindh, Pakistan

[5] Benazir School of Business, Benazir Bhutto Shaheed University, Lyari, Karachi (75660), Sindh, Pakistan

[6] Telecommunications Department, Fluminense Federal University (UFF), RJ, Brazil

[7] TerziA, Bul "Vazkresenie" 103, 1504, Sofia, Bulgaria

Agriculture is one of the vital factors in living, social, and economic stability. The current lifecycle of food safety is multifaceted and unsecured as more stakeholders are involved and sharing resources and related information. This poses serious issues for agriculture industries to provide food safety, transparency, industrial-related information integrity, reliability, trustworthiness in the supply chain, and food livestock against the evolving threats growing in Industry 4.0. This chapter bridges these divergences by enabling a mixture of the 'Internet of Things' (IoT)

* Corresponding author: abdullah.khan00763@gmail.com; abdullah.ayub@bbsul.edu.pk

and Blockchain-based novel and secure smart business agricultural transparent supply chain and forecasting architecture. Precision Business Agriculture is a blockchain-IoT-enabled FIWARE cloud-based platform to collect current and previous node transactions and deploy them to the hosts' analytics modules. In this proposed architecture, participating stakeholders create a public ledger (Ethereum) network to share and agree on distinct supply chain and business agriculture forecasting-related activities before preserving the IoT-Blockchain ledger. We have conceived and created smart (digital) contracts, implemented them using pseudo-algorithms, and represented them through sequence diagrams to handle stakeholders' interaction in the supply chain and forecasting process. This proposed solution delivers system integrity, provenance, transparency, preservation, and a robust security mechanism to store immutable agriculture-business-related information in a permissionless hash (SH-256) encrypted IoT-smart contract distributed ledger.

1. Introduction

The "Internet of Things" (IoT) and Blockchain are gaining more attention in developing "information and communication technology" (ICT) due to the success of versatile applications and distribution in nature [1, 2]. The collaboration of both technologies generates a large amount of data that can be preserved, optimized, and utilized efficiently by the business intelligence and prediction strategy [3]. The importance of such innovation from the perspective of managing proper supply chains and a new way of business forecasting smart agriculture is significant in distinct activities [4, 5]. These processes are digitalization, transparency, immutability, availability, provenance, and digital contracts [6]. The secure IoT-smart contract data of agriculture via Industry 4.0 in the food safety and supply chain sector are stored in distributed storage such as Filecoin, IPFS, Storj, PPIO, Swarm, and so forth [7]. For instance, it helps the supply chain practitioners to take advantage and also assists in evaluating the performance of the overall system of state-of-the-art technologies. So, make industrial policies and stakeholders participating rules according to the forecasting of business agriculture [4, 8].

The fourth industrial revolution (Industry 4.0 or I4) incorporated sensors, advanced machines, and wireless communication technologies to form a fully connected business [8, 9]. The evolution of the Industrial IoT (IIoT) is one of the prominent aspects considered a significant component of Industry 4.0 [10]. Moreover, the technology can collect, examine, aggregate, and analyze information about processes, system performance, orders, and supplies and trace records from suppliers, manufacturers, and customers. The proper management of food safety is the main concern nowadays; federal authorities need to focus on the policies and take action to formulate and strengthen the management of food safety as well as the production of agriculture [11, 12]. The identification and traceability of agricultural products and control operations through Cyber-Physical Systems

(CPSs). Livestock and agricultural departments manage the hunger requirements, which are growing and increasing day to day all over the globe [13]. In contrast, the industrial green agriculture revolution is robust to the general production of crops and consumption in accordance.

According to the World Food Management and Research Report, the delivery of basic nutrition leads to the cause of death of 48% of children, and almost 2.2 billion people suffer because of minerals, vitamins, and other necessary malnutrition [14]. Several research and development agencies and experts are working to enhance the crop quality and quantity of food production. Precision agriculture (PA) is a revolution in the agriculture industry that has emerged many technologies, such as ICT-based approaches, that significantly impact crop production. In this act, farmers and steering bodies can remotely monitor and control operations, for example, soil moisture condition, crop health, etc., using ubiquitous multimedia devices [15, 16]. ICT has robust effectiveness and distributed use of irrigation water control management for agriculture. It also helps in accessing and deficit data of soil moisture and weather conditions and forecasting via real-time IoT-wireless sensor-based scheduling of smart irrigation and gathering data on soil conditions of agricultural areas. Moreover, it is widely recognized for its transparency and traceability in the "food supply chain" (FSC).

Blockchain introduced innovation in economic computing, allowing people to exchange payments and related information in a reliable, cost-effective, and fast with more accurate delivery through the communication network [16]. This technology emerges as an impressive way and solution to various real-life business challenges, for example, supply chain, logistics, and forecasting security issues. On the other side, the IoT is provided with the connection of different multimedia devices and sensors for the remote collection of real-time data. The combination of blockchain and IoT provides a platform for the lifecycle of real-time data acquisition and data preservation security—time and money-saving strategies [17]. However, the minimum ratio of manufacturers and industries in food and safety uses smart production and the Industrial Internet of Things (IIoT)-based manufacturing. The blockchain industrial IoT-enabled platform with "artificial intelligence" (AI) approaches to optimize a number of parameters and iterations of production lifecycle analysis, mitigate risk, and increase productivity.

"Distributed ledger technology" (DLT) enables the security of industrial-based sensor data, such as the data within the cryptographic hash encryption [18]. It makes a novel and secure system completely trustable, transparent, traceable, provenance, and immutable. Symmetric hash (SH-256) authentication depends on two keys—for protecting node transactions in a distributed ledger, as shown in Fig. 1. The distributed application (DApp) is designed and developed according to the rule of implementing smart contracts, consensus policies, and digital signatures between participating stakeholders to trace a good through the entire supply chain [19, 20]. The node transactions are preserved in the blockchain-distributed immutable ledger (impossible to change), a perfect mechanism for storing and maintaining records of agriculture supply-chain-related information

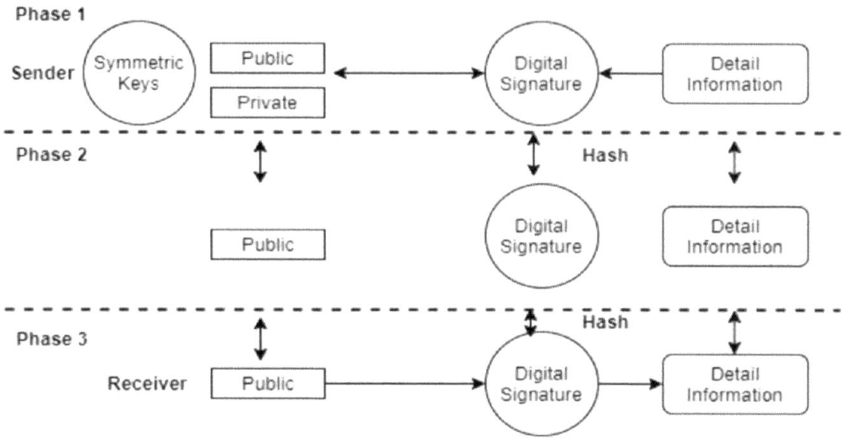

Fig. 1. The existing Symmetric-Hash (SH-256)-based authentication depends on two keys—for protecting node transactions in a Distributed Ledger.

exchange between ownerships. Moreover, blockchain permissionless public networks allow participating stakeholders to control and monitor the execution of the overall transactions and trace the previously preserved and newly created node transactions added to the ledger. In addition, it can use risk mitigation efforts to strengthen the livestock and FSC and prove the use of precision agricultural record preservation, security, and sustainability.

1.1 Motivation and Contributions

This chapter deals with the combination of Blockchain-IoT, the state-of-the-art technologies in this recent era for PA, FSC traceability, and the production of crop forecasting. We studied many related works, such as IoT-based smart agriculture and farming, livestock, and supply-chain traceability, which do not use blockchain DLT. An IoT-based coordination system in Agri-FSC: development of an efficient framework using "DEMATEL-ISM" proposed an FSC traceability framework [20], which cannot use blockchain-enabled security and privacy and without providing any benchmark framework simulation results.

However, this paper proposes a blockchain-IoT-enabled architecture for PA, FSC traceability, and crop production, allowing forecasting-related details in secure distributed storage. This architecture contains the following features: trustworthiness, transparency, provenance, immutability, assurance, control, access and availability, interpretation, traceability, and protected preservation. The traceable system ensures the privacy and protection for the overall "Precision-Agri-Ledger" (or PA-ledger) node transactions (hash SH-256 encrypted) and the storage (distributed preservation) of FSC details through the permissionless network. The contributions of this research are as follows:

- We present the design of smart business agriculture and analyses of FSC traceability and production of crops, forecasting data-related problems and challenges and security issues in IoTPA.

- A novel PA-ledger, Blockchain-enabled public permissionless network that connects IoT-based external devices to communicate (collect sensor-based data and make the decision accordingly) and provide a secure environment of FSC traceability as well as production of crop forecasting information preservation in the distributed structure architecture is presented.
- This chapter presents different smart contract (digital contract) algorithms pseudo-implementations, such as adding new nodes and updates, node transactions of the FSC and forecasting information in the distributed ledger, registration of participating stakeholders and digital signature policies, and individual node transaction verification and validation records exchange between peer-to-peer connections.
- Simulate the proposed blockchain permissionless network architecture for FSC traceability. We adopt this approach since blockchain is an open-source, decentralized distributed preservation platform employing intelligent digital contract functionality.
- At last, we discussed the privacy protection-related Blockchain-IoT-enabled implementation challenges, limitations, and issues in future open research directions.

The remainder of this chapter is structured as follows. Section 2 discusses the current concept of smart business agriculture and its features applied in the modern world; we also highlight the critical key mechanism of the supply chain in smart agriculture. Section 3 focuses on the current trend of IoT in the smart agriculture environment, and blockchain distributed ledger enables a solution (framework) for secure smart business agriculture. In addition, we discuss security measures as challenging aspects of the IoT in food safety and the agriculture supply chain. Section 4 presents a blockchain public permissionless-enabled secure architecture for smart agriculture supply chain traceability. Furthermore, we maintain crop production and forecasting information in the protected distributed ledger.

Moreover, we design and implement digital contracts for information traceability and define the event-based node transactions and storage in the distributed ledger. It also demonstrates the working operations of the proposed architecture, the process of adding new nodes, updates, transactions, and stakeholder participation verification and validation through sequence diagrams. We also presented critical blockchain-IoT-based system implementation challenges, limitations, and issues in Section 5. Finally, the conclusion appears in Section 6.

2. Smart Business Agriculture and Supply Chain Strategy

Smart agriculture is the technology that refers to IoT usage, including wireless sensors, location systems, remote monitoring and control operations systems, and AI approaches on farms. The main motivation is the robust quality and quantity of crop production while reducing human labor strength [21]. For smart business agriculture, the technologies that are involved in creating a successful business chain of events are categorized as follows: (i) precision irrigation and plant nutrition, (ii) management of climate and control monitoring, (iii) sensors (for water, irrigation, light, soil, temperature management), (iv) software platforms, (v) collaboration of GPS location systems or satellite-based location analysis, (vi) communication-based on a mobile connection, (vii) analytical and data optimization protocols.

The connectivity between all these technologies creates an IoT-smart business agriculture system—the communication mechanism connecting sensors and devices, resulting in a complex system that handles agriculture-based farming built on data received [22]. In comparison, farmers and competent business agricultural authorities can monitor the running operations on their farms and take strategic actions via controlling remotely. The lifecycle processes of smart agriculture and farming are discussed as follows [22, 23]:

- **Wireless Sensor-Based Data Collection:** In this process, the sensors are installed in different farm areas to gather field data and transmit control actions regarding temperature, soil, moisture, air, etc.
- **Examine and Analysis:** The system examines and analyzes the collected data, identifies potential problems in this process, and takes action according to the status of the monitoring process.
- **Monitoring and Decision:** The application and cyber-controlled human-managed platforms decide the actions that must be executed in this process.
- **Actions:** After the previous process is performed successfully, the actions on the sensor data are performed while analyzing the overall scenario, such as soil, air, moisture, and temperature. After that, the whole cycle starts again.

As a result, the automated smart agriculture and farming process is utilized with high precision and control monitoring. Eventually, this system is considerable and saves all resources used—lower levels of human resources, such as irrigation, fertilizers, energy, time, and cost, can be handled directly.

Supply chain management involves the entire movement of agricultural food products from a supplier to a distributor, which also records the flow of information related to food products between participating stakeholders and among supply chain processes with robust profitability [24, 25]. The supply chain management functions involve procuring raw food-related materials, developing food production, operations and controls, monitoring, marketing, finance, distribution, customer service, and representative services. The main objective of the supply chain network is to increase overall value and productivity [25]. The values are directly proportional to supply chain profitability, whereas profit is distinct between the total generated revenue from the supply chain and customer costs. The design and strategies of supply chain management are as follows:

- The structure of supply chain management decides which stages and activities are executed first;
- The selection of location and capital capacity for customer facilitation;
- One of the main aspects is to decide the food product that is to be made and the place where these can be preserved and
- The modes of transportation affect supply chain management (either positively or negatively), which also handles the source from where the information will be collected and when.

The supply chain cannot be reversed because of the nature of long-term project design, which takes in more cost so that the management decides on market account uncertainty [25].

3. Current IoT-Blockchain Solutions for Food Supply Chain

The IoT enables devices embedded with sensors to connect and communicate with different devices through the Internet [26]. In this scenario, devices may monitor and control real-time operations remotely. For instance, it can include sheds, pumps, tractors, weather stations, and digital computers to trace data directly. However, data from devices can help decision-makers by involving stakeholders and helping them make smarter and safer farming decisions; moreover, they can adapt more quickly to changing conditions [27]. The IoT-CPS can monitor fame infrastructure and conditions; by this, it frees up time, labor, and capital for investment, quality, and quantity of crops and forecasting, less human effort, increased food productivity, and smart supply chain traceability.

Blockchain distributed technology allows all parties to improve supply chain transparency and traceability and reduce administrative costs [28]. A blockchain supply chain allows participants to record prices, food-related information, quality, quantity, location, and certification more effectively to secure the food-related chain of records. The benefit of this recorded information within a blockchain distributed ledger increases the traceability of the FSC, whereas the losses are lower in the counterfeit and gray market [29]. Yet, it improves visibility and compliance over outsourced contracts and manufacturing and the strong position of the organization in the marketplace as a ledger of responsible manufacturing [28–30].

Many organizations are adopting and experimenting with blockchain IoT innovation nowadays. A public blockchain permissionless network in the FSC provides increased transparency, integrity, and provenance and reduces cost and risk across smart business agriculture [30]. So, IoT provides remote data collection, monitoring, and control facilities in the agricultural environment. This section mentions some potential benefits of applying the combination of Blockchain-IoT as follows:

- According to corporate standards, the technologies increase the transparency of the FSC;
- It calculates lower losses from the trading of the gray market;
- Enhance compliance and visibility over outsourced manufacturing contracts;
- Reduce administrative costs and increase the security of the FSC;
- It also strengthens credibility and provides a trustworthy environment while exchanging information;
- Secure participating stakeholder engagement in this distributed ledger; and
- Reduce the potential risk of FSC malpractice.

However, blockchain distributed technology provides transparency and provenance, reducing costs by enabling streamlined administrative processes and effective supply chain data auditing separately. These involved processes manually check compliance and purpose, preserve details in the distributed ledger, and accelerate through all relevant information.

3.1 Internet of Things (IoT) with 6G and Impact on Smart Precision Agriculture

The fifth-generation (5G) wireless network has initiated to spread world communities and is most expected to contribute to the increasing growth in the usage of IoT and its applications, including smart PA, transportation (e-transportation), healthcare (e-health), smart home, and city, etc. [31]. The IoT with 5G technology requires massive connectivity, a lower level of latency, a small chunk of time delay, and security. It is sufficient enough in many IoT devices to share and exchange different types of information in real-time environments [31, 32].

The ultra-low latency and security constraints promote the emergence of the sixth generation (6G) wireless networks, which support more robust functionality and faster transmission, lower latency compared with 5G, higher network capacity, and node security. Recently, the IoT with 6G promises to improve the effectiveness with no delay of new cyber-physical PAFSC traceability [32]. This collaborative technology enhances the performance of the FSC and production forecasting nodes' transactions in transparency, secure communication, validation and verification, transaction monitoring, and traceability. The technology enables wireless radio frequency (RF) channels, a contrasting family of cellular and WLAN channels. By this, the platform interoperability issue between IoT-5G vs. 6G can be achieved. It also helps maintain IoT security and privacy protocols and store information preservation in the ledger.

Still, it is essential to obtain an accurate channel to design an efficient communication pathway and network protocols to support smart PAFSC and production forecasting applications [33]. For instance, the signal processing matches industrial IoT channels, including the multi-input, multi-output, and coding of the corporative network, for example, where two nodes are used to initiate and exchange transaction information related to the PAFSC. It is observed that the system can maintain ongoing changes in the IoT paradigm (Fig. 2) with effective transactions between nodes, and it is testified by exchanging information between nodes in different environments [33].

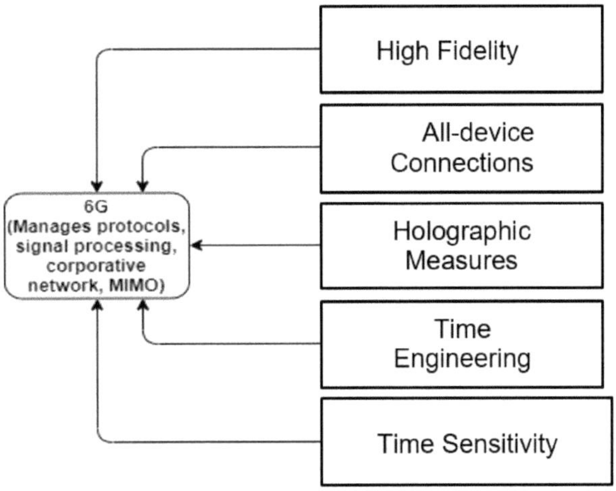

Fig. 2. Internet of Things (IoT) with 6G connectivity.

3.2 Applications of IoT with 6G

The race for the sixth-generation wireless network (6G) has started. Many experts are exploring and proposing several of the best frameworks, architectures, and state-of-the-art methods for the issues and limitations faced by previous researchers [34]. The 6G serves a wide range of new technologies, including hardware and software; this technology drives three main key factors that maintain the overall infrastructure of future networks: (i) high-fidelity holographic society, (ii) connectivity of all devices, and (iii) time engineering. High fidelity refers to those applications that simultaneously connect nature and make interactive communication, for example, holographic telepresence [34]. In contrast, the massive number of connected devices refers to connecting all IoT devices to exchange data. The communication occurs to or from the machines in the network, with little means to no human interactivity [34, 35]. Thus, applying time-engineering 6G is the most important in the latest IoT systems as this involves IIoT automation, autonomous applications, and others. Time-sensitive applications are significant in areas where people are not the data generators.

In the healthcare industry, IoT has played a vital role in providing an in-home care approach that relieves pressure by replacing the traditional reactive and hospital-centered system [35]. The increasing number of aging populations and chronically ill patients impact the current system, drastically increasing the demand for IoT for healthcare [36]. The present chapter highlighted and categorized the potential role of futuristic 6G in IoT healthcare and wireless mobile communication technology usage in this complex environment. One can improve living standards and the utilization of the vision of new technologies. The 6G IoT for healthcare is divided into two categories: (i) imperative use of human activity monitoring (for search and rescue) and (ii) assistive for the visually impaired (real-time detection for older people) [35, 36].

The increment in the number of vehicles on the road, especially the increase in the number of users in metropolitan cities, has also worsened issues and challenges [37]. The limitations and issues in transportation are as follows: (i) traffic jams, (ii) pollution, (iii) accidents, (iv) fuel consumption and scarcity, and (v) fuel and insurance costs. Many enhancements are aimed at sector productivity, carbon footprint, and safety; to improve these aspects, many experts have proposed the best solutions for making transportation systems more intelligent [38]. An intelligent system starts with smart infrastructure (highways and parking with intelligence), vehicle intelligence (AI-based reinforcement learning), and public transport intelligence [39].

A classification of e-transportation vehicles established on their capabilities, such as conventional vehicles, driver assistance, partial automation, conditional automation, high automation, and full automation, etc. [40, 41], are briefly discussed below:

- **Conventional Vehicles:** All tasks are performed by the driver;
- **Driver Assistance:** A few features assist the driver;
- **Partial Automation:** Automated features such as acceleration, etc., help the driver to monitor the complete environment;
- **Conditional Automation:** Automated, but the driver can take control at any time;
- **High Automation:** The vehicle performs the driver's task sufficiently. The driver has the choice to control or not; and
- **Full Automation:** In any condition, the vehicle performs all functions as the driver does.

4. Novel and Secure IoT-Blockchain Architecture for Agriculture Supply-Chain Traceability and Food Production Forecasting with Records Preservation

This chapter proposes a Blockchain-IoT-enabled PA secure architecture for FSC traceability and production forecasting. We used a public network structure as a blockchain distributed system in this proposed architecture; the permissionless architecture allows the food authority to manage forecasting information in a protected manner and store records in the distributed immutable ledger. This proposed architecture creates off-chain stakeholders' communication, allowing participants to establish a production file. Individuals can create files and record information for each section, such as farmers making personnel files based on products along with quantity, the warehouse being the processing unit, and processing links between other stakeholders.

The production and processing unit is responsible for the information exchange between nodes and transactions regarding production, which is gathered per the specifications and standards (regulatory). The unique identification via barcodes transfers the information to the distributor stakeholder. The distributor transfers products to the retailers, as in Fig. 3.

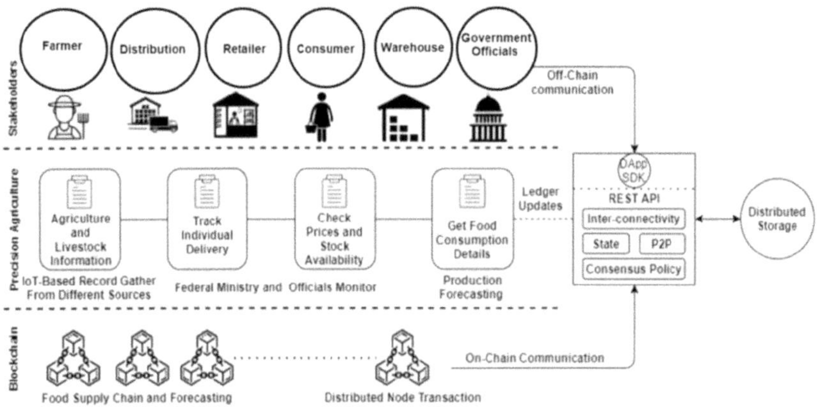

Fig. 3. The proposed IoT blockchain enabled novel and secure architecture for agriculture supply-chain traceability and preservation of food production forecasting records.

Using the IoT, one can deliver products from amid stakeholders while tracking each truck with the help of GPS. After products are transported and have reached the desired destination, the delivery is stored in the distributed ledger, which is immutable. FSC traceability is developed in the processing and product and preserved through the barcodes. In this scenario, these barcode factors are used to track location, product identity, history of transportation and preservation, delivery, expiry, and other relevant and related information of the individual products. That is completely reliable and makes it easy to trace and access records.

The blockchain distributed ledger maintains these collected records in the storage, a cryptographic hash encrypted function. We proposed an on-chain permissionless public peer-to-peer (P2P) communication network to secure and verify stakeholder registration and digital signature (a pair of keys) features. Blockchain digital contract (written in a programming language) is designed and developed to automate the FSC between parties without the involvement of any intermediary human, as shown through Contract 1.

As a tool, the blockchain composers allow building and deploying programable digital contracts and defending the node execution of the FSC, exchanging information between connected stakeholders, as shown in Fig. 4.

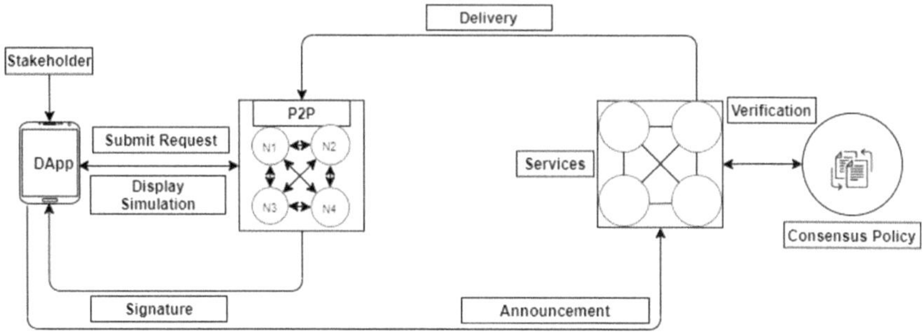

Fig. 4. The node transactions of supply-chain and agriculture-business forecasting execution.

In this architecture, only authorized stakeholders interact with this system. The person who authorizes stakeholders is the engineer of the Ministry of Food Authority. Stakeholders interact with this public network with the REST, as in Fig. 3. Moreover, this permissionless infrastructure contains different types of distributed simulation and services, including consensus verification, stakeholder announcement, execution of transaction scheduling, and more services related to digital contracts. In addition, the public network consists of several nodes, and an individual node holds an image of the information, which is immutable regarding chronological chain order.

4.1 Smart Contract

For secure FSC traceability and production forecasting record preservation, the node of supply chain information hash protected (SH-256) transaction with `trackID( )` is deployed, as shown in Contract 1. The Ministry of Food Authority initiates the system and manages addresses. A new event transaction related to FSC traceability and production forecasting details records can be added to the distributed ledger via (`CurrTransaction()` and `PreTransaction()`). We designed and implemented the digital contracts for participating stakeholder registration (`stakeholderReg()`), preserved food supply chain traceability, and updated production forecasting information in the digital ledger.

All these related transactions and preserved forecasting details are recorded and updated by Precision-Agri-Ledger in accordance with the digital contract and customized blockchain permissionless consensus policy. The `addNode transaction()` function adds a new node and executes a record update (timestamp), where the addition of node transactions and changes are shared between consensus in accordance. The digital contract preserves generic information (supply chain traceability details and production forecasting information) such as `tID`, `stakeholderReg()`, `CurrTransaction()`, `PreTransaction()`, `countTransaction()`, and `timestamp`. Once the FSC traceability and forecasting node is added to the distributed storage, the

Table 1. Precision-Agri-Ledger digital contract

Contract #1: Digital Contract for Food Supply Chain Traceability and Production Forecasting Preservation

```
   Input Constraints:"Precision-Agri-Ledger" Ministry of Food Authority
initiates and addresses the food supply chain traceability and
production forecasting record

   Tx. Int main(),
   tracking ID,(tID);
   Stakeholders of Food Supply Chain Registration and Node Transactions;

   access detail of production forecasting,
   (stakeholderReg());
   current transactions,
   previous transaction,
   (CurrTransaction);
   add new node and update ledger,
   (PreTransaction);
   timestamp(execute);

   supply chain addNode() contract (addNode transactions)
   precision-agri-ledger update contract (addUpdate transactions)
   Node transactions counter and update ledger (countTransaction)

   Food Authority (FA) is responsible for setting individual addresses
and authorize

   if Tx.int main() is FA
   then
   if node does not exist
   then
   state of counter changes, abd add new node,
   and record on the ledger;
   stakeholderReg(), CurrTransaction, PreTransaction, timestamp,
   also, send updated records notification in the
   chain ofstakeholders;
   update Precision-Agri-Ledger;
   Count is increment by unit (1);
   else
   check state errors and generate notification
   and rollback;
   break;
   system terminates;
   else
    check state errors and generate notification
       and rollback;
       break;
   system terminates;
```

blockchain public network can initiate updated transaction records from the Ministry of Food authority in the ledger and addresses among them, as shown in Contract 1.

4.2 Proposed IoT-Blockchain-enabled Food Supply-Chain Verification and Validation

IoT, conventional devices, and blockchain technology aim to build an autonomous and intelligent system for the traceability of the FSC. In a distributed environment, building trust is a complex task. For this purpose, we integrate IoT with blockchain in the FSC and production forecasting, which provides benefits in automation and preservation and saves time. The main function involved in the integration process is the registration of devices (stakeholders), role generation, allocation, and verification.

Firstly, registering the IoT-based devices with the distributed permissionless public network is shown in Fig. 5. The stakeholder service engineer (food authority engineer) verifies and registers devices. Each gets a unique identity (ID) and easily communicates with the system. By this, participating stakeholders can specify the required information, including a tracking ID, systems IDs, related information, and participation; after that, it allows them to submit a proposal of transactions per the digital contract. The blockchain-connected node network executes each transaction bid and adds the new transaction to the distributed storage. During this process, events are emitted from the nodes to communicate and inform that transactions are successfully delivered and recorded (ledger) or not among stakeholders.

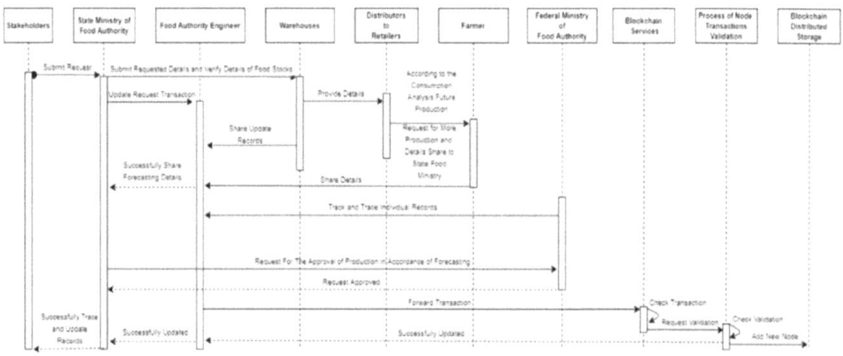

Fig. 5. Process verification and validation through sequence diagram.

5. IoT-Blockchain Related Challenges and Limitations in Precision Agriculture Supply Chain Traceability and Forecasting

A permissionless blockchain stores food production forecasting chain-of-information on a secure "Precision-Agri-Ledger" in which the involved stakeholders are registered to interact with this distributed ledger directly because of the nature of the public network. All connected nodes in this network are hash-encrypted (SH-256), as discussed in Section 4, ensuring more security where participating stakeholders publicly keep access to the ledger [42]. The distributed ledger with cryptographic hash encryption guarantees the FSC transactions and forecasting information protection in the IoT-PA environment, which cannot be tampered with and forged chain-of-information [43]. The immutable nature of blockchain makes agricultural information preservation more effective and maintains integrity and transparency between stakeholders with provenance.

The blockchain-enabled PAFSC traceability solution provides a well-organized platform for real-time transaction monitoring. Searching, selecting, and adapting the blockchain version is not easy, and most probably, blockchain-enabled public distributed ledger solutions cannot be implemented fully or adequately [44]. Scalability in the blockchain is another type of emerging challenge. More than that, many organizations are still facing the implementation of consortium systems, computational energy, and cost of operational security to enable the organization and well management of industrial IoT in PAFSC traceability [45]. The existing FSC monitoring systems and the process of production forecasting legacy are inefficient and preserve a central-based storage system, which causes more issues related to tampering, forgery, alteration, etc., and requires more maintenance costs. After applying blockchain permissionless public ledger architecture, the operation cost is reduced drastically, increasing the security of FSC trackability, maintaining the storage of forecasting-related chain-of-information records, and managing scalability and sustainability.

However, the Ministry of Agriculture and Health authority is responsible for maintaining the quality of food productivity and supply chain management; the higher authority must consider and develop different policies related to agriculture and livestock preservation under the need and consumption [46]. For this purpose, the IoT-Blockchain security implications provide a secure platform for FSC traceability and record production forecasting details in an immutable ledger, where ownership can monitor individual transactions and node deliverability. The authority of agriculture controls, systems, and supply chain stakeholders must genuinely collaborate with the existing regulatory and privacy concerns, such as real-time American Farm Bureau (AFB), for record traceability and platform interoperability [47]. The IoT-PA development enables the supply chain traceability and secure communication ecosystem to update the objectives of authority and administration.

6. Precision Agriculture, Blockchain, and IoT Ventures in Latin America (LA)

Communication and power management/harvesting relying on Wireless Sensor Networks (WSNs) must be efficient, as discussed in Ref. [48]. Technical challenges must be solved, focusing on enabling mobility and portability while reducing power consumption, latency, and acquired data dependability regarding sensor nodes to meet these criteria. In an irrigated cropland assessment study, soil moisture is measured using a WSN based on these fundamentals. This research shows and debates the elements that impact network stability and the quality of data measurements when the weather is unfavorable. LA's agroindustry is appalling and risky [49, 50]. Water, fertilizer, and arable land are among the most critical resources that must be conserved to produce more food. Farmers now, more than ever, need specific technology instruments to accomplish this aim. PA is contingent on improved agribusiness through technology. Still, there is a huge exigency for global technological ameliorations to reinforce agricultural yields. In [49], the authors systematically assessed the literature in this study and looked at several papers that advocate process improvement in agriculture. Proposals from many scholars at various latitudes helped develop a background on current methods, emphasizing irrigation optimization and pest and disease control [49–52]. Just little applied research has been done on how to increase agricultural productivity. They do this by controlling the many variables that may impact it. Studies on LA are few and far between. As a result, the LA agroindustry still has much room for evolution.

Due to its multidisciplinary constituency, Agriculture 4.0 has found infrastructural problems in handling the CPS paradigm [53–56] in general. When it comes to PA, drones have helped some vision-based frameworks, and embedded electronics have all contributed to a considerable increase in PA in recent years. Services and ways exist to help farmers utilize this new technology, such as cloud-based video systems, drones, and software. Contrary to what was alleged beforehand, these pieces of machinery are expensive and difficult to get for most farmers in Latin America (and other developing nations). A low-cost "unmanned aerial system" (UAS) envisioned for PA made up of three subsystems has been suggested in certain research. These include drones, multispectral imaging systems (that can rely on open hardware), and open-source software accountable for calculating vital information for farmers. There are four key areas of focus in our effort to build an effective drone: (1) the drone's construction; (2) its vision algorithm implementation; (3) its autonomous trajectory; and (4) its low-cost multispectral system with capabilities to handle lots of bands [49, 57, 58]. For illustration, a quadrotor's imaging structure consisting of two low-cost cameras can be amended to afford multispectral images. With UASs, it is possible to evaluate vegetation indices and other farmers' relevant data consumption/use/need. Using professional multispectral cameras and related software, one may test the competitiveness of such systems against their economic worth. Agricultural CPSs may also leverage open-source hardware and software to give an inexpensive tool for precision agriculture.

Blockchain technology implementations for Agriculture 4.0 are scarce. It is extremely difficult to obtain reports about the real Latin American scenario regarding traceability (informality is immense) and the generation of new types of rural revenue besides environmental studies employing AI in agricultural settings [59–63].

7. Conclusion

This chapter discusses the preliminary Blockchain public permissionless network-enabled IoT-based agriculture, food supply chain (FSC) and production forecasting, information preservation, confidentiality, and security-related caveats and challenges. The proposed "Precision-Agri-Ledger" enables (i) a secure architecture design meant for FSC traceability besides (ii) protected forecasting of production information in the distributed ledger. Given its decentralized and edgy design, it can help build a robust agriculture 4.0 foundation. Likewise, the Precision-Agri-Ledger interconnects different external IoT devices, encompassing several communication protocols for secure data handover between connected participating stakeholders, analysis variation in different node transactional data, and secure records in the blockchain distributed preservation.

Moreover, the main concern of this chapter is to secure a smart track-and-trace along with the overall FSC process and their related information and securely store the production forecasting artifacts in the complex IoT-smart agriculture environment. This architecture may add a new node to the permissionless network transactions and bring them up to date with the contract following the approval of the consensus. The (a) verification/validation process for an individual node transaction, in addition to (b) links between chain-of-information (i.e., production forecasting) elements, are maintained through the distributed application (DApp) by a blockchain engineer. The proposed architecture manages record integrity, transparency, provenance, immutability, and traceability for multifaceted IoT FSC environments. These traits are paramount to sanitizing and building bullet-proof agrarian structures in Latin America.

Authors Contribution

- Abdullah A. Khan has written the original draft and preparation,
- All authors have reviewed, rewrote, performed part of the literature survey, and edited,
- Mazhar Ali Dootio, Awais Khan Jumani, Mudassir Hussain, and Rukhsar Ayub have investigated and designed the architecture and explored software tools.

All authors read and agreed to the published version of this chapter.

Conflict of Interest

The authors declare no conflict of interest.

References

1. Pranto, T.H., Noman, A.A., Mahmud, A. and Haque, A.B. (2021). Blockchain and smart contract for IoT enabled smart agriculture. *Peer J. Computer Science*, 7, e407.
2. Singh, P. and Singh, N. (2020). Blockchain with IoT and AI: A review of agriculture and healthcare. *International J. Applied Evolutionary Computation* (*IJAEC*), 11(4), 13–27.

3. Waheed, U., Mansoor, Y. and Khan, M.A. (2021). BIoT-based smart agriculture: Food and crops efficiency and improvement in supply chain cycle. *Blockchain Technology for IoT Applications*, 173.
4. Khan, P.W., Byun, Y.C. and Park, N. (2020). IoT-blockchain enabled optimized provenance system for food industry 4.0 using advanced deep learning. *Sensors*, 20(10), 2990.
5. Khan, A.A., Laghari, A.A. and Awan, S.A. (2021). Machine Learning in Computer Vision: A review. *ICST Transactions on Scalable Information Systems*, 8(32).
6. Torky, M. and Hassanein, A.E. (2020). Integrating blockchain and the Internet of Things in precision agriculture: Analysis, opportunities, and challenges. *Computers and Electronics in Agriculture*, 105476.
7. Lee, S.W., Singh, I. and Mohammadian, M. (Eds.) (2021). *Blockchain Technology for IoT Applications*. Springer Nature.
8. Mistry, I., Tanwar, S., Tyagi, S. and Kumar, N. (2020). Blockchain for 5G-enabled IoT for industrial automation: A systematic review, solutions, and challenges. *Mechanical Systems and Signal Processing*, 135, 106382.
9. Khan, A.A., Shaikh, A.A., Cheikhrouhou, O., Laghari, A.A., Rashid, M. et al. (2021). IMG forensics: Multimedia enabled information hiding investigation using convolutional neural network. *IET Image Processing*. Wiley. doi: 10.1049/pr2.12272.
10. Dey, K. and Shekhawat, U. (2021). Blockchain for sustainable e-agriculture: Literature review, architecture for data management, and implications. *J. Cleaner Production*, 128254.
11. Bodkhe, U., Tanwar, S., Bhattacharya, P. and Kumar, N. (2020). Blockchain for precision irrigation: Opportunities and challenges. *Transactions on Emerging Telecommunications Technologies*, 33(1), e4059.
12. Khan, A.A., Shaikh, Z.A., Laghari, A.A., Bourouis, S., Wagan, A.A. and Ali, G.A.A.A. (2021). Blockchain-aware distributed dynamic monitoring: A smart contract for fog-based drone management in land surface changes. *Atmosphere*, 12(11), 1525.
13. Yadav, S., Garg, D. and Luthra, S. (2020, July). Development of IoT based data-driven agriculture supply chain performance measurement framework. *J. Enterprise Information Management*, 34(1), 292–327.
14. Borah, M.D., Naik, V.B., Patgiri, R., Bhargav, A., Phukan, B. and Basani, S.G. (2020). Supply chain management in agriculture using blockchain and IoT. *In: Advanced Applications of Blockchain Technology*, 227–242. Singapore: Springer.
15. Misra, N.N., Dixit, Y., Al-Mallahi, A., Bhullar, M.S., Upadhyay, R. and Martynenko, A. (2020). IoT, big data and artificial intelligence in agriculture and food industry. *IEEE Internet of Things Journal,* 99, 1–1.
16. Awan, S., Ahmed, S., Ullah, F., Nawaz, A., Khan, A. et al. (2021). IoT with BlockChain: A futuristic approach in agriculture and food supply chain. *Wireless Communications and Mobile Computing*, 2021, 1–14.
17. Sajoy, P.B. (2021). Emerging trends in the use of IoT in agriculture and food supply chain management: A theoretical analysis. *Turkish J. Computer and Mathematics Education* (*TURCOMAT*), 12(3), 3293–3297.
18. Khan, A.A., Laghari, A.A., Shaikh, A.A., Bourouis, S., Mamlouk, A.M. and Alshazly H. (2021). Educational blockchain: A secure degree attestation and verification traceability architecture for higher education commission. *Applied Sciences*, 11(22), 10917.
19. Khan, A.A., Laghari, A.A., Awan, S. and Jumani, A.K. (2021). Fourth industrial

revolution application: Network forensics cloud security issues. *In*: *Security Issues and Privacy Concerns in Industry 4.0 Applications*, 15–33.
20. Yadav, S., Luthra, S. and Garg, D. (2020). Internet of Things (IoT) based coordination system in Agri-food supply chain: Development of an efficient framework using DEMATEL-ISM. *Operations Management Research*, 1–27.
21. Farooq, M.S., Riaz, S., Abid, A., Umer, T. and Zikria, Y.B. (2020). Role of IoT technology in agriculture: A systematic literature review. *Electronics*, 9(2), 319.
22. Kassim, M.R.M. (2020). IoT applications in smart agriculture: Issues and challenges. *Proc. 2020 IEEE Conference on Open Systems* (*ICOS*), 19–24. IEEE.
23. Khan, A.A., Laghari, A.A., Liu, D.S., Shaikh, A.A., Ma, D.A., Wang, C.Y. et al. (2021). EPS-ledger: Blockchain hyperledger sawtooth-enabled distributed power systems chain of operation and control node privacy and security. *Electronics*, 10(19), 2395.
24. Wieland, A. (2021). Dancing the supply chain: Toward transformative supply chain management. *J. Supply Chain Management*, 57(1), 58–73.
25. Craighead, C.W., Ketchen, Jr. D.J. and Darby, J.L. (2020). Pandemics and supply chain management research: Toward a theoretical toolbox. *Decision Sciences*, 51(4), 838–866.
26. Laghari, A.A., Wu, K., Laghari, R.A., Ali, M. and Khan, A.A. (2021). A review and state of art of Internet of Things (IoT). *Archives of Computational Methods in Engineering*, 1–19.
27. Hajjaji, Y., Boulila, W., Farah, I.R., Romdhani, I. and Hussain, A. (2021). Big data and IoT-based applications in smart environments: A systematic review. *Computer Science Review*, 39, 100318.
28. Khan, A.A., Shaikh, Z.A., Baitenova, L., Mutaliyeva, L., Moiseev, N. et al. (2021). QoS-Ledger: Smart contracts and metaheuristic for secure quality-of-service and cost-efficient scheduling of medical-data processing. *Electronics*, 10(24), 3083.
29. Kouhizadeh, M., Saberi, S. and Sarkis, J. (2021). Blockchain technology and the sustainable supply chain: Theoretically exploring adoption barriers. *International Journal of Production Economics*, 231, 107831.
30. Wamba, S.F., Queiroz, M.M. and Trinchera, L. (2020). Dynamics between blockchain adoption determinants and supply chain performance: An empirical investigation. *International Journal of Production Economics*, 229, 107791.
31. Barakat, B., Taha, A., Samson, R., Steponenaite, A., Ansari, S. et al. (2021). 6G opportunities arising from Internet of Things use cases: A review paper. *Future Internet*, 13(6), 159.
32. Sodhro, A.H., Pirbhulal, S., Luo, Z., Muhammad, K. and Zahid, N.Z. (2020). Toward 6G architecture for energy-efficient communication in IoT-enabled smart automation systems. *IEEE Internet of Things Journal*, 8(7), 5141–5148.
33. López, O.L., Alves, H., Souza, R.D., Montejo-Sánchez, S., Fernández, E.M.G. and Latva-Aho, M. (2021). Massive wireless energy transfer: Enabling sustainable IoT toward 6G era. *IEEE Internet of Things Journal*, 8(11), 8816–8835.
34. Li, T., Liu, W., Zeng, Z. and Xiong, N.N. (2021). DRLR: A deep reinforcement learning based recruitment scheme for massive data collections in 6G-based IoT networks. *IEEE Internet of Things Journal*, 99, 1–1.
35. Nayak, S. and Patgiri, R. (2021). 6G communication technology: A vision on intelligent healthcare. *In*: *Health Informatics: A Computational Perspective in Healthcare*, 1–18. Springer, Singapore.

36. Nguyen, D.C., Ding, M., Pathirana, P.N., Seneviratne, A., Li, J. et al. (2021). 6G Internet of Things: A comprehensive survey. *IEEE Internet of Things Journal*, 9(1), 359–383.
37. Zhou, Z., Wang, M., Huang, J., Lin, S. and Lv, Z. (2021). Blockchain in big data security for intelligent transportation with 6G. *IEEE Transactions on Intelligent Transportation Systems*, 23(7).
38. Suresh Kumar, K., Radha Mani, A.S., Sundaresan, S. and Ananth Kumar, T. (2021). Modeling of VANET for future generation transportation system through edge/fog/cloud computing powered by 6G. *Cloud and IoT-Based Vehicular Ad Hoc Networks*, 105–124.
39. Hong, T. and Qi, F. (2021). Applications and Implementations of 6G Internet of Things. *6G Wireless Communications and Mobile Networking*, 208.
40. Ferrag, M.A., Shu, L., Yang, X., Derhab, A. and Maglaras, L. (2020). Security and privacy for green IoT-based agriculture: Review, blockchain solutions, and challenges. *IEEE Access*, 8, 32031–32053.
41. Mirabelli, G. and Solina, V. (2020). Blockchain and agricultural supply chains traceability: Research trends and future challenges. *Procedia Manufacturing*, 42, 414–421.
42. Zhang, X., Cao, Z. and Dong, W. (2020). Overview of edge computing in the agricultural Internet of things: Key technologies, applications, challenges. *IEEE Access*, 8, 141748 –141761.
43. Niknejad, N., Ismail, W., Bahari, M., Hendradi, R. and Salleh, A.Z. (2020). Mapping the research trends on blockchain technology in food and agriculture industry: A bibliometric analysis. *Environmental Technology & Innovation*, 101272.
44. Feng, H., Wang, X., Duan, Y., Zhang, J. and Zhang, X. (2020). Applying blockchain technology to improve agri-food traceability: A review of development methods, benefits and challenges. *J. Cleaner Production*, 260, 121031.
45. Dutta, P., Choi, T.M., Somani, S. and Butala, R. (2020). Blockchain technology in supply chain operations: Applications, challenges and research opportunities. *Transportation Research Part E: Logistics and Transportation Review*, 142, 102067.
46. Wamba, S.F. and Queiroz, M.M. (2020). Blockchain in the operations and supply chain management: Benefits, challenges and future research opportunities. *Elsevier International Journal of Information Management*, 52, 120–139.
47. Vangala, A., Das, A.K., Kumar, N. and Alazab, M. (2020). Smart secure sensing for IoT-based agriculture: Blockchain perspective. *IEEE Sensors Journal*, 21(16), 17591–17607.
48. Franciscatto, R., Shubeita, F.D., Marcon, C., Vit, A.R. and Spohn, M.A. (2012). Evaluation of a wireless sensor network applied to precision agriculture under adverse weather conditions. *Iberoamerican Journal of Applied Computing*, 2(1), 19–29.
49. Brenes, J.A. and Raventós, G.M. (2020). Research on optimization techniques applied to precision agricultural processes supported by technology in Latin America: Opportunities and a route to follow. *2020 XLVI Latin American Computing Conference (CLEI)*, 462–470.
50. Fernandez-Ordonez, Y.M. and Soria-Ruiz, J. (2017). Maize crop yield estimation with remote sensing and empirical models. *Proc. 2017 IEEE International Geoscience and Remote Sensing Symposium (IGARSS)*, 3035–3038.
51. Razmjooy, N. and Estrela, V.V. (2019). *Applications of Image Processing and Soft Computing Systems in Agriculture*. 1-300, Hershey, PA, USA: IGI Global. doi:10.4018/978-1-5225-8027-0.

52. Giacomin, J.C., Vasconcelos, F.H. and da Silva, E.J. (2007). Estimating vegetation water content with wireless sensor network communication signals. *Proc. 2007 IEEE Instrumentation & Measurement Technology Conference (IMTC)*, May.
53. Estrela, V.V., Saotome, O., Loschi, H.J., Hemanth, D.J., Farfan, W.S. et al. (2018). Emergency response cyber-physical framework for landslide avoidance with sustainable electronics. *Technologies*, 6, 42.
54. Faiçal, B.S., Freitas, H., Gomes, P.H., Mano, L.Y., Pessin, G., de Carvalho, A.C. et al. (2017). An adaptive approach for UAV-based pesticide spraying in dynamic environments. *Computers and Electronics in Agriculture*, 138, 210–223.
55. Estrela, V.V., Hemanth, J., Saotome O., Nikolakopoulos, G. and Sabatini R. (2020). Introduction to advances in UAV avionics for imaging and sensing. *In*: Estrela, V.V. et al. (Eds.), *Imaging and Sensing for Unmanned Aircraft Systems*, 1(1), 1–22. London, UK: IET. doi: 10.1049/PBCE120F_ch1.
56. Figueiredo, H.V., Saotome, O., Shiguemori, E.H., Silva Filho, P. and Estrela, V.V. (2020). Image database of low altitude UAV flights with flight condition-logged for photogrammetry, remote sensing, and computer vision. *In*: Estrela, V.V. et al. (Eds.), *Imaging and Sensing for Unmanned Aircraft Systems*, 2(5), 91–112. London, UK: IET. doi: 10.1049/PBCE120G_ch5.
57. Aroma, R.J., Raimond, K., Razmjooy, N., Estrela, V.V. and Hemanth, J. (2020). Multispectral vs. hyperspectral imaging for unmanned aerial vehicles: Current and prospective state of affairs. *In*: Estrela, V.V. et al. (Eds.), *Imaging and Sensing for Unmanned Aircraft Systems*, 2(7), 133–156. London, UK: IET. doi: 10.1049/PBCE120G_ch7.
58. Khan, A.A., Laghari, A.A, Shaikh, A., Dootio, M., Estrela, V.V. and Lopes, R.T. (2021). A Blockchain security module for brain-computer interface (BCI) with multimedia life cycle framework (MLCF). *Neuroscience Informatics*. doi: 100030. 10.1016/j.neuri.2021.100030.
59. Vargas, H.M. and Bomfim, H.M. (2014). La Geografía Latinoamericana y la Unión Geográfica Internacional (UGI): los casos de Brasil (1956) y México (1966). *J. Latin American Geography*, 13, 215–232.
60. Abreu, L.M., Hora, H.R.M., Rangel, J.J.A., Erthal, Jr. M., Razmjooy, N. et al. (2020). A multi-criteria modeling for ranking CO_2 emitting G20 countries from the Kaya Identity and their impacts on elderly health. *In*: Iano Y. et al. (Eds.), *Proc. of the 5th Brazilian Technology Symposium (BTSym'19). BTSym 2019. Smart Innovation, Systems and Technologies*. Springer, Cham, doi: 10.1007/978-3-030-16053-1_62.
61. Bocco, G. and Napoletano, B.M. (2017). The prospects of terrace agriculture as an adaptation to climate change in Latin America. *Geography Compass*, 11.
62. Faúndez-Ugalde, A., Mellado-Silva, R. and Aldunate-Lizana, E. (2020). Use of artificial intelligence by tax administrations: An analysis regarding taxpayers' rights in Latin American countries. *Comput. Law Secur. Rev.*, 38, 105441.
63. Stepanyan, A.Z. and Zaplatina, T.S. (2020). Legal regulation of robots and artificial intelligence in Latin America, the problem of human rights and AI. *Lex Russica*, 73(7).

CHAPTER

4

Constructed Wetlands: A Sustainable Technology for Wastewater Treatment and Reuse in 21st Century Agriculture

Gabriela Petroceli Mota[1*][0009-0001-9744-1173], Glacielen Ribeiro de Souza[2][0009-0003-0773-1092], Gabriel Quintanilha-Peixoto[3][0000-0002-7476-5845], Analyse Villanueva Gaete[2][0000-0003-1471-7443], Vania V. Estrela[4][0000-0002-4465-7691] and Aline Chaves Intorne[2,5*][0000-0001-8015-6926]

[1] Laboratory of Cellular and Tissue Biology, Universidade Estadual do Norte Fluminense Darcy Ribeiro, Campos dos Goytacazes, RJ, 28013-602, Brazil

[2] Laboratory of Biotechnology, Universidade Estadual do Norte Fluminense Darcy Ribeiro, Campos dos Goytacazes, RJ, 28013-602, Brazil

[3] Laboratory of Chemistry and Function of Proteins and Peptides, Universidade Estadual do Norte Fluminense Darcy Ribeiro, Campos dos Goytacazes, RJ, 28013-602, Brazil

[4] Department of Telecommunications, Universidade Federal Fluminense, Niterói, RJ, 24220-900, Brazil

[5] Laboratory of Chemistry and Biology, Instituto Federal de Educação, Ciência e Tecnologia do Rio de Janeiro, Volta Redonda, RJ, 27213-100, Brazil

The population increase and the consequent demand for food have resulted in water pollution and 25% of the world population is being affected by the water deficit. South America is the region with the greatest availability of freshwater. However, severe drought and water rationing periods are increasingly frequent. The agricultural sector is the largest water consumer, requiring the application of innovative technologies to optimize the use of the resource. Using wastewater is an alternative to reduce consumption and ensure sustainable use. In this context, biotechnological possibilities, e.g., constructed wetlands, stand out for their energy efficiency and economic and socio-environmental feasibility compared to conventional treatments. The chapter reviews the biotechnology involved in water reuse in the Agriculture 4.0 (Agri4.0 or A4) era, explores the importance of effluent treatment by phytoremediation, and presents the feasibility of harnessing

[*] Corresponding author: gabrielapetroceli@gmail.com; aline.intorne@ifrj.edu.br

water from wetlands' built-in agricultural environments. This study's importance hinges on two thought-provoking notions: A4 information technology and environmental management (sustainable growth).

1. Introduction

Food and energy come from water and other crucial living organisms [1, 2]. Climate change, population growth, and water consumption reduce freshwater availability [3]. At least one month a year, 3.6 billion people face water shortages [4]. The UN predicts that by 2050, the global population will reach 11 billion. Thus, water supply pressure will rise further (Fig. 1) [5].

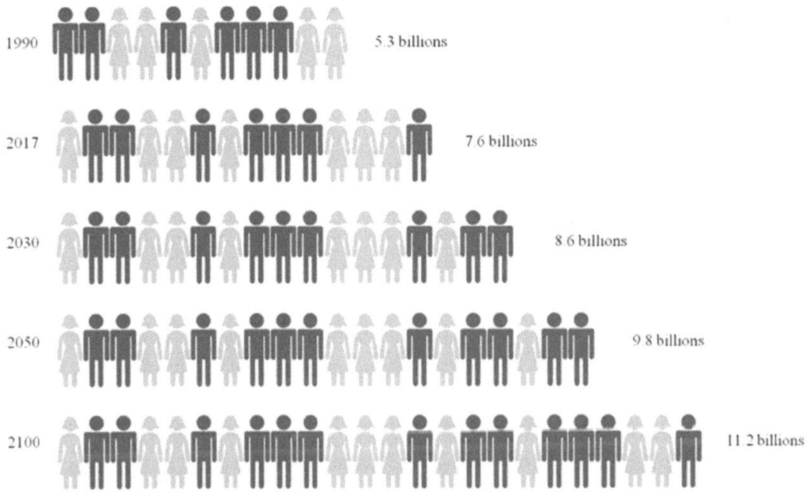

Fig. 1. World population projection until 2100. (*Source*: Adapted from UN 2017)

Water shortages are common as demand for drinking water rises [1]. South America (SA), with 26% of the world's freshwater but just 6% of the population, has experienced severe drought and water rationing in several countries. "Latin America" (LA) drinking water inequality affects 30 million people [6]. The Food and Agriculture Organization (FAO) and the World Water Council (WWC) have warned that overuse, climate change, and environmental degradation could reduce freshwater supply, especially in developing nations [7].

Evaporation and precipitation determine water resources' (WR) availability for individuals and ecosystems [8]. Climate change can disrupt the hydrological cycle and cause drought [9]. Climate projections show variations in precipitation, evapotranspiration, surface runoff, and river basin flow in SA [10]. Extreme hydrological occurrences have harmed Brazil. Studies predict an increase in the country's mean temperature and drought intensity and frequency [11]. Pollution will also affect "water quality" (WQ) in the future [12].

Worryingly low WR quality is a concern [13]. Water use climbed 600% in 100 years. Population expansion and industrial development will increase water demand for decades [14]. Anthropogenic inputs also induce water eutrophication by releasing substantial amounts of organic effluent [13].

Human activities' contaminants' sources include domestic, industrial, and agricultural effluents [15]. Industries consume large water amounts and employ varying toxicity levels in production processes. When untreated, the discharge of these effluents into water bodies represents a crucial pollution factor [16]. In Brazil, untreated effluent disposal is an environmental crime as per the National Environmental Council resolution (CONAMA 430/2011) and Federal Law 9.605/1998 [17, 18].

Due to low irrigation technology, agriculture utilizes 70% of freshwater from rivers, lakes, and aquifers worldwide [19]. Pesticides and artificial fertilizers have intensified farming to fulfill population needs [20]. These substances are often overused. Rainwater and irrigation leak leftovers to waterways, harming aquatic ecosystems [21]. Overpopulation of algae and plants, foul smell, turbidity, oxygen availability reduction (hypoxia), animal mortality, and eutrophication reduce biodiversity [22, 23]. Nitrogen fertilizers pollute surface and groundwater the most [24]. Thus, freshwater supply and quality are 21st-century challenges [25].

The 17 "Sustainable Development Goals" (SDGs) proposed in 2015 support a planet's sustainability agenda. Among the 169 goals in the 2030 Agenda, agriculture must redefine its use, conservation, and environmental protection. SDG 2 ensures food security and eradicates hunger, highlighting sustainable and resilient food production chain needs [19]. Concurrently, ODS 6 regards improving WQ, reducing pollution, eliminating effluent discharges, and minimizing hazardous chemicals and materials released to halve the proportion of untreated wastewater and substantially increase its recycling and safe reuse.

Industry 4.0 (I4.0 or I4) technologies (I4Ts) can effectively support digital transformations toward UN SDGs [26]. I4.0 adoption can help boost advanced manufacturing/processing capabilities that meet SDGs. I4.0 has become a wide-ranging research topic and brought progress in digital technologies, ultimately changing traditional agrarian practices. However, I4Ts pose challenges like novelty, corporations facing skills gaps, financial limitations, and operational I4.0 project difficulties. These caveats become overwhelming within rural contexts. "Cyber-physical systems" (CPSs), an emerging framework and paradigm to control physical and virtual systems via digital tools built on "information and communication technologies" (ICT) [26–28], can respond to intelligent wastewater and reuse management needs. CPSs' key technologies comprise sensors, actuators, wireless communications/networking, hydrodynamic modeling, data exploration, and control to assist hydric CPSs in sensing and working with water treatment plants.

Hence, "Nature-Based Solutions" (NbS) promise alternatives for WR conservation, ecological restoration, and sustainable agricultural water use [29]. The International Union defines NbS for the Conservation of Nature

(IUCN) as "actions to protect, sustainably manage, and restore natural or modified ecosystems that address societal challenges effectively and adaptively, simultaneously providing human well-being and biodiversity benefits" [30]. Therefore, "wastewater treatment" (WT) and reuse technologies often show potential to manage water availability and demand in agriculture [31], and, when added to biotechnological advances brought by Microbiology, Molecular Biology, and Bioinformatics, NbScan foster more sustainable Agri4.0.

This study's theoretical relevance focuses on two thought-provoking concepts: the information technology domain (Agri4.0), and eco-friendly management (sustainable development). Herein, constructed wetlands (CWs) emerge as a WT alternative and explore the application of wetland-treated water to foster the circular economy (CE), increasing quality and productivity.

2. More Sustainable Agriculture Reusing Water

Large wastewater volumes from homes and industries are generated daily. Europe treats 71% of wastewater. LA remediates around only 20% [32]. These waters generally contain 99% water and 1% suspended, colloidal, and dissolved solids [33]. They can hold organic matter, nutrients (mainly nitrogen and phosphorus), metals, emerging contaminants (antibiotics, hormones, personal care products, pesticides, polycyclic aromatic hydrocarbons, phenolic compounds, and volatile organic compounds), and pathogenic microorganisms (viruses, antibiotic-susceptible, and resistant bacteria, protozoan, and parasitic worms) [33–35]. The water pollution load and pollutants types may vary among sectors (urban, industrial, agriculture, etc.) and between countries and subregions [32]. Thus, wastewater requires processing before being used as a freshwater source [36].

Alternatively, wastewater recovery represents an opportunity to face the CE pressures on water supplies, which requires adopting a closed-loop system. Such a design entails reusing resources, energy, and materials several times (for the same or different purposes) with minimal processing required for subsequent use [37, 38]. The CE principles contrast the traditional linear economic model, which depends on materials and energy availability [39]. Henceforth, wastewater nutrients are reused as fertilizers, turning the production system circular, without waste, and replacing the traditional linear system.

Water reuse preserves freshwater sources by minimizing aquatic ecosystem exploitation and effluent discharge [40]. In 2015, the global water reuse capacity was 54.5 GL/day, up from 33.7 in 2010. China, the USA, the Middle East, North Africa, Europe, and West South Asia have the biggest growth [41]. Fruits, vegetables, cotton, and grains use 91% recycled water [42]. Most wastewater utilized for irrigation comes from cities/municipalities, industries, and agro-industries [43, 44]. Since effluents are reused, the potential is greater.

Wastewater production does not depend on climatic conditions. Climate change scenarios project spatial and temporal variations in the water cycle dynamics, exacerbating water supply and demand discrepancies [41, 45].

Considering this, wastewater reuse is essential in meeting the high irrigation demand [42].

2.1 Municipal Wastewaters

In some world regions, municipal water use for irrigation is already a reality [46]. About 20 million hectares of land have been irrigated with untreated or partially treated wastewater in 50 countries (approximately 10% of global irrigation practices) [36]. The reuse of municipal wastewater is a sustainable practice. It also promotes soil fertility and productivity due to the added nutritional load, reducing synthetic fertilizers' needs and decreasing environmental contaminants' discharge [36, 46]. LA produces approximately 30 km^3/year of municipal effluents. These data make estimating these wastewaters' potential nitrogen, phosphorus, and organic carbon content possible, proving agricultural applicability (Table 1) [47].

Table 1. LA municipal wastewater nutrient concentration

Resources	Concentration (Tg/year)*		
	Low	Medium	High
Total Nitrogen	0.6	1.2	2.55
Total Phosphorus	0.12	0.24	0.45
Total Organic Carbon (TOC)	2.4	4.8	8.7

* Tg = Teragram = 10^9 kg. The values obtained were the theoretical maximum that could be retrieved from urban sewage, not rural or industrial waste outside cities. The calculations use 100% resource recovery efficiencies without considering technical or economic constraints.

Source: FAO, 2017

A successful water reuse experiment occurred in Tacna, Peru, where water presenting minimum quality parameters for reuse was applied. Farmers who irrigated with treated wastewater in stabilization tanks obtained crops 50–250% larger than those who used well water and fertilizers, an advantage in the prices of the products they offer in the market [48].

2.2 Recycled Waters' Quality Parameters

Proper WT for irrigation with the desired WQ entails physicochemical and microbiological properties changes that can harm the soil while damaging human health and the environment [24, 46]. Commonly controlled physical parameters are total dissolved solids, electrical conductivity, temperature, color/turbidity, hardness, and sediment. Chemical parameters include pH, concentrations of anions and cations, calcium, magnesium, sodium, carbonates, bicarbonates, chlorides, sulfates, metals, nitrogen, phosphorus, and potassium. The microbiological parameters control bacteria like *Escherichia coli* and *Legionella*, viruses, and protozoa resembling *Giardia* and *Cryptosporidium* [49, 50].

Each country sets control mechanisms for recycled WQ in crop irrigation. Some LA countries already have specific regulations and guidelines [48, 51, 52]. There are also recommendations stipulated internationally [53]. The EU manual from [50] and the WTFAO guidelines for agriculture [49], and the "World Health Organization" (WHO) [54] exemplify references. Considering all qualitative parameters, traditional, non-ecofriendly chemical and physical WT has costly implementation and maintenance [55]. Sustainable and low-cost technologies represent a strategy to ensure quality and protect water availability [56, 57].

3. Constructed Wetlands (CWs)

CWs are artificial water treatment systems employing aquatic plants—aquatic macrophytes—as renewable resources to filter water through phytoremediation. This biotechnique has increasingly gained prominence due to its economic and socio-environmental viability, with continually improving systems [37, 56]. Literature shows the WT effectiveness by CWs using different species of aquatic macrophytes to improve water physicochemical and biological characteristics through "biochemical oxygen demand" (BOD) and "chemical oxygen demand" (COD) reductions [59, 60] and removal of:

- Suspended solids [58, 59];
- Nitrogen and phosphorus [61];
- Metals and emerging organic contaminants [58, 60];
- Pathogenic organisms (bacteria, viruses, helminths, protozoa) [60].

The bibliometric analysis presented herein used the keyword "Constructed Wetlands" in the Scopus Elsevier database, analyzing verified documents published from 1985 onwards. A significant production increase emerged, mainly in the last 20 years (Fig. 2). In total, 10,560 articles related to CWs, the year of greatest publication in 2020, with 815 productions available.

Several countries have adopted CWs, especially China, which ranks with 3,104 publications, followed by the United States (2045) and the United Kingdom (538). Among the 15 countries with the highest theme production, Brazil ranks 11th with 280 works (Fig. 2). This shows that there is potential in the region to develop such treatment systems, given the tropical climate, the water bodies' abundance, and aquatic biotechnologically explorable macrophyte species. Furthermore, agriculture is the LA economy foundation, especially in Brazil, which has large arable areas, requiring a large volume of water for irrigation [62]. Wastewater treated by CWs is an interesting alternative for irrigation. Then, considering literature since the first publication, the keywords 'wastewater' and 'irrigation' have been searched in the Scopus Elsevier databases. Between 1966 and 2021, 9,207 publications were identified, and since 2013, more than 400 theme-related publications/year appeared, reaching the highest level in 2021 with 700 publications (Fig. 3).

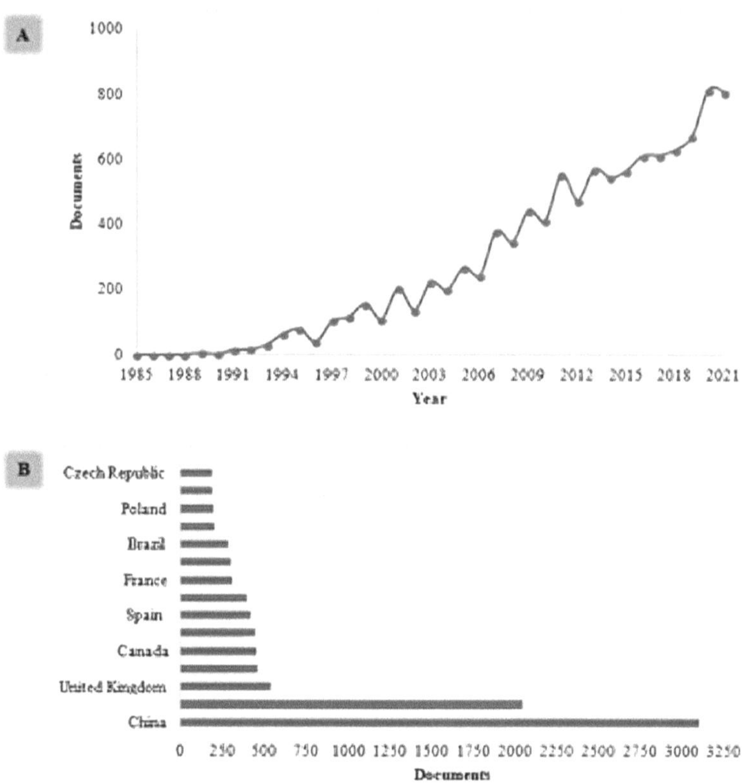

Fig. 2. Documents by year (A) and by country (B), between 1985 and 2021, for "Constructed Wetlands" in the title, abstract, and keyword. (*Source*: Scopus Elsevier, accessed on 22 November 2021)

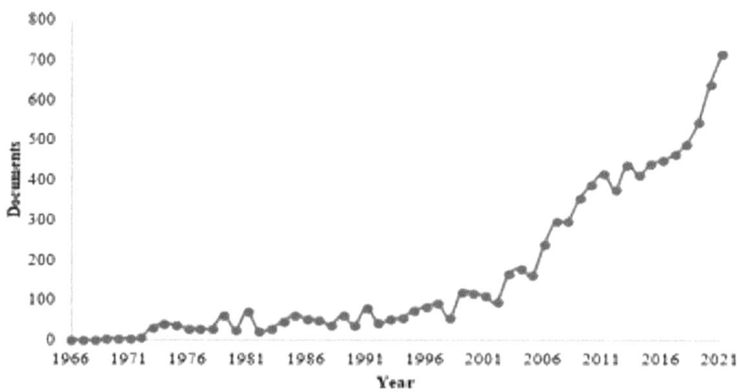

Fig. 3. Documents between 1966 and 2021 with 'wastewater' and 'irrigation' in the title, abstract, and keyword. (*Source*: Scopus Elsevier, accessed on 22 November 2021)

The publications' increase over the years reinforces the WT by CWs and their irrigation viability. Unlike conventional treatment, this low-priced technology stands out for its potential to remove pollutants, and eco-friendliness [63].

3.1 Constructed Wetlands Water Treatment

CW phytoremediation is sustainable, with process escalation, cheap installation/ maintenance costs, soil structure, fertility conservation, erosion avoidance, metal leaching prevention, and is sun-powered [64]. Disadvantages comprise a longer period than traditional treatments and reduced efficiency depending on the contaminant's concentration, toxicity, and bioavailability, which slows plant growth and decreases biomass [64]. After environmental remediation, the plant material destination might increase expenses. Food chain metals must be considered [64, 65]. Thus, biotechnology can lessen or eliminate such issues. Microbial inoculants speed treatment, break down hazardous chemicals, and create fertilizers from aquatic macrophytes [66–68]. Thus, agriculture may benefit from another high-value, nutritious product.

CWs perform preliminary, primary, secondary, and tertiary WT (Fig. 4). After grating, the effluent proceeds to the initial treatment in an aeration pond. Sedimentation ponds treat the wastewater afterward. Finally, aquatic macrophytes in plant-mediated purification [68] and free-living and related microbes [69] metabolize nutrients and pollutants in the third stage [70]. Plant species should absorb (adhere to internal tissues) and/or adsorb (adhere to the external surface) [71], incorporate nutrients and wastewater toxins [72], or decrease wind speed over the water surface, which is critical for wastewater sedimentation [73].

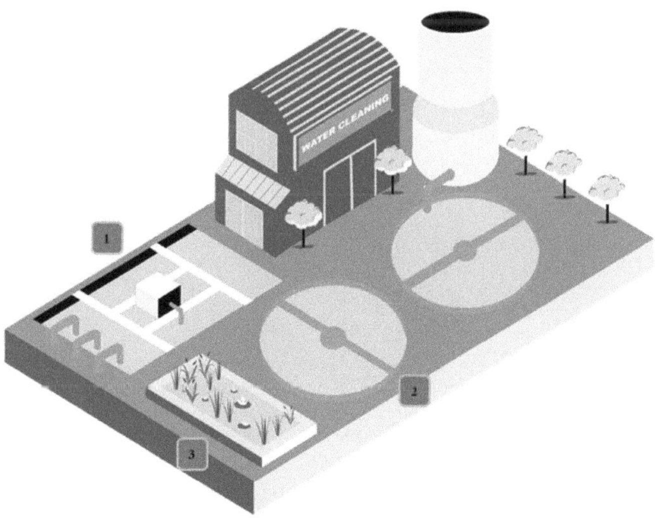

Fig. 4. Water treatment in constructed wetlands. 1: Primary treatment. 2: Secondary treatment. 3: Tertiary treatment.

Generally, the choice of CWs plants mainly includes aquatic macrophytes, meeting the requirements [74] below:

1. Tolerance to local climatic conditions, pests, and diseases;
2. Tolerance to pollutants and hypertrophic conditions;
3. Greater relative ability to absorb contaminants by direct assimilation or storage;
4. Rapid establishment, dissemination, and growth.

Aquatic macrophytes can purify effluent through direct contact with contaminated water and serve as environmental quality bioindicators [75, 76]. These plants have thin cuticles and may eliminate water pollutants over their surfaces [77]. The widely studied genera *Typha* [73], *Eichhornia*, *Salvinia*, and *Lemna* [78] have helped with contaminant removal (Table 2), differing especially regarding substrates (Fig. 5). *Typha* is rooted in the sediment. Others are free-floating macrophytes. The *Typha* species tend to outcompete other plant species in CWs and even in natural environments [73], occurring mainly as an invasive species. Thus, physical barriers between different CW treatment steps are paramount. Since *Typha* has no precise needs, [79] and [80] urge early treatment. Reduced wind speed improves particle sedimentation.

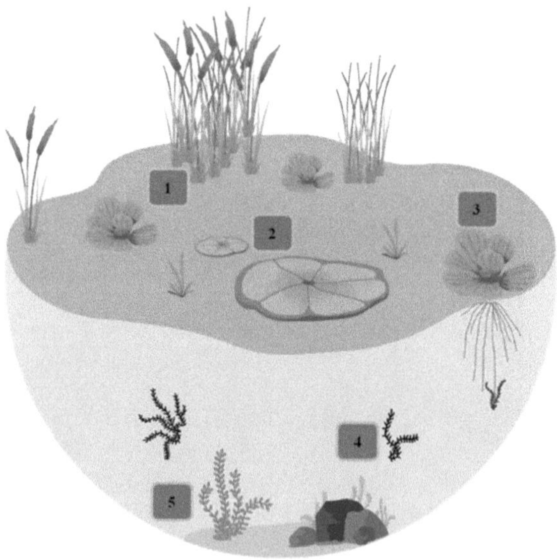

Fig. 5. Significant groups of aquatic macrophytes. 1: Emerged aquatic macrophytes: rooted in the sediment with leaves out of the water (e.g., *Typha*). 2: Emergent with floating leaves rooted in the sediment and other leaves floating on the water surface (e.g., *Nymphaea*). 3: Floating macrophytes: float freely on the water surface (e.g., *Pistia*, *Salvinia*). 4: Free submersed aquatic macrophytes: floating underwater (e.g., *Ceratophyllum*). 5: Rooted submerged aquatic macrophytes: grow entirely underwater (e.g., *Egeria*).

Table 2. Removal of contaminants by different species of aquatic macrophytes

Macrophytes	Contaminants	References
Typha sp.	Fe; Al; Mn; Zn; Co; Pb; Ni; Cr; Cd; Hg; As; Sulphates; Nitrogen; Phosphorus	Klink et al., 2013 [81] Rana et al., 2018 [82] Jeke et al., 2019 [83] Hejna et al., 2020 [84] Singh et al., 2021 [85] Bonanno, 2013 [86] Mufarrege et al., 2015 [87]
Eichhornia sp.	S; Cd; Pb; Zn; Cr; Ag; Cu; Ni; Hg; Fe; Al; Tetracycline	Mishra et al., 2017 [88] Sricoth et al., 2018 [89] Malar et al., 2015 [90] Zahoor et al., 2018 [91] Chakrabarty et al., 2017 [92] Melignani et al., 2015 [93] Suryandari et al., 2017 [94] Jayaweera et al., 2007, 2008 [95, 96]
Salvinia sp.	Mn; Pb; Cd; Ni; Cu; Cr; Drugs (Amoxicillin, caffeine, dipyrone, Ibuprofen, and Omeprazole); Phosphates; Nitrates; Ammonia	Estrella-Gómez et al., 2012 [97] Wolff et al., 2012 [98] Bizzo et al., 2014 [99] Ng et al., 2018 [100] Otomo et al., 2021 [101]
Lemna sp.	Phenol; Phosphates; Nitrates; Sulphates; Ammonia; As; Cu; Cd	Jayasri et al., 2017 [102] Ng et al., 2018 [100] Ceschin et al., 2019 [103] Radulović et al., 2020 [104] Seifi et al., 2021 [105]

Precisely for being sediment rooted, the *Typha* management is more delicate and labor-consuming, translating into a slower plant turnover. Thus, free-floating, faster-growing aquatic macrophytes can complement the WT in CWs. The *Eichhornia*, *Lemna*, and *Salvinia* are also hyperaccumulators [106]. Even though they can be applied separately in CWs, these plants might be advantageous when contaminants are not fully understood [107]. The authors consider these plants different affinities for metallic ions in suspension (tested the remediation of Zn, Fe, Cu, Cr, Ni, Cd, and Pb).

Eichhornia crassipes survives in As contaminated environments, in concentrations of up to 2 mg.L^{-1}, removing up to 20% of the metal from the environment [108]. *E. crassipes* in other metals' remediation has also been described (Table 2). *Eichhornia* has also worked for Ibuprofen and caffeine phytoremediation, two emerging pollutants that harm drinking WQ with costly chemical removal [109].

Several studies with *Salvinia* genus macrophytes have highlighted their resistance and ability to remove metals and drugs. *Salvinianatans* is capable of accumulating Al, Cr, Cu, Ni, and Zn [110, 111], with high efficiency in removing Zn and Cu (85 and 74%, respectively) [110]. *Salviniaauriculata* also tolerates a high concentration of metal viz. Cu [99, 112]. These characteristics reinforce the free-floating macrophytes' usage after *Typha*-assisted sedimentation since large particles could harm the phytoremediation of smaller contaminants. Sometimes, ornamental plants can also work in CWs, such as *Heliconia* and *Zantedeschia*, which are common LA plants [113].

3.2 Beneficial Microbes in Constructed Wetlands

Microorganisms have adapted to polluted habitats like aquatic macrophytes [55]. CWs promote wastewater cleansing and degrading substances removal. Some microbial groups boost plant development, which helps remove wastewater toxins [114, 115].

Recently, our team isolated and identified bacteria and yeasts associated with the aquatic macrophytes *S. auriculata* and *T. domingensis* [116, 117]. Studies have shown that the microbiota improves plant development, with a significant gain in "fresh biomass" (FB), even in high amounts of metal ions [116–118]. Inoculating the bacterium *Bacillus toyonensis* 3.1.3.0.X.14 into *S. auriculata*, increased FB by 12% compared to the control. Under Pb stress, the inoculated plant's FB increased by 20.3% compared to non-inoculated treatment in the presence of Pb, suggesting plant protection by the bacteria during metal stress [118]. Inoculating *Enterobacter* sp. 3.1.3.0.X.18 in *S. auriculata* created a similar result. The bacteria promoted a 29.6% FB gain. The inoculated plants showed a 51% FB increase with Cu compared to the non-inoculated treatment, showing the bacteria protected the plant from stress [117]. Plants inoculated with the yeast *Yarrowialipolytica* showed a significant increase of 20.9% in FB. In the presence of Cu, the yeast also protected the *S. auriculata*, with a 28.3% FB increase compared to the non-inoculated treatment [116]. Therefore, aquatic macrophytes microorganisms might protect these plants under stress, maintaining plant growth traits, and act as inoculants in wetlands systems.

Nonetheless, the CWs' microbial network is complex and not entirely associated with the host plants. This network includes bacteria, fungi, archaea [119, 120], protozoans, and small metazoans [121]. These organisms might possess incredibly different metabolisms, which might favor using additional carbon and nitrogen sources, the presence or absence of oxygen, and act on other steps of nutrient cycles (mainly carbon, nitrogen, oxygen, phosphorus, and sulfur) [122].

Microbial species presence empirically improves WQ when enriched with microbial inoculants [123] and other techniques [124]. Still, microbial metabolism works in an intricate network (including microorganisms in symbiosis), which is challenging, requiring modern tools involving different fields, viz., Ecology, Microbiology, Molecular Biology, and Bioinformatics.

Metagenomic studies, for example, allow knowing the genomes of 25 microbial communities in an environmental sample, including non-cultivable microorganisms [125]. In [126], the occurrence of arsenic (As) metabolism genes is verified, with resistance (*ars*R), transport (*pst*B, *glp*F, *ars*B, *ars*A, *pst* (S, A, and C), and *arc*3), and metal transformation functions (*ars*T, *aio*AB, *ars*C, and *ars*M) in bacteria associated with the aquatic macrophyte, *Hydrillaverticillata.* WT plant analyses showed the prevalence of resistance genes to 21 metals in a bacterial population predominantly belonging to the Phylum Proteobacteria (90%), especially in the genera *Pseudomonas*, *Janthinobacterium*, *Acinetobacter*, *Sphingobium*, and *Enhydrobacter* [34]. Metagenomics also characterizes enzymes likely produced by microorganisms, suitable for bioremediation processes. Enzymes might be a simpler option for microorganisms, especially in extreme environments [127]. Nonetheless, producing heterologous enzymes depends on a deep structural and functional knowledge of the given gene and respective protein.

Genomic and proteomic analyses helped identify microbes, cellular mechanisms, metabolic pathways, and key proteins causing the physiological response to environmental stresses [64, 128]. For example, when seeking potential microorganisms for metal remediation, many researchers have isolated strains from contaminated areas' samples [129]. Knowing the molecular-level relationships between aquatic macrophytes and microorganisms can control the WT process in CWs, increasing the potential to remove contaminants. This strategy can increase plant biomass, improve nutritional status, and promote stress tolerance, intensifying the macrophytes' phytoremediation potential [130]. Yet, this relationship needs additional research.

3.3 LA Constructed Wetlands

Several countries have adopted WT with CWs due to economic and environmental benefits in line with the 2030 Agenda. However, this is still not as frequent in LA and the Caribbean, even with the ideal conditions to implement this technology (especially the high availability of low-cost marginal lands, tropical and temperate climates with mild winters, and high availability of climate-adapted macrophytes) [131, 132]. It has been estimated that CWs treat only 0.22% of the total wastewater flow in this region. Simultaneously, stabilization ponds, activated sludge, and upflow anaerobic sludge blanket (UASB) reactors treat about 81% of the total flow [133].

Mexico has adopted CWs in its WT local guidelines [51] with a positive effect. These systems tend to be more resilient and require less maintenance than traditional systems–pressing issues responsible for most WT failures in developing countries [53]. However, the National Inventory of Municipal Water and Effluent Treatment Plants in Operation of Mexico [134] reports that of the 2,642 treatment plants installed in the country, 278 use CWs as a complement to other treatments such as septic, sedimentation, or Inhoff tanks, i.e., less than 10%, and only 77 stations use wetlands as the main treatment method. Interestingly, crop irrigation with untreated municipal water is common, posing health and environmental risks [135]. Conversely, irrigation of crops with recycled water is safe, adding nutrients to the soil and requiring fewer chemical fertilizers.

The Ponte dos Leites Sewage Treatment Plant, located in the Araruama municipality, Rio de Janeiro, Brazil, is a successful effluent treatment by constructed wetlands, considered the largest in LA. After its implementation, regional flora and fauna were recovered, with animal species returning to the area due to food and favorable environmental conditions. Moreover, it reduced effluent treatment operating costs, and currently, all the solid waste generated goes into composting [136, 137].

In Colombia, Chile, Peru, Argentina, Uruguay, and Paraguay, CWs for effluent treatment have been implemented in experimental or small-scale systems to purify sanitary, community, industrial, and farm effluents [132, 138–140]. Further studies and research are required about effluent reuse treated in wetlands in LA agriculture, besides a specific regulatory framework [131]. Automation, based on Agri4.0 and sustainability, could speed up this technology use.

4. Agri4.0 Water Reuse Technologies

Agri 4.0 can optimize production sustainably. Using cutting-edge technologies, e.g., the Internet of Things (IoT), transforms traditional farming practices into modern agrarian solutions. ICT advances enable producers to monitor production phases and determine the best treatments [62]. IoT is crucial in modern agriculture, with intelligent sensors processing data and using cloud storage to monitor and diagnose crop development, resources, and soil properties (pH, moisture, and salinity) in real-time for optimizing irrigation and performing water physical-chemical analysis [141, 142].

Remote sensing using satellites and/or drones is another extremely relevant approach [143], allowing for a large area's panoramic assessment and optimizing crop monitoring.

Wastewater management via these technologies can help achieve sustainable and more efficient agriculture. Hence, smart sensors can monitor recycled WQ [144, 145] to ensure it is within agricultural irrigation standards [146]. Likewise, it is possible to analyze wastewater and soil remaining nutrients for better management/application of fertilizer doses and reducing costs and environmental contamination [147].

4.1 Agri4.0

Due to unsustainable methods and insufficient utilization, agriculture globally lacks resources. International manufacturers are undertaking green initiatives to attract customers [148]. Corporate social responsibility programs help makers give back. Most companies do not meet SDGs since they need help in keeping remanufacturing, recycling, and reusing. Their inflexibility, visibility, and resilience caused these failures. Applying I4 to agribusiness leads toAgri4.0 or A4 and requires proper development to overcome challenges. Administrators can access supply-and-demand-correlated real-time data from a supply chain network through CPS dashboards [28]. This excellent opportunity for organizational learning and production lines' arrangement may utilize recycled, refurbished, and remanufactured components. I4.0 easily optimizes actions and standard operating processes, providing alternatives to business processes and significantly fewer resources and lead times.

Successful A4adoption can improve their top and bottom lines [149]. WT plants can also expect operations' cost reduction to integrate planning and scheduling activities. When applying front-end and base technologies, WT facilities with intensive I4.0 adoption will attain operational flexibility, efficiency, and effectiveness.

There is considerable apprehension for high technology-oriented firms aiming for sustainable development. Product design, services, and proper transportation selection can lower global warming while becoming internationally competitive markets. The 10-R manufacturing methodologies are refuse, rethink, reduce, reuse, repair, refurbish, remanufacture, repurpose, recycle, and recover options. These approaches can provide opportunities for a CE cleaner production model and aid agribusiness in attaining a competitive edge over rivals, thus requiring analysis of different sustainability aspects.

Large groups also confront SDG procedures for finished goods and real-time manufacturing resource scheduling. Few studies exist on 10-R scheme sustainability, and the I4.0 potential to stimulate sustainability research is premature, needing more attention from potential researchers. This study promotes the second question.

4.2 Main Aspects and Discussions

The main I4Ts that help in water reuse in agribusiness follow.

4.2.1 Sensors

They measure temperature, conductivity, and pH from inlet to outlet in drinking water and wastewater systems. Thus, waste or deviation from norms is rapidly detected and corrected, ensuring quality. Sensors report real-time wastewater volume, quality, and source. They spot water pollution by detecting ions, bacteria, and metals in real-time. Sensors reduce washing water waste, control water volume, and determine wastewater standards. Control systems can monitor wastewater treatment with intelligent sensors. Digitizing agricultural water management and reuse has many benefits. Security eliminates shortages and assures correct reuse, protecting aquatic systems, resource efficiency, and circular/green economies [27, 28], increasing supply and disposal.

4.2.2 Actuators

They (e.g., pumps, intelligent membranes, valves, alarms, drones) can regulate the water path within processes or control parameters (e.g., temperature, pH, and conductivity) from the inlet to the water reuse. So, they correct waste or abnormalities from established standards, ensuring wastewater quality. Actuators receive control information on the discarded wastewater amount, quality, and production source. Actuators can quickly remediate water pollution/waste through decision-making stages after processing real-time contaminants' sensed info, like ions, bacteria, besides metals, to solve anomalies early [27, 28].

4.2.3 Drones

They can map regions of interest, check impacted aquatic systems, and monitor crop sites. This movement begins with I4Ts already in use, e.g., for water monitoring drones with automatic functions for pumping and scrutinizing samples. These schemes can work with coupled and remotely activated peristaltic pumps to analyze the watercourse region. Drones can help insert remote sensing and surveillance into wastewater facilities while working with other machines to handle information remotely, combining smartphones and cloud databases [150].

4.2.4 Nanotechnology

Nanoparticles aligned to wetlands can clean pollutants, providing a higher quality reuse water. WT can use membrane technologies to manipulate atoms and molecules and thus treat the water. The UN perceives excellent potential in nanotechnology for water decontamination, viz., graphene-coated membranes can eliminate more than 90% of anionic dyes and reuse water in staining processes. Silver nanoparticles, through pressure pumps, help filtrate wastewater and improve the reuse water quality [151].

4.2.5 Artificial Intelligence (AI)

AI optimizes the performance of filtration systems in reuse applications [152, 153]. It will transform WT by increasing the water treatment efficiency in line with the analysis of waste or substances contained in this water and automatically activate the adequate cleaning protocol to improve quality and reduce resources and energy used. When effluent conditions vary, AI software adjusts to filtration conditions in real-time, ensuring stable performance with maximum water retrieval. AI can predict supply network water consumption by analyzing a large amount of data from past events like water usage in the last days, months, or years and use these data for decision-making. An attempt determines cyclical patterns to estimate higher water consumption or lack of water through wastewater control, reprocessing, verification, and quality.

Digital tools can benefit wastewater systems and develop a sustainable and intelligent approach. Still, one needs more studies about the economic, environmental, and social compensations resulting from adopting AI in dam maneuvers. These operations' primary results can save the effluent treatment, with quality equal to or better than houses' faucets. They may involve digitalized and automated wastewater management with real-time tracking via sensors that transmit information to visualization platforms.

4.2.6 Internet of Things (IoT)

IoT systems' optimization in the primary Internet layer (edge computing) helps monitor properties with many sensors and actuators to control waste, reuse, and check water quality. Therefore, sensors can optimize irrigation, and real-time soil diagnosis can ameliorate water expenditure and management systems [152, 153]. IoT-based applications seamlessly integrate the virtual and physical worlds, combining devices to produce data, send them to other devices, and then send them to the cloud for management resolutions and data mining. Machine-to-machine communication technology enables wireless data interchange among IoT equipment and gateways.

Further, data flows from the gateway to Internet repositories. Yet, machine-to-machine communications have different traffic features and specific problems. I4Ts, like forward-thinking engineering structures, progressive process control, and machine control can monitor and control intelligent production lines in agribusiness. The central I4T focus enablement in manufacturing enriches flexibility, efficiency, and effectiveness to augment performance.

4.2.7 Big Data (BD)

WT facilities (WWTFs) use large workforces and resources to meet regulatory limits without errors. ICT allows for collecting large data amounts from plentiful sources using sophisticated sensors. The lack of specialized tools, knowledge, operators, and engineers cannot effectively extract meaningful and valuable information from large datasets. Unfortunately, digitally stored data are often underutilized. Various data analytics techniques have been developed in the past few years. The methods are efficient for analyzing vast datasets.

Nonetheless, no wholly developed study applies these techniques to assist WT operations. Data analytics processes can immensely transform a large dataset into informative knowledge, e.g., hidden data, operational caveats, or a predictive model. Big data (BD) analytics will allow operators to clearly understand the functional status while saving operation and maintenance costs and reducing the human resources required. Ultimately, the method can enhance the operating WT framework performance [154, 155]. Unstructured data from various sources, such as intelligent digital sensors (temperature, pressure, flow, weight, density, and power utilization) connected via wireless networks and mobile devices with special applications, can generate a high data volume for analysis to extract key knowledge for decision-making. This can improve the accuracy of forecasts, supplying a greater degree of visibility, higher resource efficiency, asset utilization, and enhanced throughput times.

Decision-making is one function of data quality and stakeholder commitment. BD research is gaining prominence as supply chains are increasingly complex. Methods for scrutinizing and managing data quality in supply chain management are essential while investigating the BD role and analytics in sustainable agriculture enhancement. BD, predictive analytics, and social/environmental performance can be combined.

4.2.8 Cloud/Fog/Edge computing

Cloud Computing (CC) is a deep-rooted model for service-centric systems. Nevertheless, ultra-low latency, high bandwidth, safety, and real-time analytics limit CC when examining and outputting results for a large data amount. Fog (FC) and Edge Computing (EC) offer solutions to CC restrictions. The number of agricultural applications combining CC, FC, and EC increased in the last decades in the Agri4.0 domain. Identifying all relevant research on new computing paradigms for intelligent agriculture is vital to better work with CC, FC, and EC blends. Likewise, it also analyzes and examines the agricultural application domains, research methodologies, and combinations of them. Additionally, the architecture models' components can explore the communication protocols between layers and those relating to one layer. Finally, new challenges in smart agriculture, the intense usage of mobile devices, and future research directions will require access to remote databases [156].

4.2.9 Blockchain

Blockchain is a technology that is independent of central authorities. Blockchain has continued evolving in public and private models because it can permit transparent, fraud-proof, and auditable digital asset handling. The evolution became more significant by mixing smart contracts with blockchain, like Ethereum. Smart contracts are the most transformative application of blockchain in the current digital world. Research showed how blockchain could fuel the shift toward a smart city, i.e., a town using ICT to enable a premium lifestyle at a reduced cost and resource utilization. Blockchain and smart contracts can lead to decentralized, automatic, transparent, and secure smart city management [28, 157, 158].

A reward-based blockchain framework can afford an efficient flow of liquid waste to the proper disposal areas. Such expedient helps maintain adequate environmental cleanliness and hygiene. Controlled and auto-enforced activities utilize smart contracts for system integrity, sustainability, and real-time management, thus creating open, transparent, and decentralized surroundings. Wastewater management (WWM) can be fourfold: (i) storage, (ii) collection, (iii) disposal, and (iv) recycling. The suggested smart WWM system is a step toward synchronizing these four steps.

I4.0 research in operations management is gaining momentum as corporations prioritize digitalization. Agri4.0 technologies would lead to more competitive global supply chain operations. I4Ts can integrate all critical functions to share common data, records, and knowledge in real time throughout the supply chain network. They can also be applied to automate necessary operations activities, permitting enriched visibility and mitigating risks.

4.3 Sustainable Development

The UN 2030 agenda for SDGs focuses primarily on people, the planet, and prosperity. There are 169 goals and 231 indicators where each country establishes its objectives on a national level to achieve a better quality of life for its population.

As the current sources of competitive advantages deplete, agribusiness must focus increasingly on new sources of competitive advantage for sustainability. This concept integrates environmental and social parameters to allow long-term financial viability. The literature lacks sustainability-related studies about an integrated approach. Cultural and economic factors also considerably influence any country's sustainability developments.

Therefore, identifying key sustainable supply chain drivers and understanding their inter-relationships can help managers integrate sustainability aspects in business operations. Supply chain transparency can improve social and ecological parameters, reinforcing "Environmental, Social, and Governance" (ESG) ideas and compliance practices.

Lean practices can help directly and indirectly achieve supply chain sustainability per literature. Businesses are gradually adopting CE principles by recycling and remanufacturing. Building trust is essential in a sharing economy. Firms will perish on this planet without complying with SDGs.

5. Conclusion

Considering the world population and the projected 2100 growths by the UN, water supplies will become even more critical, and there will also be greater demand for food production. However, agriculture already consumes (and wastes) the largest quantity of water. In addition, the decrease in water quality is another concern, and it is urgent to adopt sustainable measures for decontamination and optimization of water use, seeking to conserve WR.

Therefore, large wastewater volumes generated daily can be reused in agriculture to protect resources and save on fertilizers costs, stimulating the circular economy (CE) and meeting high demands. Agricultural recycled water application requires adequate treatment to meet the physical-chemical parameters according to each country's legislation to guarantee the aimed quality.

The WT process by CWs is a promising, sustainable, and low-cost technology already in use, bringing positive results. This technique can decontaminate wastewater, even with high loads of nutrients and contaminants. Macrophytes and microorganisms have efficient survival mechanisms and strategies even in eutrophic or severely contaminated environments. Together, these organisms can be a biotechnological tool to promote effluent purification reactions and remove organic and inorganic water contaminants.

Some LA countries employ Wetland-mediated effluent purification. Its implementation reduces operating costs compared to conventional methods and feeds the CE since solid waste is destined for composting, not landfills. Instead, treated wastewater usage in CWs for irrigation is an underused strategy with great potential.

Literature indicates different 10R-based primary- and advanced-level I4Ts to explore sustainable development outcomes in WT with agricultural applications.

Wastewater monitoring technologies offer greater productivity and economic, social, and environmental advantages.

This research indicates that the more robust the delivery system is, the better the degree of implementation and the more encouraging the operational results. The I4.0 delivery system comprises senior management support, preparation, design resources, provision to research organizations and universities, and simplifying I4Ts'adoption. The collected works indicate significant ICT demand among small, medium, and large companies. I4Ts coupled with biotechnology can fill this gap by connecting the process phases and consolidating CE.

References

1. Srivastava, J., Gupta, A. and Chandra, H. (2008). Managing water quality with aquatic macrophytes. *Rev. Environ. Sci. Biotechnol.*, 7(3), 255–266.
2. Nations, U. (2018). *Nature-based Solutions for Water*. UN World Water Development Report.
3. Yoon, J., Klassert, C., Selby, P., Lachaut, T., Knox, S., Avisse, N. et al. (2021). A coupled human-natural system analysis of freshwater security under climate and population change. *Proc. Natl. Acad. Sci.*, USA, 118(14).
4. Boretti, A. and Rosa, L. (2019). Reassessing the projections of the World Water Development Report. *NPJ Clean Water*, 2(1).
5. United Nations. (2019, Dec.). *Population Facts, How Certain are the United Nations Global Population Projections*? No. 6. Dept. of Economic and Social Affair.
6. Silva-Rodríguez de San Miguel, J.A. (2018). Water management in Europe and Latin America. *Manag. Environ. Qual. An. Int. J.*, 29(2), 348–367.
7. FAO & WWC. (2015). *Towards a Water and Food Secure Future: Critical Perspectives for Policy-makers*. Rome and Marseille.
8. Konapala, G., Mishra, A.K., Wada, Y. and Mann, M.E. (2020). Climate change will affect global water availability through compounding changes in seasonal precipitation and evaporation. *Nat. Commun.*, 11(1), 1–10.
9. Liu, J., Yang, H., Gosling, S.N., Kummu, M., Flörke, M., Pfister, S. et al. (2017). Water scarcity assessments in the past, present, and future. *Earth's Future*, 5(6), 545–559.
10. Brêda, J.P.L.F., de Paiva, R.C.D., Collischon, W., Bravo, J.M, Siqueira, V.A. and Steinke, E.B. (2020). Climate change impacts on South American water balance from a continental-scale hydrological model driven by CMIP5 projections. *Clim. Change,* 159(4), 503–522.
11. Zilli, M., Scarabello, M., Soterroni, A.C., Valin, H., Mosnier, A., Leclère, D. et al. (2020). The impact of climate change on Brazil's agriculture. *Sci. Total Environ.*, 740, 139384.
12. Niemi, L., Landová, P., Taggart, M., Boyd, K., Zhang, Z. and Gibb, S. (2022). Spatiotemporal trends and annual fluxes of pharmaceuticals in a Scottish priority catchment. *Environ. Pollut.*, 292 (October 2021).
13. Wang, B., Stirling, E., He, Z., Ma, B., Zhang, H., Zheng, X. et al. (2021). Pollution alters methanogenic and methanotrophic communities and increases dissolved methane in small ponds. *Sci. Total Environ.*, 801, 149723.

14. Wada, Y., Flörke, M., Hanasaki, N., Eisner, S., Fischer, G., Tramberend, S. et al. (2016). Modeling global water use for the 21st century: The water futures and solutions (WFaS) initiative and its approaches. *Geosci. Model Dev.*, 9(1), 175–222.
15. Rycewicz-Borecki, M., McLean, J.E. and Dupont, R.R. (2016). Bioaccumulation of copper, lead, and zinc in six macrophyte species grown in simulated stormwater bioretention systems. *J. Environ. Manage.*, 166, 267–275.
16. Abbas, S.Z., Rafatullah, M., Ismail, N. and Lalung, J. (2014). Isolation, identification, characterization, and evaluation of cadmium removal capacity of *Enterobacter* species. *J. Basic Microbiol.*, 54(12), 1279–1287.
17. Brasil, Lei nº. 9.605, de 12 de fevereiro de 1998, art. 33. Available from: http://www.planalto.gov.br/ccivil_03/leis/l9605.htm.
18. CONAMA, Resolução nº 430/11, 2011. Available from: http://conama.mma.gov.br/?option=com_sisconama&task=arquivo.download&id=627.
19. Jägermeyr, J. (2020). Agriculture's historic twin-challenge toward sustainable water use and food supply for all. *Front. Sustain. Food Syst.*, 4(April), 5.
20. Kebede, E. (2021). Competency of rhizobial inoculation in sustainable agricultural production and biocontrol of plant diseases. *Front. Sustain. Food Syst.*, 5(August), 1–22.
21. Sharma, S., Rana, V.S., Pawar, R., Lakra, J. and Racchapannavar, V.K. (2021). Nanofertilizers for sustainable fruit production: A review. *Environ. Chem. Lett.*, 19(2), 1693–1714.
22. Khan, M.N. and Mohammad, F. (2014). Eutrophication: Challenges and solutions. *In*: Ansari A. and Gill, S. (Eds.). *Eutrophication: Causes, Consequences and Control 2*, 1–262. Springer Link.
23. Scherer, L. and Pfister, S. (2021). Modelling spatially explicit impacts from phosphorus emissions in agriculture. *Int. J. Life Cycle Assess.*, 20(6), 785–795.
24. Khan, N. (2018). Natural ecological remediation and reuse of sewage water in agriculture and its effects on plant health. *Sewage*. doi: 10.5772/intechopen.75455.
25. Harguinteguy, C.A., Cofré, M.N., Fernández-Cirelli, A. and Pignata, M.L. (2016). The macrophytes *Potamogeton pusillus* L. and *Myriophyllum aquaticum* (Vell.) Verdc. as potential bioindicators of a river contaminated by heavy metals. *Microchem. J.*, 124, 228–324.
26. Monteiro, A.C., França, R.P., Estrela, V.V., Fernandes, S.R., Khelassi, A. et al. (2020). UAV-CPSs as a test bed for new technologies and a primer to Industry 5.0, in imaging and sensing for unmanned aircraft systems. *In:* Estrela V.V. et al. (Eds.), *Deployment and Applications*, Volume 2. IET.
27. Estrela, V.V., Saotome, O., Loschi, H.J., Hemanth, J.D., Farfan, W.S. et al. (2018). Emergency response cyber-physical framework for landslide avoidance with sustainable electronics. *MDPI, Technologies*, 6(2), 42.
28. Kerin, M. and Pham, D.T. (2019). A review of emerging industry 4.0 technologies in remanufacturing. *J. Clean Prod.*, 237.
29. Vörösmarty, C.J., Osuna, V.R., Cak, A.D., Bhaduri, A., Bunn, S.E. et al. (2019). Ecosystem-based water security and the Sustainable Development Goals (SDGs). *Ecohydrol. Hydrobiol.*, 18(4), 317–333.
30. Cohen-Shacham, E., Walters, G., Janzen, C. and Maginnis, S. (Eds.) (2016). *Nature-based Solutions to Address Global Societal Challenges*, 1–32. IUCN.
31. Slobodiuk, S., Niven, C., Arthur, G., Thakur, S. and Ercumen, A. (2021). Does irrigation with treated and untreated wastewater increase antimicrobial resistance in soil and water: A systematic review. *Int. J. Environ. Res. Public Health*, 18(21).

32. Turcios, A.E., Miglio, R., Vela, R., Sánchez, G., Bergier, T. et al. (2021). From natural habitats to successful application: Role of halophytes in the treatment of saline wastewater in constructed wetlands with a focus on Latin America. *Environ. Exp. Bot.*, 190(July).
33. Ungureanu, N., Vlăduț, V. and Voicu, G. (2020). Water scarcity and wastewater reuse in crop irrigation. *Sustain.*, 12(21), 1–19.
34. Gupta, S.K., Shin, H., Han, D., Hur, H.G. and Unno, T. (2018). Metagenomic analysis reveals the prevalence and persistence of antibiotic- and heavy metal-resistance genes in wastewater treatment plant. *J. Microbiol.*, 56(6), 408–415.
35. Melendez-Marmolejo, J., Garcia-Saavedra, Y., Galvan-Romero, V., de Leon-Martinez, L., Vargas-Berrones, K. et al. (2020). 14 Emerging contaminants. Environmental problems associated with antibiotic use. New detection and remediation techniques and legislative perspectives in Latin America. *Rev. Salud. Ambient*, 20(1), 3–61.
36. Jaramillo, M.F. and Restrepo, I. (2017). Wastewater reuse in agriculture: A review about its limitations and benefits. *Sustain*, 9(10).
37. Masi, F., Rizzo, A. and Regelsberger, M. (2018). The role of constructed wetlands in a new circular economy, resource oriented, and ecosystem services paradigm. *J. Environ. Manage.*, 216, 275–284.
38. Sgroi, M., Vagliasindi, F.G.A. and Roccaro. P. (2018). Feasibility, sustainability, and circular economy concepts in water reuse. *Curr. Opin. Environ. Sci. Heal.*, 2, 20–25.
39. Fortunati, S., Morea, D. and Mosconi, E.M. (2020). Circular economy and corporate social responsibility in the agricultural system: Cases study of the Italian agri-food industry. *Agric. Econ.* (*Czech Republic*), 66(11), 489–498.
40. Colella, M., Ripa, M., Cocozza, A., Panfilo, C. and Ulgiati, S. (2021). Challenges and opportunities for more efficient water use and circular wastewater management. The case of Campania Region, Italy. *J. Environ. Manage.*, 297(July), 113–171.
41. Voulvoulis, N. (2018). Water reuse from a circular economy perspective and potential risks from an unregulated approach. *Curr. Opin. Environ Sci. Heal.*, 2, 32–45.
42. Shoushtarian, F. and Negahban-Azar, M. (2020). World-wide regulations and guidelines for agricultural water reuse: A critical review. *Water* (*Switzerland*), 12(4).
43. Janeiro, C.N., Arsénio, A.M., Brito, R.M.C.L. and van Lier, J.B. (2020). Use of (partially) treated municipal wastewater in irrigated agriculture: Potentials and constraints for sub-Saharan Africa. *Phys. Chem. Earth*, 118–119, 102906.
44. Libutti, A., Gatta, G., Gagliardi, A., Vergine, P., Pollice, A. et al. (2018). Agro-industrial wastewater reuse for irrigation of a vegetable crop succession under Mediterranean conditions. *Agric. Water Manag.*, 196, 1–14.
45. Goonetilleke, A. and Vithanage, M. (2018). Water resources management: Innovation and challenges in a changing world. *Water* (*Switzerland*), 9(4), 1–5.
46. Fito, J. and Van Hulle, S.W.H. (2021). Wastewater reclamation and reuse potentials in agriculture: Towards environmental sustainability. *Environ. Dev. Sustain.*, 23(3), 2949–2972.
47. FAO: *Reutilización de aguas para agricultura en América Latina y el Caribe: Estado, principios y necessidades*. (2017). Santiago de Chile.
48. Moscoso, J. (2016). *Manual de buenas prácticas para el uso seguro y productivo de las aguas residuales domésticas*. Autoridad Nacionaldel Agua (ANA) y Food and Agriculture Organization of the United Nations (FAO).
49. FAO (1992). *Wastewater Treatment and Use in Agriculture*. FAO Irrigation and Drainage Paper 47. Rome.
50. Alcalde-Sanz, L. and Gawlik, B.M. (2017). Minimum quality requirements for water

reuse in agricultural irrigation and aquifer recharge: Towards a legal instrument on water reuse at EU level. EU Science Hub – European Commission. 57 p.

51. CONAGUA (Comisión Nacional del Agua) (2018). Manual de agua potable, alcantarillado y saneamiento. Diseño de plantas de tratamiento de aguas residuales municipales: Zonas rurales, periurbanas y desarrollos ecoturísticos. México D.F.: Secretaría de Medio Ambiente y Recursos Naturales.
52. Pinzón, G.L.A. (2017). El reúso de aguas residuales en Colombia. Derechoaguas Tomo VII.
53. Salgot, M. and Folch, M. (2018). Wastewater treatment and water reuse. *Curr. Opin Environ. Sci. Heal.*, 2, 64–74.
54. World Health Organization (WHO) (2006). Planificación de la seguri daddel saneamiento. Manual para el uso y la disposición seguros de aguas residuales, aguas grises y excretas. ISBN 978 92 4 354924 8 OMS.
55. Abbas, S.Z., Rafatullah, M., Hossain, K., Ismail, N., Tajarudin, H.A. and Khalil, H.P.S.A. (2018). A review on mechanism and future perspectives of cadmium-resistant bacteria. *Int. J. Environ. Sci. Technol.*, 15(1), 243–262.
56. Liu, F., Zhang, S., Luo, P., Zhuang, X., Chen, X. and Wu, J. (2018). Purification and reuse of non-point source wastewater via Myriophyllum-based integrative biotechnology: A review. *Bioresource Technology*, 248, 3–11.
57. Vymazal, J. (2011). Constructed wetlands for wastewater treatment: Five decades of experience. *Environ. Sci. Technol.*, 45(1), 61–69.
58. Riggio, V.A., Ruffino, B., Campo, G., Comino, E., Comoglio, C. and Zanetti, M. (2018). Constructed wetlands for the reuse of industrial wastewater: A case study. *J. Clean Prod.*, 171, 723–732.
59. Omidinia-Anarkoli, T. and Shayannejad, M. (2021). Improving the quality of stabilization pond effluents using hybrid constructed wetlands. *Sci. Total Environ.*, 801, 149615.
60. Odinga, C.A., Swalaha, F.M., Otieno, FAO, Ranjith, K.R. and Bux, F. (2013). Investigating the efficiency of constructed wetlands in the removal of heavy metals and enteric pathogens from wastewater. *Environ. Technol. Rev.*, 2(1), 1–16.
61. Ruan, W., Cai, H., Xu, X., Man, Y., Wang, R. et al. (2021). Efficiency and plant indication of nitrogen and phosphorus removal in constructed wetlands: A field-scale study in a frost-free area. *Sci. Total Environ.*, 799, 149301.
62. Grimblatt, V. (2020). IoT for agribusiness: An overview. *Proc. IEEE 11th Lat. Am. Symp. Circuits Syst.* (LASCAS), 1–4.
63. Zheng, Y., Sun, Z., Liu, Y., Cao, T., Zhang, H. et al. (2021). Phytoremediation mechanisms and plant eco-physiological response to microorganic contaminants in integrated vertical-flow constructed wetlands. *J. Hazard. Mater.*, 424, 127611.
64. Azubuike, C.C., Chikere, C.B. and Okpokwasili, G.C. (2016). Bioremediation techniques–classification based on site of application: Principles, advantages, limitations, and prospects. *World J. Microbiol. Biotechnol.*, 32(11), 1–18.
65. Pandey, P.K., Sharma, S.K. and Sambi, S.S. (2015). Removal of lead(II) from waste water on zeolite-NaX. *J. Environ. Chem. Eng.*, 3(4), 2604–2610.
66. Imran, M., Zulfiqar, S., Saeed, H. and Shakoori, F.R. (2021). Multi metal resistant *Klebsiella pneumoniae* KW is an efficient copper accumulator and bioremediator of industrial wastewater maimoona. *Pakistan J. Zool.*, 53(6), 2175–2182.
67. Han, S., Li, J., Zhou, Q., Liu, G. and Wang, T. (2019). Harmless disposal and resource utilization of wastes from the lake in China: Dewatering, composting and safety evaluation of fertilizer. *Algal Res.*, 43, 101623.

68. Gonçalves, E.C.B.M., Moura, F.J. and Teixeira, M.A. (2021). *Cyperus giganteus* pruning residues from constructed wetlands: Potential for energy production. *J. Clean Prod.*, 325, 129319.
69. Sultana, M.Y., Akratos, C.S., Pavlou, S. and Vayenas, D.V. (2014). Chromium removal in constructed wetlands: A review. *Int. Biodeterior. Biodegrad.*, 96, 181–190.
70. Wang, Q., Hu, Y., Xie, H. and Yang, Z. (2018). Constructed wetlands: A review on the role of radial oxygen loss in the rhizosphere by macrophytes. *Water*, 10(6).
71. Sharma, S., Singh, B. and Manchanda, V.K. (2015). Phytoremediation: Role of terrestrial plants and aquatic macrophytes in the remediation of radionuclides and heavy metal contaminated soil and water. *Environ. Sci. Pollut. Res.*, 22(2), 946–962.
72. Bi, R., Zhou, C., Jia, Y., Wang, S., Li, P. et al. (2019). Giving water bodies the treatment they need: A critical review of the application of constructed floating wetlands. *J. Environ. Manage.*, 238, 484–498.
73. Vymazal, J. (2013). Emergent plants used in free water surface constructed wetlands: A review. *Ecol. Eng.*, 61, 582–592.
74. Sundaravadivel, M. and Vigneswaran, S. (2001). Constructed wetlands for wastewater treatment. *Crit. Rev. Environ. Sci. Technol.*, 31(4), 351–409.
75. Dhote, S. and Dixit, S. (2009). Water quality improvement through macrophytes: A review. *Environ. Monit. Assess.*, 152(1–4), 149–153.
76. Pang, S., Zhang, S., Lv, X.Y., Han, B., Liu, K. et al. (2016). Characterization of bacterial community in biofilm and sediments of wetlands dominated by aquatic macrophytes. *Ecol. Eng.*, 97, 242–250.
77. Rezania, S., Taib, S.M., Md Din, M.F., Dahalan, F.A. and Kamyab, H. (2016). Comprehensive review on phytotechnology: Heavy metals removal by diverse aquatic plants species from wastewater. *J. Hazard. Mater.*, 318, 587–599.
78. Ansari, A.A., Naeem, M., Gill, S.S. and AlZuaibr, F.M. (2020). Phytoremediation of contaminated waters: An eco-friendly technology based on aquatic macrophytes application. *Egypt J. Aquat. Res.*, 46(4), 371–376.
79. Stefanakis, A.I. (2020). Constructed wetlands for sustainable wastewater treatment in hot and arid climates: Opportunities, challenges, and case studies in the Middle East. *Water*, 12(6).
80. Li, X., Yang, L., Xu, K., Bei, K., Zheng, X. et al. (2021). Application of constructed wetlands in treating rural sewage from source separation with high-influent nitrogen load: A review. *World J. Microbiol. Biotechnol.*, 37(8), 1–8.
81. Klink, A., Macioł, A., Wisłocka, M. and Krawczyk, J. (2013). Metal accumulation and distribution in the organs of *Typha latifolia* L. (cattail) and their potential use in bioindication. *Limnologica*, 43(3), 164–168.
82. Rana, V. and Maiti, S.K. (2018). Municipal wastewater treatment potential and metal accumulation strategies of *Colocasia esculenta* (L.) Schott and *Typha latifolia* (L.) in a constructed wetland. *Environ. Monit. Assess.*, 190(6).
83. Jeke, N.N., Zvomuya, F., Cicek, N., Ross, L. and Badiou, P. (2019). Nitrogen and phosphorus phytoextraction by Cattail (*Typha* spp.) during wetland-based phytoremediation of an end-of-life municipal lagoon. *J. Environ. Qual.*, 48(1), 24–31.
84. Hejna, M., Moscatelli, A., Stroppa, N., Onelli, E., Pilu, S. et al. (2020). Bioaccumulation of heavy metals from wastewater through a *Typha latifolia* and *Thelypteris palustris* phytoremediation system. *Chemosphere*, 241, 125018.

85. Singh, S. and Chakraborty, S. (2021). Bioremediation of acid mine drainage in constructed wetlands: Aspect of vegetation (*Typha latifolia*), loading rate and metal recovery. *Miner. Eng.*, 171, 107083.
86. Bonanno, G. (2013). Comparative performance of trace element bioaccumulation and biomonitoring in the plant species *Typha domingensis*, *Phragmites australis* and *Arundo donax*. *Ecotoxicol. Environ. Saf.*, 97, 124–130.
87. Mufarrege, M.M., Hadad, H.R., Di Luca, G.A. and Maine, M.A. (2014). Metal dynamics and tolerance of *Typha domingensis* exposed to high concentrations of Cr, Ni and Zn. *Ecotoxicol. Environ. Saf.*, 105(1), 90–96.
88. Mishra, J., Singh, R. and Arora, N.K. (2017). Alleviation of heavy metal stress in plants and remediation of soil by rhizosphere microorganisms. *Front. Microbiol.*, 8, 1–7.
89. Sricoth, T., Meeinkuirt, W., Pichtel, J., Taeprayoon, P. and Saengwilai, P. (2018). Synergistic phytoremediation of wastewater by two aquatic plants (*Typha angustifolia* and *Eichhornia crassipes*) and potential as biomass fuel. *Environ. Sci. Pollut. Res.*, 25(6), 5344–5358.
90. Malar, S., Sahi, S.V., Favas, P.J.C. and Venkatachalam, P. (2015). Mercury heavy-metal-induced physiochemical changes and genotoxic alterations in water hyacinths [*Eichhornia crassipes* (Mart.)]. *Environ. Sci. Pollut. Res.*, 22(6), 4597–4608.
91. Zahoor, A., Ahmad, F., Hameed, M. and Basra S.M.A. (2018). Structural and functional aspects of photosynthetic response in *Eichhornia crassipes* (Mart.) Solms under cadmium stress. *Pakistan J. Bot.*, 50(2), 489–493.
92. Chakrabarty, T., Afrin, R., Mia, M.Y. and Hossen, M.Z. (2017). Phytoremediation of chromium and some chemical parameters from tannery effluent by using water Hyacinth (*Eichhornia crassipes*). *Res Agric. Livest. Fish*, 4(3), 151–156.
93. Melignani, E., de Cabo, L.I. and Faggi, A.M. (2015). Copper uptake by *Eichhornia crassipes* exposed at high level concentrations. *Environ. Sci. Pollut. Res.*, 22(11), 8307–8315.
94. Suryandari, M.K., Hariati, A.M. and Mahmudi, M. (2017). Removal of lead from water by *Eichhornia crassipes* (Mart.) Solms. *Imp. J. Interdiscip. Res.*, 3, 2387–2392.
95. Jayaweera, M.W., Kasturiarachchi, J.C., Kularatne, R.K.A. and Wijeyekoon, S.L.J. (2007). Removal of aluminum by constructed wetlands with water hyacinth (*Eichhornia crassipes* (Mart.) Solms) grown under different nutritional conditions. *J. Environ. Sci. Heal. – Part A Toxic/Hazardous Subst. Environ. Eng.*, 42(2), 185–193.
96. Jayaweera, M.W., Kasturiarachchi, J.C., Kularatne, R.K.A. and Wijeyekoon, S.L.J. (2008). Contribution of water hyacinth (*Eichhornia crassipes* (Mart.) Solms) grown under different nutrient conditions to Fe-removal mechanisms in constructed wetlands. *J. Environ. Manage.*, 87(3), 450–460.
97. Estrella-Gómez, N.E., Sauri-Duch, E., Zapata-Pérez, O. and Santamaría, J.M. (2012). Glutathione plays a role in protecting leaves of *Salvinia minima* from Pb^{2+} damage associated with changes in the expression of SmGS genes and increased activity of GS. *Environ. Exp. Bot.*, 75, 188–194.
98. Wolff, G., Pereira, G., Castro, E., Louzada, J. and Coelho, F. (2012). The use of *Salvinia auriculata* as a bioindicator in aquatic ecosystems: Biomass and structure dependent on the cadmium concentration. *Brazilian J. Biol.*, 72(1), 71–77.
99. Bizzo, A.L.T., Intorne, A.C., Gomes, P.H., Suzuki, M.S. and Esteves, B.S. (2014). Short-term physiological responses to copper stress in *Salvinia auriculata* Aubl. *Acta Limnol Bras.*, 26(3), 268–277.

100. Ng, Y.S. and Chan, D.J.C. (2018). Phytoremediation capabilities of *Spirodela polyrhiza*, *Salvinia molesta* and *Lemna* sp. in synthetic wastewater: A comparative study. *Int. J. Phytoremediation*, 20(12), 1179–1186.
101. Otomo, J.I., de Jesus, T.A. Coelho, L.H.G., Monteiro, L.R., Hunter, C. et al. (2021). Effect of eight common Brazilian drugs on *Lemna minor* and *Salvinia auriculata* growth. *Environ. Sci. Pollut. Res.*, 28(32), 43747–43762.
102. Jayasri, M.A. and Suthindhiran, K. (2017). Effect of zinc and lead on the physiological and biochemical properties of aquatic plant *Lemna minor*: Its potential role in phytoremediation. *Appl. Water Sci.*, 7(3), 1247–1253.
103. Ceschin, S., Sgambato, V., Ellwood, N.T.W. and Zuccarello, V. (2019). Phytoremediation performance of *Lemna* communities in a constructed wetland system for wastewater treatment. *Environ. Exp. Bot.*, 162, 67–71.
104. Radulovic, O., Stankovic, S., Uzelac, B., Tadic, V., Trifunovic-Momcilov, M. et al. (2020). Phenol removal capacity of the common duckweed (*Lemna minor* L.) and six phenol-resistant bacterial strains from its rhizosphere: In vitro evaluation at high phenol concentrations. *Plants*, 9, 5991–6017.
105. Seifi, A. and Dehghani, M. (2021). Influence of nano-titanium dioxide particles (TiO_2 NPs) on improving phytoremediation efficiency of As/Cu/Cd from copper mine wastewaters using *Lemna minor*. *Arab. J. Geosci.*, 14(6).
106. Anand, S., Bharti, S.K., Kumar, S., Barman, S.C. and Kumar, N. (2019). Phytoremediation of heavy metals and pesticides present in water using aquatic macrophytes. *Phyto and Rhizo Remediation*, 89–119.
107. Kodituwakku, K.A.R.K. and Yatawara, M. (2020). Phytoremediation of industrial sewage sludge with *Eichhornia crassipes*, *Salvinia molesta* and *Pistia stratiotes* in batch fed free water flow constructed wetlands. *Bull. Environ. Contam. Toxicol.*, 104(5), 627–633.
108. Jasrotia, S., Kansal, A. and Mehra, A. (2017). Performance of aquatic plant species for phytoremediation of arsenic-contaminated water. *Appl. Water Sci.*, 7(2), 889–896.
109. de Oliveira, M., Atalla, A.A., Frihling, B.E.F., Cavalheri, P.S., Migliolo, L. and Filho, F.J.C.M. (2019). Ibuprofen and caffeine removal in vertical flow and free-floating macrophyte constructed wetlands with *Heliconia strata* and *Eichornia crassipes*. *Chem. Eng. J.*, 373, 458–467.
110. Dhir, B. and Srivastava, S. (2011). Heavy metal removal from a multi-metal solution and wastewater by *Salvinia natans*. *Ecol. Eng.*, 37(6), 893–896.
111. Mandal, C., Bera, S., Dey, N. and Adak, M. (2016). Physiological alterations of *Salvinia natans* L. exposed to aluminium stress and its interaction with polyamine. *Plant Sci. Today*, 3(2), 195.
112. Panfili, I., Bartucca, M.L., Ballerini, E. and Del Buono, D. (2017). Combination of aquatic species and safeners improves the remediation of copper polluted water. *Sci. Total Environ.*, 601–602, 1263–1270.
113. Sandoval, L., Zamora-Castro, S.A., Vidal-Álvarez, M. and Marín-Muñiz, J.L. (2019). Role of wetland plants and use of ornamental flowering plants in constructed wetlands for wastewater treatment: A review. *Appl. Sci.*, 9(4), 1–17.
114. Ohore, O.E., Qin, Z., Sanganyado, E., Wang, Y., Jiao, X. et al. (2022). Ecological impact of antibiotics on bioremediation performance of constructed wetlands: Microbial and plant dynamics, and potential antibiotic resistance genes hotspots. *J. Hazard. Mater.*, 424, 127495.
115. Mothé, G.P.B., Quintanilha-Peixoto, G., Souza, G.R., Ramos, A.C. and Intorne, A.C.

(2021). Overview of the role of nitrogen in copper pollution and bioremediation mediated by plant-microbe interactions. *Soil Nitrogen Ecology*, 249–264. Springer

116. Barroso, L.M. (2018). Interação de planta aquática e microrganismos promotores do crescimento vegetal resistentes a cobre. Dissertação de mestrado - Universidade Estadual do Norte Fluminense Darcy Ribeiro. Campos dos Goytacazes, RJ.
117. Silva, J.T.S. (2019). Promoção do crescimento de *Salvinia auriculata* Aublet mediada por *Enterobacter* sp. sob estresse por cobre. Dissertação de mestrado - Universidade Estadual do Norte Fluminense Darcy Ribeiro. Campos dos Goytacazes, RJ.
118. Orozimbo, K.B. da S. (2020). Bactérias promotoras do crescimento vegetal isoladas da planta aquática *Salvinia auriculata* com resistência a chumbo. Dissertação de mestrado - Universidade Estadual do Norte Fluminense Darcy Ribeiro. Campos dos Goytacazes, RJ.
119. Wang, D., Bai, Y. and Qu, J. (2021). The *Phragmites* root-inhabiting microbiome: A critical review on its composition and environmental application. *Engineering*, 9, 42–50.
120. Knapp, S., Schmauck, S. and Zehnsdorf, A. (2019). Biodiversity impact of green roofs and constructed wetlands as progressive eco-technologies in urban areas. *Sustain.*, 11(20).
121. Yozzo, D.J. and Osgood, D.T. (2013). Invertebrate communities of low-salinity wetlands: Overview and comparison between *Phragmites* and *Typha* marshes within the Hudson river estuary. *Estuaries and Coasts*, 36(3), 575–584.
122. Meng, P., Pei, H., Hu, W., Shao, Y. and Li, Z. (2014). How to increase microbial degradation in constructed wetlands: Influencing factors and improvement measures. *Bioresour. Technol.*, 157, 316–326.
123. Qiu, Z., Egidi, E., Liu, H., Kaur, S. and Singh, B.K. (2019). New frontiers in agriculture productivity: Optimised microbial inoculants and *in situ* microbiome engineering. *Biotechnol. Adv.*, 37(6).
124. Lee, D.J., Show, K.Y. and Wang, A. (2013). Unconventional approaches to isolation and enrichment of functional microbial consortium: A review. *Bioresour. Technol.*, 136, 697–706.
125. Datta, S., Rajnish, K.N., Samuel, M.S., Pugazlendhi, A. and Selvarajan, E. (2020). Metagenomic applications in microbial diversity, bioremediation, pollution monitoring, enzyme and drug discovery. A review. *Environ. Chem. Lett.*, 18(4). 1229–1241.
126. Zhen, Z., Yan, C. and Zhao, Y. (2020). Influence of epiphytic bacteria on arsenic metabolism in *Hydrilla verticillata*. *Environ. Pollut.*, 261, 114232.
127. Pandey, A., Tripathi, P.H., Tripathi, A.H., Pandey, S.C. and Gangola, S. (2019). Omics technology to study bioremediation and respective enzymes smart bioremediation technologies: Microbial enzymes. Elsevier Inc., 23–43.
128. Chandran, H., Meena, M. and Sharma, K. (2020). Microbial biodiversity and bioremediation assessment through omics approaches. *Front. Environ. Chem.*, 1, 1–22.
129. Marzan, L.W., Hossain, M., Mina, S.A., Akter, Y. and Chowdhury, A.M.M.A. (2017). Isolation and biochemical characterization of heavy-metal resistant bacteria from tannery effluent in Chittagong city, Bangladesh: Bioremediation viewpoint. *Egypt J. Aquat. Res.*, 43(1), 65–74.
130. Sessitsch, A., Kuffner, M., Kidd, P., Vangronsveld, J., Wenzel, W.W. et al. (2013). The role of plant-associated bacteria in the mobilization and phytoextraction of trace elements in contaminated soils. *Soil Biol. Biochem.*, 60, 182–194.

131. Rodriguez-Dominguez, M.A., Konnerup, D., Brix, H. and Arias, C.A. (2020). Constructed wetlands in Latin America and the Caribbean: A review of experiences during the last decade. *Water*, 12(6).
132. Maine, M.A., Sánchez, G.C., Hadad, H.R., Caffaratti, S.E., Pedro, M.D.C. et al. (2016). Humedales construidos para tratamiento de efluentes de industrias metalúrgicas en Santa Fe, Argentina. *Tecnol y Ciencias del Agua*, 7(1), 5–16.
133. Noyola, A., Padilla-Rivera, A., Morgan-Sagastume, J.M., Güereca, L.P. and Hernández-Padilla, F. (2012). Typology of municipal wastewater treatment technologies in Latin America. *Clean – Soil, Air, Water*, 40(9), 926–932.
134. CONAGUA (Comisión Nacional del Agua). (2019). Inventario nacional de plantas municipales de potabilizacióny de tratamiento de aguas residuales en operación. México D.F. Secretaria de Medio Ambiente y Recursos Naturales.
135. FAO. (2014). Reutilización del agua en la agricultura: ¿Beneficios para todos? Inform esobretemashídricos. Roma, Italia.
136. Curty, A.F. (2021). Recovery of urban landscape and biodiversity through phytopurification. *2021 International Conference: 27th World Congress of Architects*, 61–67.
137. de Souza, F.P. and Azevedo, J.P.S. (2020). An overview of urban lagoons in Rio de Janeiro: Relevant aspects in the management of Rodrigo de Freitas and Araruama lagoons and the Jacarepaguá lagoon system. *Eng. Sanit e Ambient*, 25(1), 197–204.
138. García, J.A., Paredes, D. and Cubillos, J.A. (2013). Effect of plants and the combination of wetland treatment type systems on pathogen removal in tropical climate conditions. *Ecol. Eng.*, 58, 57–62.
139. Avellán, C.T., Rousseau, D.P.L. and Lens, P.N.L. (2011). Adapting to socioeconomic, operational and environmental challenges of dairy farm effluent purification in Uruguay through the use of surface flow constructed wetlands. *Water Pract. Technol.*, 6(3).
140. Maine Hrhma. (2016). Memorias de la III conferencia panamericana de sistemas de humedales para el tratamiento y mejoramiento de la calidad del agua. *In*: Memorias de la III Conferencia Panamericana de Sistemas de Humedales para el Tratamiento y Mejoramiento de la Calidad del Agua.
141. Boursianis, A.D., Papadopoulou, M.S., Gotsis, A., Wan, S., Sarigiannidis, P. et al. (2021). Smart irrigation system for precision agriculture: The AREThOU5A IoT Platform. *IEEE Sens. J.*, 21(16), 17539–17547.
142. Gagliardi, G., Lupia, M., Cario, G., Gaccio, F.C., Angelo, V.D. et al. (2021). An Internet of Things Solution for smart agriculture. *Agronomy*, 11(2140), 1–18.
143. Winkler, R. (2021). MeteoMex: Open infrastructure for networked environmental monitoring and agriculture 4.0. *Peer J. Comput. Sci.*, 7, 1–23.
144. Sharma, N., Chandrasekar, S. and Sundar, K. (2018). Suitability of ANN and GP for predicting soak pit tank efficiency under limited data conditions. *MATEC Web Conf.* 203.
145. Nanda, I., Shukla, V.K., Dhanekula, V.J., Gadipudi, M., Penugonda, V. et al. (2021). Design and real-time implementation of smart water management using LabVIEW and IoT. *Proc 2nd IEEE Int. Conf. Comput. Intell. Knowl. Econ.*, 524–528.
146. Ramesh, M.V., Nibi, K.V., Kurup, A., Mohan, R., Aiswarya, A. et al. (2017). Water quality monitoring and waste management using IoT. *GHTC 2017 – IEEE Glob., Humanit. Technol. Conf. Proc.*, 1–7.

147. Rotondi, D., Straniero, L., Saltarella, M., Balducci, F., Impedovo, D. and Pirlo, G. (2019). Semantics for wastewater reuse in agriculture. *Conf. Proc. – IEEE Int. Conf. Syst. Man. Cybern.*, 598–603.
148. Ar, I.M. (2012). The impact of green product innovation on firm performance and competitive capability: the moderating role of managerial environmental concern. *Procedia Soc. Behav. Sci.*, 62, 854–864.
149. Muthukrishnan, L. (2021). An overview on the nanotechnological expansion, toxicity assessment, and remediating approaches in agriculture and food industry. *Environ. Technol. Innov.*, 102136.
150. Estrela, V.V., Hemanth, J., Saotome, O., Nikolakopoulos G. and Sabatini, R. (2020). *Imaging and Sensing for Unmanned Aircraft Systems: Deployment and Applications*, 2. ISBN 978-1-78561-6440.
151. Pal, P. (2017). Nanotechnology in water treatment. *Industrial Water Treatment*, 513–536.
152. Razmjooy, N. and Estrela, V.V. (2019). *Applications of Image Processing and Soft Computing Systems in Agriculture*. IGI Global.
153. Hemanth, J. and Estrela, V.V. (2017). Deep learning for image processing applications. *In*: Advances in Parallel Computing. Amsterdam, Netherlands, 31: IOS Press. ISSN 978-1-61499-822-8.
154. Wongburi, P. and Park, J.K. (2021). Big Data analytics from a wastewater treatment plant. *Sustainability*, 13, 12383.
155. Estrela, V.V., Hemanth J., Loschi, H.J., Nascimento, D.A., Iano, Y. and Razmjooy, N. (2020). Computer vision and data storage in UAVs. *Imaging and Sensing for Unmanned Aircraft Systems*, 2(1), 23–46. IET, London, UK.
156. Estrela, V.V., Andreopoulos, N., Sroufer, R., de Jesus, M.A., Mamani, W.D.H. and Peixoto, A. (2021). Transmedia ecosystems, quality of experience, and quality of service in fog computing for comfortable learning. *2021 IEEE Global Engineering Education Conference (EDUCON)*, 1003–1009. IEEE.
157. Sen Gupta, Y., Mukherjee, S., Dutta, R. and Bhattacharya, S. (2021). A blockchain-based approach using smart contracts to develop a smart waste management system. *Int. J. Environ. Sci. Techno.*, 1–24.
158. Khan, A.A., Laghari, A.A., Shaikh, A.A., Dootio, M.A., Estrela, V.V. and Lopes, R.T. (2022). A blockchain security module for brain-computer interface (BCI) with multimedia life cycle framework (MLCF). *Neurosci. Informatics*, 2, 100030.

CHAPTER

5

Herdade do Freixo do Meio: A Case Study of Mediterranean Agroforestry in Portugal Analyzed from Place to People Perspective

Albino Lopes[1][0000-0001-8836-0024], Patrícia Araújo[2,3,4*][0000-0002-4211-9638], Carla Gomes da Costa[5][0000-0003-3711-5621] and Lígia Nascimento[2,3,6][0000-0002-6285-1631]

[1] Centro de Administração e Políticas Públicas (CAPP), ISCSP, Lisboa, Portugal
[2] ISMAT, Intrepid Lab, Portimão, Portugal
[3] CETRAD, Porto, Portugal
[4] Unidade de Investigação e Desenvolvimento do ISLA Santarém, Santarém, Portugal
[5] CETRAD – Universidade Europeia – Lisboa, Portugal
[6] Centro de Investigação em Organizações, Mercados e Gestão Industrial (COMEGI), Universidade Lusíada de Lisboa, Portugal

This study uses in-depth interviews, observation, document analysis, and future-based investigative exercises to examine the 'Herdade do Freixo do Meio' (HFM), a State, manor, or grange in the South of Portugal (Mediterranean). The sustainable, integrative practices implemented include agroforestry studies, tourism, electronic commerce, archeology, general management, business practices, and technology and research's role in the "big picture" intervention. The Syntropic farming concept inspired the study, despite the term only emerging in 2013. According to some authors, a participatory business structure increases the value of an agroforestry activity, creating the conditions for implementing a smart, ecological, locally based, economically sustainable, and generalizable new production paradigm. Results indicate biodiversity and a sustainable economy. Besides recovering the 'montado' (cork and holm oaks), HFM has a dozen continuous differentiation-integration experiences: roughly 15 distinct fruit tree species arranged in the 'community' were planted in degraded

* Corresponding author: pattaraujo@gmail.com

property regions. A four-quadrant cognitive framework (opposed two by two) originated from ideas raised in interviews and validated through documents and people's activity observation. It points to future effort paths and aspects of the transformation already undertaken: (i) the rules regarding main routines of primary work and transformation/packaging/marketing are being stabilized; (ii) both individual and group initiatives are still insufficient; (iii) the team spirit and collaborative culture is shown, and all parties desire its deepening; and (iv) the high added-value products organization for customers is advanced, but lacking the professional work regarding cost control. After three years, biodiversity destruction has reversed. Business-wise, financial recovery was achieved with an increasing profit.

1. Introduction

Nowadays, "the extraction of global resources is paralleled by expulsions of living beings from social, economic, and physical environmental spaces" [1], and finally, humans are understanding and striving for the emergence of new paradigms, including green governance models [2] and other nature-based solutions, to try to give voice to a new revolution, one that takes all the ecosystem truly into account.

The term nature-based solutions, "in the context of the exclusionary and predatory projects that gather behind it, reveals something fundamental. Western civilization has based much of its domination on its troubled and hostile relationship with nature: an exploitable resource and something to venerate as untouched and 'pure' [3].

As Suša states [1], describing the overall societal feeling:

> "Short-term interests and exploitation devastating social and environmental conditions on this planet drive most threats: forests are scalped, rivers run dry, species are going extinct, and humans are changing the climate. Institutional systems are programmed to squeeze more resources from the planet. Societies fail to recognize hidden threats, and the ability to survive is questioned [...]. It seems evident that, in the absence of a struggle for revolutionary transformation, alternative civilizational rules, and responsible global regulations, which could be both political and moral, the (relatively) long-term future of capitalist self-market-regulation-through-destruction may be connected with the failure of humankind to respond."

In response, Agriculture 4.0 (Agri4) emerges as an analogy to Industry 4.0 (aka the Fourth Industrial Revolution or I4) and perspectives on the improvement of income, sustainability, and working conditions, as well as the challenges related to improving food production and avoiding biodiversity loss [4].

Technology increases agricultural sector competitiveness through specialization and new technological demands. However, these authors state that the current specialization of agriculture, monoculture, is not as efficient as diversified agricultural systems in supporting biodiversity, maintaining soil quality, carbon sequestration, water retention, energy efficiency, resistance, and resilience to climate change.

According to Libinsky et al. [4], among the multiple challenges to be overcome, four of them stand out, which deserve attention from research, political leaders, rural producers, and technology developers: (i) the need to increase food and energy production sustainably; (ii) the need for investments in infrastructure and connectivity; (iii) the risk of crop concentration and loss of biodiversity; (iv) specialization risk and economic competitiveness based in few products.

To Rose et al. [5], three tenets of sustainable intensification should guide Agri4: people, production, and the planet. Many Agri4-related narratives regarding environmental benefits have been predominately framed, with little attention on social sustainability. Authors contribute with a framework of multi-actor co-innovation to guide responsible socio-technical transitions. In this sense—multi-actor co-innovation—we will describe a case study in Portugal involving multiple actors and people management.

So, humans have been searching for nature-based solutions, although there is somehow a paradox within this search, as Schmidlehner [3] notices:

> "Nature – objectified as 'resource' – can be aggressively exploited without moral scruples; living beings can be crammed into monocultures or industrial livestock farms and genetically manipulated to maximize production. This predatory attitude is contrasted with the romantic veneration of an ideal distant 'motherly' nature, bringing forth images of untouched natural landscapes and 'virgin forests'."

Regarding Portugal's situation, since the 1960s, agricultural activity has faced a paradox worthy of careful consideration when analyzed from a management perspective. While management studies had begun to discover the differentiation strategy virtuosities and the consequent need for integration [6], an intensification of standardized production in fields is seen as monoculture systems (agricultural and/or forestry).

An intensified agricultural economy, based on standardized production is translated into an intensive and mechanized monoculture system, having also dissociated the branches of agriculture and forest with serious ecological consequences. This specialized, subsidized, and globalized agricultural economy, also destructive of small and medium-sized properties, voracious in energy and phytosanitary products, has been assumed the aura of a "green revolution" promoting population and metropoles growth, appearing to public opinion—against all warnings of multiple worldwide research [7, 8]—as the 'solution' for the world. However, a minority of people became aware that a serious ecological crisis was underway, which had been previously announced for a long time, evidencing that the center of all possible and widespread revolutions is always moved by people and their mindset, as the case of the newborn blue revolution.

This chapter aims to present the Herdade do Freixo do Meio (HFM) case study. HFM is in the municipality of Montemor-o-Novo, the location where an experience was initiated four years ago. HFM is a State, Manor, or Grange, a big, diverse farm in the South of Portugal (Mediterranean).

Over the past 50 years, the landscape, agro-industrial work, and economy have changed drastically. It has been a troubled period characterized by a succession of events: (i) from the creation to the closing of an agricultural cooperative, which was later seen as having no viability, thus worsening the latifundium structure deficiencies (580 acres destined for cork oak forest/'montado', sheep grazing and wheat cultivation areas); (ii) afterwards, a business ecosystem of diversified food production succeeded the previous agricultural cooperative (currently with 500 IDs/references, sold in an association system of regular consumers, physical and e-commerce stores); (iii) from this business ecosystem, a set of seven industrial micro-units have been developed for a long time, which process the agroforestry production; (iv) thereafter, an ecological experimentation was started in 10 fruit tree groves and vegetable production trials; (v) various formulas for reversing the biodiversity degradation continue to be tested in order to determine which would be the best; (vi) finally, a new entrepreneurial-cooperative experiment was recreated (a 'mixed' solution of work and consumption), involving about 30 workers, some partnerships, and about 200 consumers, i.e.,

> "From a technical point of view, the target area is a system characterized as agroforestry in which three types of strata coexist: the dispersed arboreal, shrub, and creeping plant strata. According to the literature review from [12], in Portugal, the cork oak forest, namely, the holm oak, is in regression, and, as a result, the soils of this agroforestry region (about one million ha) are in an advanced state of aridity."

At HFM, located in the southern region of Portugal (the most markedly Mediterranean), from a strictly technical point of view, in addition to having recovered the space designated as 'montado' of cork oaks and holm oaks, experiments are effectively underway that we can read as differentiation-integration, in the light of the reflection of the authors cited above, i.e., around 15 different species of fruit trees have been planted. The radical novelty is the fact that groves of a new type, whose experience we describe below, are organized into natural 'communities' in the image of teams of people with differentiated profiles [9] and have been planted in degraded areas of the property, according to a method adapted from [10]. The first succession/syntropic agroforestry experiment began four years ago. Verifying a situation of practically a complete reversal of biodiversity degradation is possible. In addition, from a managerial point of view, the people in charge of HFM seem to be trying new experimentation socially relevant to transforming the agricultural enterprise into a community (which could configure a second-order cooperative model), integrating activities organized in micro-enterprises, and could aspire to achieve 'lean' management [11].

The management model to be tested (if the occasion and conditions are favorable) would follow the strict principles of total quality management and cooperation, thus opting for an organizational formula that is eminently participative and oriented towards the greatest possible autonomy and integration. The idea pursued is to propose to the cooperative's management that HFM could initiate a process of Organizational Development (OD) in two phases: (i) as a business incubator; (ii) to be able to evolve, afterward, into a Multifunctional Cooperative formula, integrating a plurality of productive micro-units. The businesses to be developed would lose nothing with the autonomy, as proven by a plurality of studies [13]. The authors study three strategies to achieve success: partnerships that can be activated; a communication strategy based on proximity or other values; and the choice of one or more specialists in that type of business.

The leadership/management dyad could be separated from each other: (i) current leadership could start reserving itself for the function of animating the global experience, presiding over the 'cooperative' in transformation and the institutional development of the project; (ii) a manager (professional?) could, in the short/medium term, start taking over the strictly operational management of the business, in articulation with the people in charge of the various units that would be stabilized, under the superior direction of the multifunctional cooperative.

Looking ahead, we intend to find an alternative to the grain or forest monocultures. Suppose this agroforestry experimentation can take advantage of all its potential, from a production capable of feeding 1,200 people in the medium term. In that case, we can move to a total of 12,000 while guaranteeing three objectives that have been long postponed in this region that has suffered so much since the infamous wheat campaigns of the 1940s: (i) an agile and participative organization; (ii) endowed with a production that consumers value; (iii) capable of guaranteeing better salaries; (iv) generating much more and better employment; (v) and above all, in a time of dramatic environmental stress, to produce 'syntropic', nature in balance and which is sustainable. To generate gains in terms of arable land, that is, to achieve the inverse of the entropy to which nature has been subjected to since the so-called cereal revolution of the Neolithic period, when the sedentarization of the human species happened.

Approached from this global sustainability perspective, agroforestry can be the "lever that lifts the world" that Archimedes dreamt of. Agroforestry differentiates and integrates where other organizational systems have systematically failed in one or both dimensions. This type of experimentation can be conducted without needing more resources than communities can mobilize.

The fight against climate change seems more complex and simpler than has been admitted at a time when everything seems to indicate that the future is heading towards continuing the same solutions as in the past, based on 'interested' solutions. Sustainable solutions will be postponed as long as the disconnection introduced by the idea of monoculture is maintained, even if more space is allocated to so-called wilderness areas.

The example of Amazon reported, and so celebrated by the media in the autumn of 2021 and named "Partners for the Amazon Platform (PPA)", an institution placed under the sponsorship of the American USAID, is an example of the recipe—an "aspirin for cancer"—that is, it would have achieved a collection of about $800,000 to support small businesses of the indigenous communities, designated as the guardians of the forest.

Although this apparent reconciliation with nature may seem appealing, the large wilderness areas will probably be managed by lobbies specialized in taking advantage of new 'environmental' financial resources models. A not-at-all-reassuring index is the resounding failure of fossil fuel consumption, which has not fallen for several decades (80% of total energy).

Humankind, knowledge, and nature meet in the complexity and simplicity of procedures in this Organizational Development case study; in turn, it seems to be the challenge insufficiently made aware of by stakeholders at the most diverse levels.

2. Research Problem

What above all motivates and justifies scientific research is not the search for a possible consensus among experts, but rather the search for observable data, for facts accepted by the plurality of the various interested parties (stakeholders), validating them by confronting them with the hypothesis of reproducible experimentation. In September 2021, Germany was on the stage for demonstrations against "intensive agriculture" in the Algarve and Alentejo (Portugal), such as the one organized by Friederike Huer, an activist who organizes tourist trips.

A case study, within the limits of its apparent modesty, has the merit of searching for data and its validation through experimentation. In addition, the research progress consists, according to K. Popper's well-known proposal [14], of the search for the small detail that allows the so-called scientific consensus to be invalidated rather than in its confirmation.

The present case set out above seeks a double confrontation, even if, in terms of the wording of this text, we hope to combine the two: (i) that of the opinion generally favorable to the industrial management model of arable land; (ii) and that of the effectiveness of participative management by agricultural workers, organized in small autonomous enterprises (in reality they would be micro-units), integrating a second-order cooperative.

The need to find a way to enhance environmental and economic sustainability does not seem to be compatible with the current situation of disconnection (agroforestry sustainability, accompanied by a persistent economic imbalance), as noted by [15], a conclusion which deserved the endorsement of the HFM leader, the co-supervisor of the referencing thesis.

Could the core of the environmental problem be more of the productive unit's leadership/management order, in general, and of strategic human resources management (SHRM) in particular, than of a technical order?

Figure 1 summarizes how everything flows until it reaches the final two squares concerning the two main stakeholders: producers and consumers, and how this project might change those people's lives for a new sustainable paradigm.

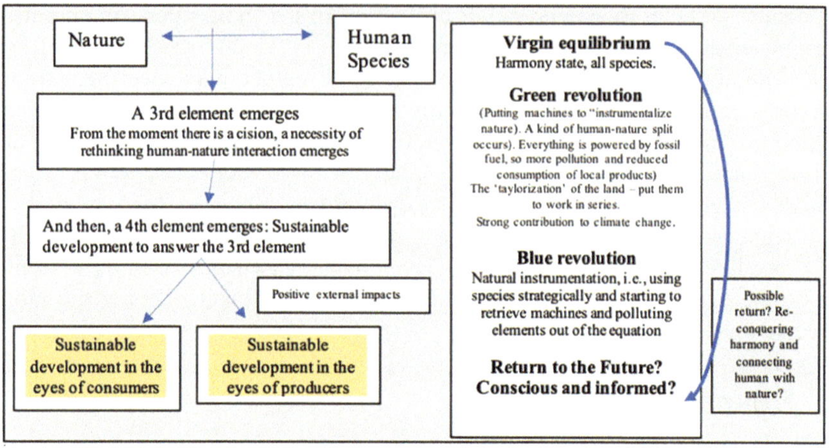

Fig. 1. Diagram flow: From virgin to integrated nature, to the return to the 'future' and its impacts on people.

3. Theoretical Approach

The elimination of forests to disseminate intensive agricultural or agroforestry production has led to a situation in which the following are combined and enhanced, if we refer, among others, to Pasini's literature review [16]: (i) the degradation of agricultural soils; (ii) the loss of cellulosic raw materials; (iii) the need for large amounts of fertilizers, water, pesticides, and fuel for the machinery involved in the production process; (iv) the destruction of an integrated system of the agroforestry type, capable of storing huge amounts of carbon (or CO_2), suspended in the atmosphere, and which would be associated with the climate change that affects us (involving a high risk of lack of environmental control). It should be noted that only a small part of the CO_2 emitted is dissolved in the atmosphere, most of it going into the oceans. Climate change, caused by CO_2, is today the third cause of biodiversity loss, but it could soon overtake the first two (human activity and deforestation in favor of agriculture). In turn, the oceans are becoming more acidified so calcium fixation will be affected. The rise in sea level, which risks being dramatic in the next two centuries, is linked to what is being done today. This data from UN bodies (namely the GIEC) also speaks of inequality, with the populations that emit the least CO_2 suffering the most.

These four very complex issues, with which humankind is confronted, underlie this so-called "green revolution" deemed necessary to satisfy the growing need for food. There are also other associated issues, namely: (i) the economic and nutritional importance for populations; (ii) the quantification of the positive externalities in terms of biodiversity, as well as the respective remuneration in favor of the populations which could have an interest in safeguarding the same forests. These difficulties, which have existed for several years, seem to perpetuate.

From the farm management (and, even more specifically, SHRM) point of view, the issues are no less complex. Indeed, the difficulties regarding effective coordination of collective action are long-standing and difficult to resolve, neither by the market nor the state. In the expectation of a solution through action by the state, easier and more immediate, people seem trapped in an overwhelming dilemma: (i) there is an urgency to find a solution; (ii) the state has powerful means; (iii) the disconnection between environmental and economic sustainability persists. The public powers seem to be even less capable of equating the terms of an efficient solution than the private ones by failing to find a safe course to a necessary collaboration of the actors involved in power games, every day more complex (in terms of authors such as [17-19]. Meanwhile, public opinion begins to perceive the situation as dramatic, and science struggles with the natural difficulty in producing evidence without reaching conclusions acceptable to all stakeholders [20].

The studies carried out on the specific region of the Mediterranean cork oak forest, as [21] states, identify a set of issues to be addressed, which can be summarized as follows, seemingly articulated around the need for the whole of society (read, above all, consumers) to become aware of the environmental sustainability problem: (i) the desirable maintenance of the cork oak forest model cannot be left solely to farmers, although they are, as direct managers of ecosystems, the key to the solution; (ii) like any economic activity, multifunctional agriculture depends on support markets, hence on the society as a whole; (iii) it would be necessary to create more efficient mechanisms that remunerate and encourage regulation and support services; (iv) in a special way, the most conscious consumers, who value food security and environmental sustainability, would provide the pillar of economic sustainability to multifunctional agriculture to be developed.

Assuming the theoretical position is based on this 'awareness' centrality, which we call the majority thesis, farms involved in a complex network of "meeting points" between organic production and conscious consumers who bought these goods, would be sustainable, benefiting health and the environment. This solution implementation is complicated, but one that is easy to understand for all interested parties. It is curious to note that the social representation of the proposed solution is equivalent to the Taylorist-Fordist model (produce to sell), abandoned by the Western industry following the 1973/1975 crisis [22], in the wake of which companies turned to the Japanese solution (sell to produce, according to [11]. From this paradigm, technical issues would be complicated but easy to execute. Yet, the problem becomes more complex from the perspective of people's suitability regarding this production model.

To sum up: in agriculture and forestry, especially in fragile areas such as the Mediterranean cork oak forest, the transition from mass production supported by sophisticated technical systems and simple-to-understand human skills (the green revolution) to a complex articulation between supply and demand requires managing between the two sides of the economy, based on organized people involved in an organization open to improvisation [23] and the permanent construction of provisional solutions that are always being updated.

4. Reformulated Problem Based on the "Think Local" Approach and Exploring a New Working Hypothesis

This work explores the two listed questions (the technical and the managerial) and the intricacies of an action project grounded in a social nature in the form of a (mixed) cooperative that conciliates production and consumption, from which the prospects of private profit and centrally imposed directive action would, by definition, be excluded.

But does it make sense to study a particularly complex issue from a specific case in a relatively arid area of the Mediterranean forest when the global issues are so serious? The doubt relates to the widespread conviction that, to solve local and global problems, it is necessary to start by "thinking globally": in practical terms, the awareness of the whole society.

"Thinking local" may be a better method to prepare for global action than "thinking globally". As per Voiron, C. et al. [24], this is especially true in the unknown Mediterranean.

Since the 1960s, the eco-farming world in Portugal has faced a situation that, when analyzed from the people management angle, reveals a paradox that should be carefully considered. This paradox is that agricultural and forestry management contradicts the best internationally established practices. Indeed, at the same time, research into the management related to large companies (particularly industrial companies) is discovering the virtues of combining differentiation and integration [6] on the one hand and the small autonomous or semi-autonomous business dimension as a space for freedom and creativity [25] on the other, in the agricultural economy management we are witnessing the increasing dominance of standardized production, limited to a few species only. There would be about 7,000 edible plant species in temperate climates, according to what can be seen in the video about permaculture in [26]. This has been going on for some time, but the systematic error of monocultures has been translated into an increasingly advanced system of 'monoculture', intensive and progressively mechanized. Machines have their place, as the Centre for Research on Syntropic Agriculture (CEPEAS) video shows [27].

The agriculture and forestry branches, traditionally interconnected in a 'syntropic' way, were also disassociated, with the most serious ecological consequences, completing the cycle of estrangement between human culture and nature. At times, the so-called eco-functionality (the idea that some species should not even be allowed to exist) has been radically abandoned. It is as if we wanted to find another (entropic) basis for thinking about the laws of nature, challenging the "divine power" of which the biblical parables speak, or the Greek fables of

Esopus enlighten us. Scientists who have written about climate issues in *Nature* magazine, for example, consider that the conclusions of the 1972 Club of Rome report (*The Limits to Growth*) about predictions on energy consumption, pollution, and population growth were deemed alarmist (and "irresponsibly absurd"). Today the conclusions would be more tailored and alarming than ever [28, 29].

Götsch, a pioneer who has studied these themes, says that it was as if humankind had been punished by being condemned to find ways of resisting the plagues it unleashes [10]. It will have to do so, he presumes, either through phytosanitary measures or through genetic modifications or genetically modified organisms (GMOs), but without achieving the stability lost by giving up sustainable agriculture (in a situation of sharing with other living beings). In fact, as in the myth of Sisyphus, when the pharmaceutical industry seems to have found the "effective poison" against a given pest, the latter strengthens again after the inevitable adaptation and the resulting resistance capacity acquisition.

The 'new' agricultural/forestry economy, prevailing after the Second World War, which is markedly industrial, seems to have acquired several characteristics that we could try to systematize. It has become specialized, subsidized, globalized, far removed from the balance inherent in biodiversity (with the massive loss of trees, shrubs, pastures, and other green elements of a natural fertilizing type) and dependent on large-scale or chemical fertilization and phytosanitary treatments. From a socioeconomic point of view, it has taken the form of a primary production that is averse to employment and to the defense of small and medium-sized properties (see the case of a medium-sized property Happen Filns (2021), Syntropic Forest with 17 years old [32]), voracious in water (which should be preserved and not consumed in irrational irrigation systems), extremely demanding in energy, or even permeable to experiments leading to genetic transformations and drastic reductions in traditional varieties, among many other multiple aspects.

In the meantime, it has assumed the mythical mantle of the "green revolution", which has led to population growth (consider independent India, which in 1947 had 360 million and today's population has jumped to about 1,300 million), especially in large metropolises. This agricultural model has finally become visible to public opinion, against all the warnings of leading researchers [30, 31] among other critical consciences [7, 8], as the 'solution' for feeding the world, starting with the poorest and most populous countries.

In this context, it is important to explore the generic hypothesis that it was not traditional agriculture that impoverished the people but rather the lack of adequate organization, in conjunction with the introduction of the idea of collective intelligence, effective in the productive system, per the intuition of [6], when these authors looked at what hindered the creation of value in the industrial-based economy (the deficit of articulation between the contradictory movements of differentiation and integration).

For all that, it seems certain that a critical consciousness that would allow innovation has not been created. Even today, there is no concerted action against the currents that have led to funding energy, production, and the commercialization of products such as coffee or soya. The debate on the possibilities of action by local authorities to consolidate an eco-economic paradigm has not been fruitful. Above all, more global policies have been left to the powerful from all ideological backgrounds.

Thus, in the agroforestry world, there was widespread adherence to the industrial paradigm at a time when the paradigm known as techno-services [38], anticipated the consequences of the revolution by the internet in terms of organizational skills centrality. This rationale consists of bringing together the ability to produce more food (by combining the study of the 'intelligent' action of natural species in ecosystems and conducting experiments guided and enhanced by the action of the knowledge thus obtained) and, at the same time, improving arable land. Only the result of this triple combination of interactions: the action of nature, the study of this 'intelligent' action, and human experimentation that respects the laws of ecosystems. The succession of plantations, for example, cannot be well achieved without this triple conceptualization. This will allow the emergence of the new economy, also called the circular or blue economy by Gunter Pauli, which means using the land economically, preserving it, and improving it.

Relying on the same author, Lopes [9] advocated the interactive character of the New Information and Communication Technologies (NICT) in the work organization, as in social life. Since the 1990s, NICT has powerfully influenced a whole series of major consequences, among which are: (i) the productive reorganization; (ii) the elimination of hierarchical and bureaucratic models, in particular with the valorization of the organizational culture management; and (iii) the redefinition of political power, which would no longer have the ability to control activities and regulatory instruments fully; (iv) or, the markets globalization.

The consequences of this revolution have been read extremely, and it is necessary to reflect on it once again regarding the so-called "green revolution". The latter, left to its dynamic, subjected to the techno-services revolution, would allow the big monopolies to take advantage of the first two consequences, to impose a globalization of the financialized economy, and to limit or dominate the regulatory power of national states.

Meanwhile, the Crozerian perspective met the anticipation conditions of what was being prepared, i.e., the technical-economic bases to demonstrate that the former space of tangible goods would be occupied by collective knowledge/intelligence. This would irreversibly, more than the investments in tangible means, be converted into value [39].

According to the same authors, there would be eight critical success factors (of intangible nature) capable of leading technologically competent economies to act in the complexity and to take advantage of the produced knowledge. Adapting the reasoning to the agroforestry farms situation and leading them to compete effectively against those that bet on standardized processes, the agroforestry farms should be excellent at (i) generating passion in the workers; (ii) constantly reinventing the organization; (iii) developing the teams' concentration and focus;

(iv) creating quality service supporters; (v) adapting the teams to change; (vi) being quick in terms of responsiveness; (vii) betting on flexibility; (viii) generating confidence.

Together with the paradigm shift in the business economy comes, thus, a new Gestalt of concepts that, if applied to the agricultural economy, is likely to lead to an awareness, even if by a small minority of people, that a serious ecological crisis, long announced, is underway, and that action must be taken. The priority is, therefore, to show to society that there are examples of considering the cause-and-effect relationship and, above all, that it is possible to get out of the impasse by competing, in the area of agroforestry, based on the idea of leading a truly assumed culture of implementing gradual procedures in search of greater added value [40].

5. Research Questions

The story that interests us begins in 1990, with the return of a 580 acres estate to the previously expropriated owners, after a 15-year 'adventure' under the banner of the so-called "Agrarian Reform". "Agrarian Reform" consisted of expropriating the large estates in the country's south (latifundia zone) and giving them to rural workers organized in cooperatives, who would continue to exploit the land and take over its management. The "Red Carnation" cooperative (one of the flagships of that reform) left an ecologically heavy heritage supported by chemical-mechanical agriculture. It cultivated, almost exclusively, wheat (still very dependent on heavy machinery—soil industrialization, using phytosanitary products and fertilizers), sheep farming, and cork production (which paid the bills). Nature would work like a clock: we would want it to work by eliminating parts. The hours it ticks may be equivalent to the end of the species experimenting against natural laws.

The paradox to be highlighted is related to the unemployment raging in the village of Foros de Vale de Figueira. It is the property's new management that will foster job creation. They begin immediately with poultry farming. The diversification and eco-functionality, which attributes a specific function to each being, breaking with the theory of specialization, will effectively coincide with the creation of wealth and the multiplication of employment [33]. This new orientation gave rise, in the meantime, as mentioned, to seven micro-factories involving more than 500 biological references/products, instead of concentrating on the production of raw materials: (i) olive oil mill; (ii) acorn (a 'superfood', as Alfredo Sendim emphasizes [34]) husking and flour production unit, which, in turn, allows advancing towards the production of sausages, and notably those of the 'vegan' type; (iii) poultry slaughterhouse; (iv) meat processing unit; (v) winery; (vi) bakery; (vii) vegetable preservation unit (dried tomatoes, pickles, peppers, etc.), or of aromatic herbs.

This context hypothesizes that there is a positive relationship between participatory business structure and value creation (in line with [41]), of an agroforestry activity, creating, from there, the conditions for implementing a new productive paradigm that meets the 'intelligence' of nature, that is ecological (understanding of the same nature), locally based (to optimize production and consumption), and economically sustainable and generalizable, according to reference authors such as [10].

6. Case Study

The relative backwardness of Portugal has preserved large areas where the reversal of biodiversity degradation seems to have met good conditions to be tested in favor of "mother nature". The agroforestry approach conceived the tree as the 'technology' that forms par excellence, arable land, and, logically, one of a biodiverse nature [35]. Deforestation for the mass production of food is an attack on nature's most precious resource: topsoil.

A living, arable soil of one centimeter in the Mediterranean area conditions, which could take hundreds of years, probably a thousand, to form, could be formed in about 10 times less time through the tree's 'work'. They are, therefore, the true engines of nature, the most efficient being to transform energy into food, as stated by Alfredo Cunhal Sendim, the engineer who has led this experience and named it an "agroforestry succession system" [15].

Herdade do Freixo do Meio is located on the geological boundary between Alto Alentejo and Ribatejo, near the village of Foros de Vale de Figueira, in the municipality of Montemor-o-Novo, about 8 km from the urban center of Montemor-o-Novo, and it is approximately 580 acres (Fig. 2).

Fig. 2. Herdade do Freixo do Meio: geographical limits and aerial view [15].

7. Method

In a historical context like the present one, in which the need to respond to the "Covid-19" pandemic from a solid scientific consensus is continuously repeated, we think it is necessary to evoke the thoughts of [42]. This author, one of the great epistemologists in the history of science, liked to recall the bases for establishing the principles of knowledge creation: the advancement of knowledge can only be effectively achieved by adopting a systemic approach based on the questioning of knowledge accepted as acquired. The author positions himself in line with the central idea of the "discourse of method" or methodical doubt explored in this founding work of modernity. As can be seen, according to the famous philosopher, that idea presupposes an exercise of reflective re-elaboration regarding the theses, the procedural rules arising from them, and the established practices accepted as consistent, with the same views and regulations, by the reference authors of a given scientific area. In science, turnkey solutions are not serious. As Kuhn termed it, scientific knowledge is subject to uninterrupted questioning aiming to detect the slightest flaws, present even in solidly established paradigms. Therefore, we propose to analyze the experiment underway at HFM based on the epistemology of Popper, Bachelard, and Kuhn [44].

This case study was developed based on in-depth interviews conducted with the leader of HFM, as well as on direct observation and document analysis. Researchers used informal note-taking to register interactions.

Furthermore, we draw on the influence of future research methods to imagine future scenarios, following recommendations by [43].

8. Organizational Data Analysis

Based on a selection of the notions and concepts put forward in the interviews and validated based on document analysis and observation of people's activities, a "researcher's notebook" was built, based on which it was possible to reach a descriptive analysis of the organizational system and develop some "cognitive frameworks" [23], an instrument for analyzing the representations shared by the social actors in question, which we consider to be robust from a theoretical and pedagogical point of view.

This theoretical framework can organize many representations (both social and scientific [22]), given that it has the essential characteristics for the communications sharing by most relevant social actors within a systemic approach to reality.

This cognitive framework technique enables the identification of a typology from which four quadrants emerge, in general terms, sometimes potential and sometimes paradoxical.

8.1 Organizational Context

HFM, with all its assets (land, facilities, brands, and machinery), is a family property and is loaned to a mixed or user cooperative society, as it is called, which operates it (comprising around 30 producers and 200 consumers). This cooperative is managed collectively by an elected board of directors, and the leader of the legal entity that owns the property is, coincidentally, also the cooperative's president. For ease of reading, we only refer to the cooperative when discussing organizational reality.

8.2 Data Presentation

Our analysis will consider two parameters in detail, starting with biodiversity. There are subsets of about 30 different species of fruit trees, organized in 'communities', that have been planted in degraded areas of the property, according to methods adapted from [10]. Portuguese researchers have contributed to this theme. There are 10 kinds of wood with different configurations at HFM. These are plantations of rows of fruit trees (spaced every two meters), adapted to the dry Mediterranean climate and subject to a wide temperature range. There are about 30 interacting species, starting with those of the Quercus family (holm and cork oaks)/climbing vines/olive trees and wild olives/fig trees/lemon trees and other citrus fruits/stone pines/walnut trees/almond trees/pomegranate trees/quince trees/mulberry trees/carob trees/banana trees/or loquat trees. All this immense variety is combined with other trees, such as strawberry and elderberry trees, or woody trees, such as eucalyptus, plane trees or willow, and pyrrhic or cypress trees. To complete the diversification of species, there are also a considerable number of shrubs (such as raspberries, rosemary, mastic trees, cactus, rock roses, and brambles, among others). The combination of all this aims to occupy all the available space at altitude, from the ground plants to the tallest trees, optimally managing growth and sunlight penetration. In a forest of this type, which has existed for three years, as we have seen, the temperature dropped 5 degrees centigrade concerning the temperature outside (from 32 degrees, it dropped to 27 when we took about 20 steps inland).

In this context, about three years after the beginning of the first fully consistent agroforestry experience, it is possible to verify, by direct observation, a situation of almost complete reversal of biodiversity degradation.

The dream behind this agroforestry model is to have all the possible strata occupied in height, from the 50 meters of walnut trees to the holm and cork oaks (the cork oaks and holm oaks seek water about 80m deep, providing shade without which life would hardly be possible in the HFM region) immediately below, then the olive trees, the citrus trees, and finally the lower layers with the vines wrapped in willows and even lower down, the peach trees, before reaching the lower plants such as the vegetables. This stratification in height, says Sendim, will be conjugated according to a certain diversification of species, in the meantime, under experimentation, to reach the best solutions. This stratification is considered decisive so that the soil is not subject to very high temperatures that prevent, among other undesirable effects, the penetration in the rainwater depth. The water cycle in the soil would itself benefit from the forest replica, in the subsoil, in parallel with the roots' growth, in proportion to the trees' height, simultaneously protecting each other through the exchange of microelements (mushrooms) and creating the conditions for an "abundant life" in the proximity of the superficial or medium-depth roots.

Could one live only from producing vegetables and fruit in a woodland system? It does not seem possible at the present knowledge stage. There is the alternative of composting. In HFM, about 4 out of 580 acres have been set aside to produce vegetables and fruit in a forced system (greenhouses) and the open air from composted 'humus'. This alternative does not fit in with soil correction. Still, it does not degrade it and, above all, remains a sustainability solution that allows a better articulation between production and consumption, contributing to improving the economic conditions of the whole.

Regarding animal life, the pig and the other animals (including wild ones) in the territory of the 'montado' are not just one more element in the food chain. They represent, among many other circumstances, the emergence of determining functions, e.g., those of the "carbon transformation" (incorporated into plant elements) into "nitrogen and phosphorus", as explained by our interviewee. Like other animals, cows, sheep, chickens, or turkeys, are all part of a holistic activity of aerating and fertilizing the land, or controlling insects and biomass excess, continues Sendim.

Therefore, these animals' role in the cork oak forest areas is very relevant in every respect, concludes our interlocutor, comforted by the collaborators with whom he interacted. There is even a foreign doctoral student developing his empirical research at HFM and with whom a guided visit was made to the main experimental agroforestry woodland of the estate: (i) the animals living freely in the 'montado' control the biomass and insects, improve the soil, accelerate the diversification of plant species, and logically increase overall biodiversity; (ii) their meat, as a result of genetics, constant outdoor exercise and a diet based on the richness of the plant cover and especially acorns, provides not only minerals and proteins of high nutritional value, but even cholesterol considered good for the body (HDL).

This victory over the degradation of biodiversity and the connection to a group of consumers involved and committed to experimentation has made it possible to achieve a remarkable set of around 500 product references for sale.

8.3 Data Analysis and Results

The data collected allows us, thus, to conclude that the results found in the analysis of HFM management points toward a high biodiversity production and an economic base close to unsustainability. In fact, despite having recovered the space designated as 'montado' (a true "ancestral agroforestry system", where cork oaks and holm oaks interact with vegetable gardens and pastures), HFM is having a dozen experiences underway that can be interpreted through the reflection of the authors cited, as differentiation-integration, we are forced to conclude that differentiation is perhaps exaggerated and integration insufficient.

For HFM, the axes and quadrants configure a paradoxical model, as they are systematically opposed, two by two. The first points to the path of greater or lesser effort to be made, while the second refers to aspects related to the transformation of nature to be implemented or already started in the meantime.

The quadrants thus drawn could be presented as follows: (I) there is a lack of effort to stabilize the rules concerning the main routines of primary work and processing, packaging, and commercialization of organic food products, in accordance with consumer needs, as well as some work methods supported by technology without the use of fossil energy, i.e., **organic production, which is technologically sustainable, but insufficient from a social, organizational and economic point of view**; (II) in turn, the initiatives of the associated producers and groups of loyal consumers, despite the efforts of all, are insufficient in terms of technological, economic, organizational, and social sustainability, i.e., **a developed group culture, but lacking creativity and clarification of objectives**; (III) team spirit, collaborative culture, and association with consumer groups is evident, and their continued deepening and moderate extension to other mutually winning partnerships is desired, i.e., **there are creative teams with a collaborative culture, associated with satisfied consumer groups**; (IV) finally, the organization for high added value products for customers is at an advanced stage, but lacks all professional work for brand creation, for effective presence in demanding local markets or for cost control and efficiency in general, i.e., **objectives that are insufficiently designed in terms of demand and systematic pursuit**. The synthetic cognitive framework presents the configuration in Fig. 3.

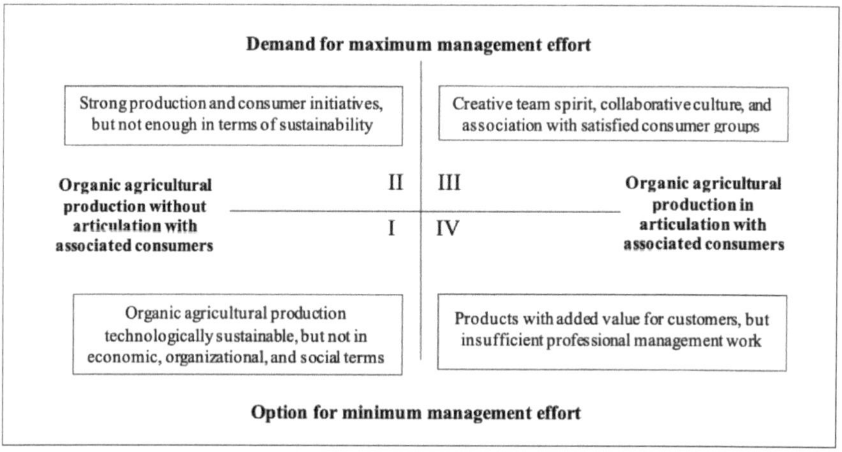

Fig. 3. HFM environmental efficiency and sustainable management cognitive framework.

According to the content analysis we have been proposing, HFM seems to be in what we have defined as quadrant IV, having distinguished itself as a case of excellence in the experimental domain in terms of sustainable agroforestry, but which runs the risk of dramatic unsustainability at the organizational and management level, which may configure an unviable economic situation.

Indeed, on the side of the organization and producers' participation and the consumers' involvement, a lack of professional management would position

HFM in quadrant III. Thus, involving all social actors in an effort of incremental creativity allows to take advantage of the achieved agroforestry disruption and can be seen as a model to be theorized and disseminated as a truly successful alternative to industrialized agriculture, the population balance of the territory and the depletion of rural areas.

9. Discussion

The analysis of the results from this experimentation in the Mediterranean cork oak forest area that we have been presenting and commenting on, projected onto a problem approach on a global scale, can be summarized, for better understanding by readers, according to two new cognitive maps that we will discuss below. The maps in question are, in turn, conceived based on two complementary and enhancing axes rather than antagonistic: nature's action (with its limitations) and man's action (which must deal with finite resources).

We often hear 'echnicians' defending the preservation of significant spaces from man's action, as if the human species were necessarily the enemy of nature. Man is desirably present, but, according to Sendim, only if he respects "nature's project", about the role of "managing the light, the succession of vegetables throughout the year and the system's evolution (from the herbs around the tree, as if it were a placenta, to the pruning phase)".

When humans wanted to impose their project on nature [36], the result ended systematically in entropy (in this case, soil erosion or desertification by about 80%, with all its entourage of undesirable externalities such as the contribution of about 30% to climate change, contamination of aquifers, etc.). Figure 4 illustrates this idea.

Sendim draws a heavy balance of the current situation of the planet, which "gets worse" every year without us knowing, even today, how to act to meet the intelligence of nature again.

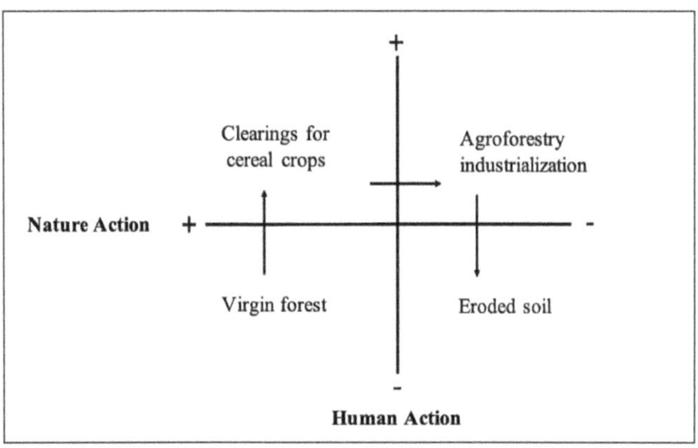

Fig. 4. Forest and artificial (Natural Forest versus Agroforest resulting from human action).

The economic model followed in the world was inspired by the industrial revolution and led to a low and fragile vegetation cover, with unpredictable consequences for the future of humankind.

Given its resilience, life would certainly return to a virgin forest after thousands of years. Still, the future of humanity requires a solution based on the inspiration of the so-called "Intelligence of Nature", and therefore, the agroforestry of the planet should be chosen.

A determining aspect of this results presentation is that it could constitute an interesting alternative to the dominant cognitive model centered on climate change.

Indeed, the issue of agroforestry concerns everyone. People living in cities, owning small amounts of land or gigantic estates, everyone can contribute to the dominance of spaces dedicated to agroforestry and contribute to the decrease of air temperature. Conversely, the climate change discourse leaves people distressed and far from the will to act, thinking about the uselessness of their efforts. In third place, the problem would be placed in terms of citizenship (internal responsibility) and not politics (attribution of external responsibility). Finally, by letting nature do its work, just guiding its action (since it is truly intelligent), fewer fossil energy resources would be needed, and food would be closer to the consumption areas instead of traveling halfway around the world to reach our tables.

The world would already face a politically paradoxical issue in the current situation: the unfeasible and inevitable immediate halt of economic growth [45, 46] to effectively control greenhouse gas emissions before development has reached so many people. In recent and widely publicized conferences, the issue has meanwhile been translated by the even more dramatic inevitability of the decrease of GDP at the global level since local carbon neutrality is not conceivable. Let us consider what appears to be a propaganda effect, which consists of deciding to immediately close the two coal-fired thermal power stations in Portugal, which ensure continuous production, replacing them with intermittent solar and wind energy, supposedly to decarbonize the national economy. However, data from the International Energy Agency (cited in a lecture by Prof. J.-M. Jancovici) state that between 2000–2017, global coal consumption increased 14 times more than solar energy and six times more than wind energy. To call this a decarbonization program is like holding back the force of the wind with the lift of a finger. What is certain is that the public authorities are utilizing these 'utopias' instead of reserving them. According to the authors, the GDP loss could have reached a reduction of 5% per year. In other words, the equation would undoubtedly be a choice between GDP and the planet's ecological health. Those authors conclude that if governments do not stop growth, nature will do it.

We believe that, instead, with agroforestry, the commitment to development could be achieved without loss of wealth, shared in global terms, and with a generalized increase in employment. Leadership/management action would be in line with the sustainable human resources management view [47], which amplifies the desired results beyond the economic/financial interests, fostering the social and environmental dimensions [48] over a long-term time horizon and contributing to the common good, notably to the achievement of the 17 sustainable development goals defined by [49, 50].

In the eyes of common sense, moving forward with the agroforestry solution may seem to be a utopia.

The case study shows that the agroforestry solution is a powerful one since it combines, among other factors, the circumstance that, on the one side, it is in our hands to achieve it; on the other side, it can create the foundations of a collaborative, locally based solution that can easily be envisaged at a global level; and also, it can be effective for a coherent strategy to combat climate change by getting to the root of the problem of loss of biodiversity; or, finally, it can involve all stakeholders in the solution (from urban gardens to school gardens, or smallholdings to large estates). Above all, the case shows how, in the new conditions of the digital revolution, people can reconcile themselves to a less stressed life in harmony with mother nature.

What world might result from effectively combating the trend for the industrialization of food production and abandoning the financialization of agricultural and forestry production? We may not yet know how to quantify the change, but the case seems to provide the basis for the constitution of a viable alternative without having, to its disadvantage, the repetitive and distressing evocation of the usual sterile catastrophism about climate change, which leaves no room for a pedagogy of the future.

Figure 5 illustrates our view of the intelligent human action impact on the future of natural forests and agroforests.

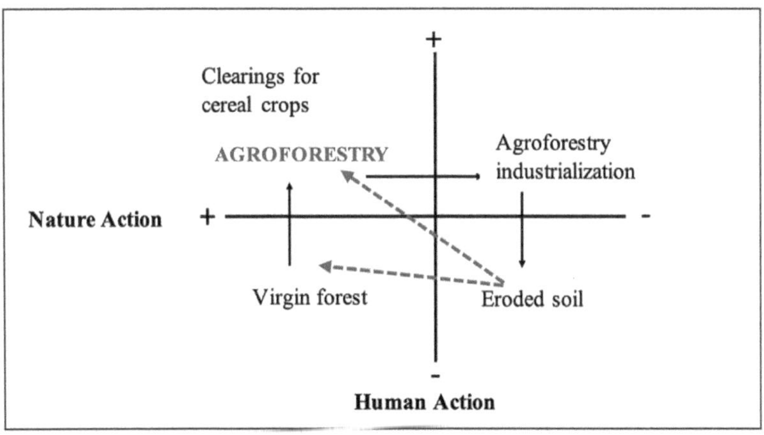

Fig. 5. Forest and mankind: Natural Forest versus Agroforest resulting from intelligent human action.

As for the continued dominance of the financial model of economic management, it only seems to reserve a violent encounter with the wall of reality for future generations.

10. Future Paths

The organization, as the poor relative of such an ambitious project and the bearer of the future, has been the main artifact of our industrial civilization since the last decade of the 19th century. It has guided our capacity for collective action in such a way as to enhance it, making compatible the tendentially divergent individual behaviors, channeling our freedom towards the idea of project and creativity, and allowing us to obtain results previously thought impossible, to paraphrase [18]. The university should perhaps be called into this organizational discussion, for although it discursively sponsors interdisciplinarity, it has continued developing limited frames of reference insensitive to complexity. In particular, the training evolution regarding professionals open to technical and managerial integration is very poor, leading organizations to a management deficit that forces an effort without results and without sponsoring an effective leadership of business projects, as shown in Warren Bennis' pioneering studies. The theoretical advances of the two universes to be made compatible (technical and managerial) have not benefited from the complexity theory approaches, particularly from a work modeling as a system of actions, of which Edgar Morin dreamt.

As we, humans, partake and co-create in this syntropic approach, we can integrate these approaches with other related philosophies that also aim for the ecosystem equilibrium within the psychology of sustainability and transduce those reflections upon the human development perspective, passing through mindfulness living frameworks, ethical and sustainable leadership, and naturalistic intelligence promotion—the last addition of Howard Gardner's theory of human intelligence [51].

The outcomes of this research could be anticipated on two distinct levels: local and global. First, on the estate level (local), HFM aims to transit from a local production, entirely biological, capable of feeding 1,200 human beings, to a total of 12,000 in a mid-term future, whereas remunerating workers are clearly above market values, based on an organization that properly explores the four quadrants of the last of the models mentioned above.

Second, on a global level (or 'glocal', since the aim is to act locally but think globally), this project aims to share the science (in its trans, cross, and multidisciplinary view of the community and the planet) and bring the people parcel to the mix: to achieve a syntropic view, we must reach the people, the people managers, the students, and the educators. Therefore, we end this case study presenting a final product to share: a guide, a framework, a new paradigm to reach "people professionals" worldwide to spread the science and start the mindset change to start applying nature-inspired innovation (Biomimicry) and achieve a life lived in syntropy.

Suppose there is an opportunity to adequately model the vast system of actions compatible with the HFM experiment. In that case, the current biological production is intended to be significantly expanded soon. Otherwise, the solution that seems to be within reach of this extraordinary work collection (in more than one capacity) may go into entropy, and the opportunity to highlight the conditions for the HFM future and of all of us, may be lost.

11. Conclusions

In-depth knowledge of the Mediterranean forest cycles has proved to be the key to success for agroforestry. This is a succession of three cycles of 9,000 years each (the pioneer trees cycle, the transition trees cycle, and then the climax cycle). The novelty, says Sendim, is that the various species specific to each cycle could be simultaneously planted [37]. In the HFM estate, a man seems to experience the project of integral cooperation with nature. In HFM, nature will have, from now on, arrange a meeting with the human community. A real willingness is being established between both parties to experience a dialogue between the 'intelligence' of nature and the collective intelligence of all those who cooperate (through production or consumption).

HFM property management is rehearsing this new, socially relevant experiment of transforming a traditional agricultural enterprise in the south of Portugal into a community (a so-called second-order cooperative), integrating activities organized into micro-enterprises in terms of 'lean' management [11]. In the first phase, the idea is that HFM would function as an incubator and then as a Multifunctional Cooperative, integrating a plurality of productive micro-units based on a strong idea of participation of all the actors involved. The leadership/management will soon be separated, with the current leader dynamizing the whole experience and presiding over the 'cooperative' in transformation (ending the current phase of managerial centralization) and with a professional manager taking over the operational business management, in articulation with those responsible for the various units that are being established.

References

1. Suša, O. (2019). Global dynamics of socio-environmental crisis: Dangers on the way to a sustainable future. *Civitas—Revista de Ciências Sociais*, 19(2), 315–336. doi: 10.15448/1984-7289.2019.2.31969.
2. Hba, R. and Manouar, A.E. (2017). *ICT Green Governance: New Generation Model based on Corporate Social Responsibility and Green IT*. arXiv preprint arXiv:1701.08714.
3. Schmidlehner, M. (2021). Analysing the discourse of 'green' capitalism: The meaning of nature in 'nature-based'. *World Rain Forest Movement, Bulletin*, 55. 18 May. Retrieved from: https://wrm.org.uy/articles-from-the-wrm-bulletin/section1/analysing-the-discourse-of-green-capitalism-the-meaning-of-nature-in-nature-based/
4. Lisbinski, F., Mühl, D., Oliveira, L. and Coronel, D. (2020). Perspectives and

challenges of Agriculture 4.0 for the agricultural sector. *In*: VIII Simpósio da Ciência do Agronegócio 2020: Inovação e empreendedorismo no Agronegócio, 422–431. CEPAN - Centro de Estudos e Pesquisas em Agronegócios: Universidade Federal do Rio Grande do Sul, Brazil.

5. Rose, D.C., Wheeler, R., Winter, M., Lobley, M. and Chivers, C.-A. (2021). Agriculture 4.0: Making it work for people, production, and the planet. *Land Use Policy*, 100, 104933. https://doi.org/10.1016/j.landusepol.2020.104933.
6. Lawrence, P. and Lorsch, J. (1967). Différenciation et intégration dans les organisations complexes. *Administrative Science Quarterly*, 12(1), 1–47.
7. Sorman, G. (1989). *Les vrais penseurs de notre temps*. Fayard.
8. Visvanathan, Sh. (2003). From the Green Revolution to the Evergreen Revolution: Studies in discourse analysis. *IDS Seminar on Agriculture Biotechnology and the Developing World*. 1–2(October), 1–20.
9. Lopes, A. (2012). *Fundamentos de Gestão de Pessoas – Para uma síntese epistemológica da iniciativa, da competição e cooperação*. Sílabo.
10. Götsch, E. (1995). *O Renascer da Agricultura*. Centro Sabiá.
11. Womack, J.P, Jones, D.T. and Roos, D. (1992). *A máquina que mudou o mundo*. Campus.
12. Abreu, M.I.A. (2020). *Caraterização de bolotas de Quercus ilex subsp. rotundifolia ao longo de um gradiente climático no Alentejo*. Mestrado em Biologia dos Recursos Vegetais, Instituto Superior de Agronomia, Universidade de Lisboa, Portugal.
13. Cheriet, F. and Maurel, C. (2020). Être petit et réussir à l'export: étude de dix cas d'entreprises vitivinicoles françaises. *Management international/Gestiòn Internacional*, 24(6), 114–126. https://doi.org/10.7202/1077353ar.
14. Popper, K.R. (1959). *The Logic of Scientific Discovery* (5th Edn.), 46, 40–41. London: Hutchinson.
15. Pinto, F.G.T. (2020). *Sistemas agroflorestais de sucessão: Regresso ao passado que nunca existiu*. Estudo de caso, em Portugal. Dissertação de Mestrado em Engenharia Agronómica. Instituto Superior de Agronomia, Universidade de Lisboa, Portugal.
16. Pasini, F.S. (2017). *A agricultura sintrópica de Ernst Götsch: História, fundamentos e seu nicho no universo da agricultura sustentável*. Ph.D. thesis, Universidade Federal do Rio de Janeiro, Brazil.
17. Simon, H.A. (1947). *Administrative Behavior: A Study of Decision-making Processes in Administrative Organization*. Macmillan.
18. Crozier, M. and Friedberg, E. (1977). *L'acteur et le système*. Seuil.
19. Mintzberg, H. (1983). *Power in and Around Organizations*. Prentice Hall College Div.
20. Demaze, M.T. (2010). Éviter ou réduire la déforestation pour atténuer le changement climatique: le pari de la REDD. *Annales de Géographie, 119e Année, N° 674 (jul-août)*, 338–358.
21. Auriault, V. (2012). *Agricultura Multifuncional no Montado Alentejano -Estudo de Caso da Herdade do Freixo do Meio*. Tese de Mestrado em Ecologia Humana e Problemas Sociais Contemporânos, Faculdade de Ciências Sociais e Humanas, Universidade Nova de Lisboa, Portugal.
22. Lopes, A. (2016). Liderança da cultura e o problema da eficácia organizacional em Portugal. *In*: *Valorizar a tradição – Orações de sapiência no ISCSP*. Edições de Aniversário.
23. Weick, K.E. and Bougon, M.G. (1986). Les organisations en tant que cartes cognitives: Tracer les voies du succès et de l'échec. *L'organisation pensante*, 102–135.

24. Voiron-Canicio, C., Dubus, N., Loubier, J.C. and Liziard, S. (2009). Évaluer les impacts du changement climatique sur le fonctionnement d'une aire urbaine littorale: Outils d'aide à la décision existants. *Fifth Urban Research Symposium.*
25. Schumacher, E.F. (1979). *O negócio é ser pequeno: Um estudo de economia que leva em conta as pessoas.* Zahar Editores.
26. https://www.youtube.com/watch?v=uhESenZyrT0.
27. https://www.youtube.com/watch?v=Q9UtFSu7-Fs.
28. https://www.youtube.com/watch?v=iuUk1yu-A0o.
29. https://www.youtube.com/watch?v=kYKf8f_KByI.
30. Sen, A., Smit, B. and Wandel, J. (1985). *The Standard of Living.* Cambridge: Tanner Lectures on Human Values Retrieved from http://tannerlectures.utah. edu/_documents/a-to-z/s/sen86 Adaptation, adaptive capacity and vulnerability, *Global Environmental Change*, 16, 282–292.
31. Swaminathan, M.S. (1987). The Promise of Agroforestry for Ecological and Nutritional Security, Agroforestry. *In:* A Decade of Development, 25–42. International Council for Research in Agroforestry, Nairobi, Kenya.
32. https://www.youtube.com/watch?v=YBPLrr9Hph0.
33. RTP (2019). Estrada Nacional. Available at: https://www.youtube.com/watch?v=keGPLXJy8Vk.
34. Sendim, A. (2019). Interview to the leader of the Herdade do Freixo do Meio, Alfredo C. Sendim. Available at: https://www.youtube.com/watch?v=UcbGq_Bi1yI.
35. RTP2. (2019). O Plano de Florestação da Herdade do Freixo do Meio - Reportagem Biosfera (Farol de Ideias/RTP2). Available at: https://www.youtube.com/watch?v=D5TQgj9FThA
36. https://www.youtube.com/watch?v=QS5hhBzkrWY.
37. https://www.youtube.com/watch?v=oDVOR_BBJfg.
38. Crozier, M. (1989). *L'Entreprise à l'écoute.* InterEditions.
39. Martins, A.E. and Lopes, A. (2012). *Capital intelectual e gestão estratégica de recursos humanos: Fundamentos e estrutura da gestão e do conhecimento nas organizações.* Editora RH.
40. Martinaityte, I. (2014). *Leveraging employee creativity through high performance work systems: A multilevel perspective.* Ph.D. thesis, Universidade de Aston.
41. Milano, P. (2017). VIII. Paul Roger Lawrence Jay William Lorsch – Environnement, organisation, adaptation: La contingence structurelle. Dans: Sandra Charreire Petit (éd.), *Les Grands Auteurs en Management*, pp. 113–128. EMS Editions.
42. Bachelard, G. (1965). *La formation de l'esprit scientifique, contributions à une psychanalyse de la connaissance objective.* J. Vrin.
43. Inayatullah, S. (2007). Questioning the future methods and tools for organizational and societal transformation. Tamkang University. http://www.metafuture.org/library1/FuturesStudies/The-Study-of-the-Future-questioning_the_future_ch_1-2007.pdf.
44. Mayr, E. (2005). *Biologia, Ciência Única: Reflexões sobre a autonomia de uma disciplina científica* (Leite, M., Trad.). São Paulo: Companhia das Letras, Brazil.
45. Jancovici, J.-M. (2000). Le réchauffement climatique: Réponse à quelques questions élémentaires. Les Cahiers du M.U.R.S. n° 39 - 3ème trimestre.
46. Jancovici, J.-M. and Le Treut, H. (2009). *L'Effet de serre: Allons-nous changer le climat*? Champs.

47. Ehnert, I., Parsa, S., Roper, I., Wagner, M. and Müller-Camen, M. (2016). Reporting on sustainability and HRM: A comparative study of sustainability reporting practices by the world's largest companies. *The International J. Human Resource Management*, 27(1), 88–108. doi: 10.1080/09585192.2015.1024157.
48. Stankevičiūtė, Z. and Savanevičienė, A. (2018). Designing Sustainable HRM: The core characteristics of emerging field. *Sustainability*, 10(12). doi:10.3390/su10124798.
49. Stahl, G., Brewster, C., Collinngs, D. and Hajro, A. (2020). Enhancing the role of human resource management in corporate sustainability and social responsibility: A multi-stakeholder, multidimensional approach to HRM. *Human Resource Management Review*. https://doi.org/10.1016/j.hrmr.2019.100708.
50. United Nations. (2015). *Transforming our World: The 2030 Agenda for Sustainable Development*. Available at https://sdgs.un.org/2030agenda
51. Gardner, H. (2006). *Multiple Intelligences: New Horizons*. Basic Books.

CHAPTER

6

Agrobiodiversity and Agriculture 4.0 Challenges with Seed – Existing Improvements, But How about Erosion?

E.G. Rocha*[0000-0002-6100-7551], R.O. Souza[0000-0002-5004-2669], J.V.D. Fonseca[0000-0002-0257-5903], C.C. Nascimento[0000-0002-7801-5865], J.L. Nunes[0000-0002-9327-710X] and P.F. Alves[0000-0001-8924-7082]

Federal University of Goiás (UFG), UFG Law School. Campus I - Praça Universitária - Av. Universitária Esquina com 5ª Avenida s/n - Setor Leste Universitário, Goiânia, Goiás, CEP 74605-220, Brazil

Seeds are at the heart of agriculture. Preserving them and making them more diverse are major challenges for guaranteeing food sovereignty, as well as agrobiodiversity. Climate change, the emergence of new pathogens, and catering to different food crops require us to think more and more about seed variety. However, what has been observed is an accelerated process of genetic erosion. This chapter aims to introduce the promotion of agrobiodiversity as one of the main challenges of Agriculture 4.0. In terms of methodology, wide-ranging research with documents was conducted, including a bibliographical review of international agreements and interviews. As a result, we verified an intense legal and industrial process that favors genetic standardization by removing from the farmer his historical role in selecting and adapting seeds to local realities. The conclusion is that the 21st-century farmer cannot be just a seed consumer but also an active breeder.

1. Introduction

This chapter was developed by the group "Sementes e Soberania Alimentar

* Corresponding author: eduardofdufg@ufg.br

(Seeds and Food Sovereignty)", which is linked to the postgraduate Agricultural Law program at the Federal University of Goias (UFG). Between 2019 and 2020, research financed by the Brazilian government (Brazilian acronym: CNPq, which stands for National Council for Scientific and Technological Development) was carried out concerning the challenges that the modernization of agriculture has posed for agrobiodiversity, the environment, and the guarantee of food sovereignty.

In terms of methodology, wide-ranging research with documents was conducted, including a bibliographical review of Brazilian, Latin American, and international literature on the subject of agrobiodiversity; an assessment of international agreements involving intellectual property about seeds was carried out, as well as the legislation of Brazil, Argentina, and Bolivia. Finally, interviews were conducted with 72 people, farmers, agribusiness agents, and government representatives from the three countries were subjected to investigation.

This chapter will therefore present some of the conclusions obtained via this research, counting the concept of agrobiodiversity as one of the major challenges of Agriculture 4.0. International agreements, as per the (i) Agreement on Trade-Related Aspects of Intellectual Property Rights (TRIPS), and (ii) convening of the International Union for the Protection of New Varieties of Plants (UPOV), has directly impacted national legislation. This has led to the stimulating of a narrow view of technological advances, which have privileged uniformity, homogeneity, stability, and intellectual property to the detriment of agricultural biodiversity. Reports from the Food and Agricultural Organization of the United Nations [22], the United Nations Development Programme (UNEP 2020), as well as specialized literature point to a significant loss in agrobiodiversity created by the current agricultural model.

Therefore, the 21st-century's precision agricultural technology is faced with an immediate challenge of reversing the losses in accentuated genetic and plant varieties that have occurred since many of these species are adapted to different climatic realities and crops. For these reasons, an epistemological shift will be defended, in which the technologies of Agriculture 4.0 engage in dialog, learning, and teaching with the traditional farmer, smallholders, and local producers–historical agents for adaptation and local improvements concerning seeds. Precision agriculture must also be the agriculture of solidarity, recognizing that they are multiple actors in the fields with different needs and distinct roles, among them the small farmer and local peoples and communities.

The course of this chapter is divided into three parts. The first part discusses the agricultural and technological paradigm through which modern agriculture is passing – that of the Green Revolution. The second part of the chapter will consider the role of farmers in improving seeds, highlighting what the specialized literature has pointed out: that small farming has historically played a role in promoting agrobiodiversity. Next, the role that international legislation regarding intellectual property and national legislation has played in creating problems related to the uniformization of seeds will be discussed. The final part of the chapter will consider the need for informed dialog in which precision technology and traditional knowledge are taken into consideration and in which their respective functions are respected.

2. Green Revolution, Technology Package, and Improved Seeds

Contemporary agriculture results from a series of innovations, scientific discoveries, and technological processes. However, behind the appearance of a standardized model, contemporary agriculture is, at its essence, a complexity of knowledge, practices, and related (social and environmental) elements.

Additionally, there are various ways of 'doing' agriculture, some more dependent on industrial components and others based on communal and agroecological techniques. However, changes experienced by the sector in the last few decades significantly favor the agribusiness model. This is a reflection of a predominant economic logic that has valued the 'business' component of agriculture more than the 'ecological' component.

A large part of the methods currently employed in what is known as Agriculture 4.0 was developed to increase production to meet market supply demands. Despite important advances, such as exploiting drones, software, and even gains in production, the preservation of agrobiodiversity and guaranteed food security are not always considered.

Therefore, saving seeds, an aspect that guarantees biological diversity and food quality, represents one of the major challenges of Agriculture 4.0. Therefore, understanding the Green Revolution phenomenon is essential to mapping out this issue and how it has come to include legal and political aspects.

The strategy of the Green Revolution originated in the 1960s and was largely brought about by discoveries in plant genetics. The Green Revolution is characterized by profound agricultural transformations, such as intensive mechanization, a strong dependency on industrial inputs, the adoption of chemical fertilizers and agrotoxins, and improved seeds defended as high yield. It is large-scale mechanized agriculture based on the uniformity of crops and means of cultivation and the standardization of farm products [4, 18].

The Green Revolution was purported to be a great political and technological achievement, a strategy for achieving peace and abundance and overcoming hunger through increased food production. Years later, it was realized that the 'miraculous' aspirations of the Green Revolution were never truly reached [21].

At the heart of the Green Revolution lay what was known as High Yielding Variety (HYV). The term High Yielding Variety refers to the seed varieties created in international agricultural research centers. It suggests that this seed possesses a higher yield when compared to local varieties [21].

With the arrival of the Green Revolution in periphery countries, what was before presented as a solution to eradicating hunger soon proved to be limited. Through disseminating its technology package, the Green Revolution consolidated industrial agriculture, characterized by investments in technologies and machinery, chemical inputs, and uniformity in the seed. On the one hand, the Green Revolution increased profit margins for large corporations that could render large-scale agriculture feasible. On the other hand, as pointed out by innumerable studies in rural sociology, it generated a greater dependency for farmers on factors that are exogenous to the farm, resulting in debt, crises in the fields, and rural exodus [3, 4, 12].

With agriculture shifting to an industrial base, seeds were also modified to be adapted to a cultivation system based on standardization and large-scale production. These changes favored uniform, stable seeds and aggravated genetic erosion. Agricultural diversity is based on plurality in cultivation, use, and crossing, something that was compromised under the standardization perpetuated by the Green Revolution. At the same time, the farmers were displaced from their historical role as an on-farm conserver of seeds and promoter of improvements to adapt them to different climatic and cultural realities. This function was relegated to specialized companies and research centers, and the farmer came to be seen as merely a consumer of seeds.

Changes in agricultural systems brought about by advancing components of the Green Revolution in the fields led to the mischaracterization of the "cultural, ecological, and environmental identity" of communities [11]. Local varieties and farmer knowledge became increasingly delegitimized by technical-scientific knowledge, diminishing the importance of local farming in maintaining agrobiodiversity. The legal and economic persecution of local varieties accompanied this epistemic delegitimization.

According to international rules established by UPOV, which impact nearly all the world's countries, a seed designated for commercialization must be different, homogenous, and stable. These rules have led to farmers experiencing great difficulty in commercializing, selling, or purchasing traditional seeds in formal markets [18].

It became clear from statements made by local farmers and leadership during the interviews carried out as part of this research project that legislation makes access to traditional seeds difficult by prohibiting them from circulating freely. It was reiterated consistently during interviews that difficulty accessing seed is an important factor in local farmers abandoning them and beginning to exclusively plant certified seed.

Seed improvement made in an industrial context has been accompanied by a rapid process of appropriation, through which these resources have ceased to be characterized as being a "common good" and have come to be controlled by financially powerful business groups that are generally also responsible for the commercialization of chemical inputs. The traditional practice of sharing seeds by farmers has become increasingly restricted despite offering the potential for generating agricultural diversity.

The rich genetic material and plant diversity resulting from historical efforts and the diffuse and continuous improvement, adaptation, and conservation of varieties carried out by farmers is used as a base for developing high-yield seed varieties. Yet, at the same time, and in a contradictory fashion, traditional practices for the exchange and conservation of seed are restricted or even prohibited under national legislation, which has compromised informal seed systems and, as a result, agrobiodiversity.

3. The Role of Farmers in Seed Improvement

Seeds are a central component of agriculture; everything begins and generally ends with them [23]. Traditional seed is the result of collective efforts that have accumulated over hundreds of generations of local farm workers cultivating, growing, adapting, selecting, conserving, and exchanging them, using different forms of management and designating them for a variety of uses (cultural, religious, commercial, human and animal consumption, among others) [24]. People and communities across a variety of countries and regions have improved and adapted seed varieties for a wide range of environments, climatic, sanitation, and soil conditions, and different social and production needs [23].

These processes involve the management of seeds in which the majority of these functions take place locally, within a family or between families of local farmers, through interpersonal social networks that involve the exchange and/or purchase of seeds in local markets, forming what is known as informal or local seed systems [25, 26].

For Gill et al. [27], informal seed systems are characterized by multiple components: the saving of seed by local farmers, indigenous annual and perennial crops, informal seed markets, seed networks and "germplasm gatekeepers", informal mechanisms for seed storage, and the preservation of knowledge associated with the local system. These systems "are called informal because they are farmer organized and led, flexible and dynamic, indigenous to a community or set of communities, local or regional in scope and scale, and typically undocumented".

The local and consistent reproduction of seeds by local farm workers is a process that is essential to the function of local systems. This reproduction allows varieties "to continuously interact with each other and their agroecological and socioeconomic environment" [28].

This source of seed supply is characterized by the local reproduction of seed by the farmers themselves using 'local' practices for selecting, producing, and conditioning seed. It includes saving plant material for the next planting and informal diffusion of seed to other farms, often within the community. In most cases, it is a non-specialized seed production integrated into producing grains, roots, and tubers for consumption and marketing [28].

Informal seed systems play a crucial role in ensuring food security since they are the main source of crops in local communities and farmers and are responsible for diversity in food production. Informal seed systems also provide plant species that are commonly neglected and under-utilized for food consumption and which provide these communities with essential nutrients [27].

The conservation of biodiversity is also at the center of informal seed systems, which expand the genetic base of production through the adaptability of multiple species and varieties of crops to specific production systems and microclimates [27]. The *in situ* or on-farm conservation of biodiversity or, in other words, the improvement, adaptation, and maintenance of varieties of seed and plants carried out by farmers, traditional communities, and indigenous peoples on their territories ensure the adaptive dynamism of agricultural biodiversity [18].

The *in situ* conservation of locally adapted varieties reduces risks to local agricultural systems, such as disease, low soil fertility, and severe weather conditions like drought [27]. This adaptability of germplasm to local conditions, as well as the diversity and genetic variation of crops, the result of selection and experimentation techniques developed by communities, guarantee greater resistance to adverse conditions in local seed systems.

As per Fredenburg et al. [29], in informal seed systems, farmers exchange traditional or Creole seed varieties and use improved varieties from the formal system, using seed saved from their harvest. The exchange of seeds is made through discussion within the community or purchases made on local markets. The seed networks constructed by local farmers make up an integral part of informal seed systems, allowing individuals access to seed through connections between territories and the exchange of seed and knowledge with which they are associated [30].

Farmer seed networks are open and dynamic systems. New varieties are created by managing systems adapted to environmental conditions and local markets, exchanging knowledge, practices, and diversity between populations in different regions and ecosystems [30]. This system provides access and allows for the propagation of seed together with the knowledge, practices, and innovation associated with the management of agrobiodiversity, promoting seed security, the subsistence of agricultural communities, the conservation of agricultural diversity, as well as farmer food security, and autonomy [26, 31].

In the greater part of the Global South, informal systems, which include networks for the exchange and saving of local and non-certified seeds, are a preferred source for most farmers and Indigenous groups [31] since, depending on the country's crops between 60% and 100% of seeds planted are produced and traded by these people [28]. In Latin American and Caribbean nations, the Food and Agriculture Organization (FAO) of the World Health Organization (WHO) estimates that around 75% of seed used is provided through local or informal seed systems [18].

Even formal seed systems, which are highly regulated and composed of public institutions and private industries that work with a genetic enhancement of plants in laboratories, are facilitated and dependent on germplasms created, improved, and maintained by farmers, traditional communities, and indigenous groups in informal systems, which are generally the means through which new commercial varieties are developed.

4. The Restricting of Seed through Law and Uniformity

Despite the importance of informal seed systems, international intellectual property agreements, such as the rules of the UPOV, have increasingly restricted or even prohibited the existence of informal systems. The research that guided this chapter concluded that national legislation regarding cultivars, as a rule, are linear internalizations and offer few differences from international requirements. The seed and knowledge with which informal seed systems are associated, including the crossing of varieties, adaptation to the environment, exchange, management practices, planting, multiplication, drying, and storage, created and performed by farmers, have been affected by processes expanding the reach of intellectual property rights.

The expansion of the reach of intellectual property rights with regards to plant varieties through means of patent laws and protection laws for cultivars limit the rights of farmers to save seed for their use, prohibiting such a practice or determining that farmers may only save seed for replanting in the following harvest or on their properties, which also affects the exchange of seed between farmers. Intellectual property rights, in how they have been conceived and implemented, thereby represent a limitation to the free utilization and circulation of seed, which have been maintained by farmers for thousands of years, resulting in a loss of plant varieties and associated practices and knowledge.

The loss of seed varieties and associated knowledge was a recurring concern in the interviews conducted with farmers and representatives from rural social movements in Brazil, Argentina, and Bolivia. The following excerpt, taken from an interview with an Argentinian farmer and activist, summarized what was stated in other interviews:

> "Consequently, our reliance on seed purchases has increased, and the knowledge and range of kinds we formerly had have dwindled. When plant variety is reduced in fields, insect and microbial diversity are reduced in the soil. Everything, even the plants themselves, has a role in the fields. To have a single variety for which seed saving is prohibited implies that all of this knowledge, this culture, is gone, leaving us reliant on oral transmission."
>
> "[...] the result is that we have become increasingly dependent on the purchase of seed, and the knowledge and diversity of varieties we once had are being lost. When there is less plant diversity in the fields, there is less diversity of insects and microorganisms in the soil. Everything, each plant itself, serves a purpose in the fields (...) To have a single variety for which the seed is not allowed to be saved means that all of this knowledge, this culture is lost, leaving us dependent (oral information)."

These quotations are significant in revealing the concern shared by farmers and activists with a growing dependency on formally certified seed and how this has compromised genetic diversity and diversity concerning knowledge and culture.

There is an expropriation of seed created and maintained by farmers and a reappropriation by companies and public and private institutions that transform the seed into goods protected by intellectual property rights and supported by legislation that confers and consolidates these rights.

This process of expropriation and appropriation of seed can be understood through two pillars: agrarian or scientific persecution shaped by the Green Revolution and legal persecution, represented by laws surrounding the seed, intellectual property rights, and agricultural contracts [13].

This agrarian or scientific persecution occurs through the epistemological disqualification of traditional knowledge, its cultivation techniques, and its seed. Non-scientific knowledge is considered antiquated, archaic, and damaging to agriculture, even though this is not confirmed in practice. This results in epistemic scorn, which makes the exchange of knowledge difficult, and places technical-scientific knowledge generated by the Green Revolution in a position of privilege. Therefore, the series of techniques and technologies developed with a base in farmer knowledge that is sometimes thousands of years old and has ensured diversity, food security, and balance with nature are not valued or simply disregarded [19].

Meeting market demands is almost an exclusive parameter of agricultural production, which determines increases or decreases in productivity and demands for quality controlled by food systems in which large international corporations increasingly dominate the commodity chain. The multifunctionality of agriculture and the variety of purposes that farms serve in communities is thereby overlooked. Food and its objectives are as diverse as the cultures that produce it [9].

In addition to being epistemically disqualified, traditional seeds are excluded due to standardized technologies developed to meet the demands of large-scale agriculture brought about by the Green Revolution, which demands intense uniformization throughout its processes. Traditional non-uniform seeds are thereby seen as a problem for technology based on the presupposition of the existence of homogeneity in the fields [4].

The control exerted by large multinational companies on seed, as well as technologies associated with their plantings, such as chemical fertilizers and pesticides, make the demand for genetic uniformity increasingly apparent. Therefore, improvements made to varieties carried out by farmers over millennia and supported by their sharing practices are sometimes disregarded.

The legal persecution of seed takes place through seed laws that require obligatory registration and certification for production and commercialization, bilateral contracts between companies and producers, which are developed to ensure control of the agricultural production chain, and intellectual property rights consisting of the rights of plant breeders (cultivar protection laws) and patent systems [13].

The development of these legal mechanisms, which represent an increase in regulation of access to phytogenetic resources, accompanies a global trend in recent decades towards advancing marketing-oriented logic in agriculture. The majority of current seed legislation in a variety of countries (there are 77 members of UPOV, including the United States, countries of the European Union, Russia, and China) offer significant support for a model which includes the appropriation of these resources for economic purposes (UPOV, 2021).

With the advance of the appropriation and persecution of farmer seed varieties by policies and legislation aligned with agro-industrial standards disseminated under the Green Revolution, the phenomenon of genetic erosion has become a concerning reality. The wide-scale loss of diversity in plant genetics presents irreversible risks to food quality and, consequently, human health, thereby creating a problem with environmental and social dimensions [10, 22, 32, 33].

According to the FAO [22], the main causes of genetic erosion are deforestation, overgrazing, population pressures, and changes in agricultural practices. International and national norms that favor uniformity and stability as a condition for registration and open commercialization have supported this process. Scientific and agricultural persecution spurred by the Green Revolution, which has developed with a base in technologies that depend on the homogeneity and standardization of crops, has also favored this loss of diversity.

New biotechnologies have reconfigured the meaning attributed to biodiversity in that what was once a base element in the sustenance of life has become raw material dominated by large companies [20]. This change in perspective reveals an important component of the Green Revolution's political rationale: profit interests accompanied by financing and requirements (such as the homogenization of seeds and the creation of legal instruments protecting improved varieties).

It did not take long to consolidate seed domination by the international agroindustry, and a large part of the seed trade came to be controlled by private multinational companies. This power, which soon came to control the entire food system, represented a new phase of the Green Revolution [9, 10].

In this fashion, at the start of the 1960s, a strong campaign was observed in Europe seeking to protect new varieties like patents. On the 2nd of December 1961, in Paris, the International Convention for the Protection of New Varieties of Plants was created. It constituted an important global framework for private plant improvers to receive payments of patent rights for varieties they created [10].

The development of a plant variety protection system, such as that which occurred, is based on the idea that genetic improvement is carried out by professional plant breeders in specific sectors, denying the role of farmers in innovation and promotion of diversity in agriculture [18]. The challenges farmers face in maintaining their agricultural practices are well-documented even today since, despite being the subject of certain legal provisions (to a greater or lesser extent, depending on the country), legal apparatuses remain linked to the defense of industrial seed and certification.

A perception that national legislation concerning cultivars has heavily favored agribusiness to the detriment of farmers was frequently mentioned during interviews with individuals associated with the issue, as can be seen in the following excerpts:

> "And this question of rights is one that we have been watching carefully from the start since, first and foremost, all Law relating to seed and cultivars is written to benefit huge firms' agribusiness, the Rural Bench [6, 7]. The draft legislation is tailored toward the interests of major corporations; when you come in with work that involves diversity and agroecological management, you run up against legislation that quickly reveals that it does not comply with seed regulations. A provision in the Brazilian seed legislation refers to "uniformity, stability, and distinguishability". This demonstrates unequivocally that the Law is solely concerned with very homogenous materials. If you pick a variable variety and a diverse setting, the Law makes no provisions, and oral information is employed."
>
> "And this issue of rights is something that we have been following closely since the beginning because, first of all, all legislation related to seed and cultivars is created to serve the agribusiness of large companies, the Rural Bench. They put together legislation geared towards the interests of large companies; when you come in with some work in which you deal with diversity, with agroecological management, you come up against legislation and soon see that it does not meet the requirements of seed laws. The Brazilian seed law contains an item regarding "uniformity, stability, distinguishability." This makes it readily apparent that the Law is only interested in highly uniform materials. If you take a variety that has variability and if you take an environment of diversity, the Law does not offer any provisions (oral information) [6, 7]."

Most national legislation is an internalization of the requirements set forth by the agreement known as TRIPS, from 1994, by virtue of the obligatory nature of its ratification as a condition for a country becoming a member of the World Trade Organization (WTO). As TRIPS determined that member states must protect plant varieties, the UPOV System, established under the UPOV Convention, was seen as an appropriate alternative for conferring this protection. Therefore, as previously mentioned, many countries (77 in total) opted for this model, adhering to the resolutions from 1978 or 1991 or the referred Convention [18].

The UPOV system, intellectual property rights, and associated seed legislation are vital mechanisms for understanding contemporary agribusiness. There are reflections of global commercial relationships and the activities of powerful business groups in the agricultural sector, indicative of the existence of a corporate food regime.

The monopolistic appropriation of phytogenetic resources, and the imposition of barriers to the traditional methods practiced by farmers, have proven to be problematic. This is because they have favored genetic erosion and negatively affected the achievement of food security. Confronting this issue is, therefore, one of the key challenges of Agriculture 4.0 that, despite relying on support from cutting-edge technologies and involving professionals from the most diverse areas of science, requires an introspective consideration of its primordial elements: agrobiodiversity and the farmer.

5. Intellectual Property Legislation on Plants in Brazil and Argentina and the Seed Uniformization

At the end of the 1990s, advances in research into biotechnology reinforced agribusiness strategies in Brazil. Multinational corporations came to control the seed market, raw materials for agroindustry, and the food supply, receiving support from government policies oriented towards the guidelines for free trade from the World Trade Organization and International Monetary Fund [1].

Consolidating the multinational seed industry and agrobiotechnology created pressure to impose global Intellectual Property Rights (IPRs) onto plant varieties. In the Brazilian context, specific legislation that regulates IPRs with regards to plant varieties are Federal Law no 9.456/1997 (Brazilian Plant Variety Protection Law), inspired by the UPOV Convention of 1978, and Federal Law no 10.711/2003 (Law of Seeds and Seedlings).

Brazil signed the TRIPS Agreement, approved in 1994 by the WTO, which obligated member states to adopt a system for the protection of intellectual property with regards to plant varieties. In 1996, Brazil passed Federal Law n° 9.279 of the 14th of May 1996, which governs rights and obligations in relation to industrial property (Brazilian Industrial Property Law), and in 1997, approved internal plant variety protection legislation, Federal Law n° 9.456, of 04/25/1997 (Brazilian Plant Variety Protection Law), which governs the granting of intellectual property rights on plant varieties to so-called breeders, through plant variety protection certification.

In 1999, Brazil began adhering to the UPOV Resolution of 1978. The 1978 version of UPOV establishes that even when breeder rights protect a determined variety, it may still be used for replanting. In 2003, Brazil approved the Law of Seeds and Seedlings Law (Federal Law n° 10.711 of 2003), establishing a formal system for improving, producing, storing, commercializing, and distributing seeds and seedlings.

In Argentina, the agro-industrialization process was propelled by the Import-substituting Industrialization (ISI) period, which occurred between 1930 and 1970. Agroindustry was concentrated in the Pampean Region, where food and raw materials of an agricultural or cattle-rearing origin were processed for internal markets and exportation. In 1935, the "Ley de Granos y Elevadores" (Grain and Elevator Law – Federal Law n° 12.253/1935) was created, which sought to stimulate the adoption of improved seed and organized the market through means of a regulatory system for production and commercialization. In 1956, the

Argentinian National Agricultural Technology Institute (INTA) was created to stimulate an agricultural and cattle-rearing sector that was increasingly marked by incorporating technologies and policies focused on development. In 1960, authorities involved in agricultural policy, INTA, and seed companies began to demand and demonstrate the need for modern legislation regulating the seed market.

During this period, the Argentinian state constructed a political line and regulatory frameworks related to seed [15], such as the "Ley de Semillas y Creaciones Fitogeneticas" (Federal Law nº 20.247/1973–Law on Seeds and Phytogenetic Developments) [5], which governed the production, certification, and commercialization of seed in the country and established intellectual property rights for plant varieties or cultivars. This Law created the National Registry for Seed Trade and Regulation (RNCyF–the acronyms in Spanish) and the National Registry of Cultivars (RNC), which were aimed at regulating the production, certification, and commercialization of seed, and the Argentinian National Registry of Cultivar Ownership (RNPC) which granted intellectual property protection to plant breeders.

In 1981, the first varieties were registered in the National Registry of Cultivars. Still, the use of intellectual property rights for cultivars only entered into effect at the end of the 1980s through means of judiciary acts carried out by wheat and soybean breeders [2]. In 1994, Argentina ratified the UPOV Convention's Resolution of 1978 through Federal Law nº 24.376, which approved the international agreement for plant protection. In 1995, the Patent Law (Federal Law nº 24.481/1995) was passed in response to the requirements of the TRIPS Agreement, allowing patents to be taken out for genes and transgenic microorganisms.

Brazilian and Argentinian seed laws reflect these countries' adherence to the UPOV system. In accordance with the regulations established by the UPOV Resolution of 1978, both countries established homogeneity, stability, and distinguishability as requirements for the registration and subsequent commercialization of new plant varieties. Traditional varieties that did not meet these requirements were generally excluded from commercial activities. Brazilian legislation opened up exceptions allowing Indigenous groups, beneficiaries of agrarian reform, and traditional communities to trade and commercialize seed among one another, an exception that does not exist in Argentinian legislation. The result is a process of standardization and uniformization of plant varieties and the loss of seed that does not fulfill these criteria, such as local seed, due to difficulties in access and conservation.

Criticisms of the requirements for registration in the formal system were recurring during the interviews conducted with farmers and representatives from rural social movements. The following interview excerpts represent well that which was repeated by many different individuals:

> "We have seen a decline in genetic diversity. Private seed concentration results in the uniformization of a few high-yielding cultivars bred primarily for

> commodity production. We lack the nutritional content and quality necessary to provide healthy, safe, and sovereign foods for the people, which we politically protect via the food sovereignty paradigm. It seems critical to protect freedom and the greatest possible variety of seed, as well as unlimited access by any farmer to any seed reserve or the ability to exchange seeds with other farmers (oral information)."
>
> "[...] we have had an erosion of genetic biodiversity (...) the private concentration of seed results in the uniformization of few high yield varieties intended for the production of, above all, commodities. The concern is not us having the highest nutritional value or quality to produce healthy, safe, and sovereign foods for the people, which is what we defend with the paradigm of food sovereignty (...) politically, to me, it seems essential to defend freedom and the largest diversity of seed possible, unrestricted access to any seed reserve by any farmer, or the power to trade them with other farmers (oral information)."

Registration excludes local plant varieties from formal commercialization in an unjustified manner. This problem is largely a result of the diversity and variability of traditional seed-making patenting impossible. For this reason, any plant that does not meet the formal system's criteria for uniformity is excluded as a matter of principle without a deeper examination of the quality of the varieties that meet these criteria. For example, a more profound analysis may prove that a local variety of corn is better adapted to specific climatic conditions and, therefore, more productive, requiring fewer fertilizers in the location where it was adapted. As another example, a determined local variety may be more in demand in a region, not due to its productivity, but because of its flavor or its relationship with certain cultural traditions. However, in both cases, the seed would be prohibited from entering into the formal seed systems in countries that are signatories of UPOV for not being registered and, therefore, unable to be openly commercialized.

The characteristics required for the registration and commercialization of seed and seedlings are only found in seed that is said to be improved. As a result, Creole seeds, which are heterogeneous and have a high variability index, are kept on the margins of the formal system for the commercialization of seeds. This seed cannot be commercialized because it is not registered, except in exceptional cases where Brazilian legislation allows the sale and trade between smallholder farmers, indigenous groups, traditional communities, and agrarian reform beneficiaries.

Seed laws have served to impose the formal system, which requires registration, as the only legal system available. This legislation prohibits or imposes severe restrictions on informal systems. These laws have, therefore, contributed to a reduction in biological diversity and to greater genetic homogeneity since, by prohibiting or limiting informal systems, they exclude smallholders, farmers, and traditional communities from being seed-adapting assets and promoters of diversity. Restrictions to informal systems, in addition to genetic erosion, drive a total loss of knowledge with which such systems are associated, such as alternative cultivation techniques and related technologies [16, 32, 33].

Under the auspices of increasing quality and security in agricultural production, as well as intellectual property rights, a model has been adopted that promotes increased homogenization and genetic erosion, reducing the biological diversity that forms the basis for the resilience and sustainability of food systems [32, 33].

6. Conclusion

Agriculture 4.0 uses methods employed in Industry 4.0, bringing new elements of precision, automation, and robotic agriculture into agriculture and cattle rearing, such as high-performance computational methods, networks of sensors, machine-to-machine (M2M) communication, connectivity between mobile devices, cloud computing, analytic methods, and solutions for processing large volumes of data and building support systems for management decision-making [8].

The promise is that Agriculture 4.0 will reduce water, fertilizer, and pesticide use by implementing sophisticated technology. Agriculture 4.0 will thereby help to increase productivity, and efficiency in the use of inputs, reduce labor costs, improve work quality and worker safety. It will also lessen environmental impacts in soil preparation, seeding, fertilization, irrigation, and harvesting [8].

Applying these production techniques presents a significant capacity for adaptation in the agricultural sector, offering solutions for the wide array of emerging issues (excessive soil use and global warming). This makes it possible to reduce waste, which occurs in large-scale proportions, resulting in more efficient agriculture [17].

Despite the achievements and promise of Agriculture 4.0, one of its great challenges will be breaking down the epistemic hierarchy and trying to learn and teach other forms of agriculture, acknowledging that they also have their specific technologies, knowledge, and limitations. Because of epistemic prejudice, the modern sees what is different as merely being behind the times. At the same time, imagining that technology 4.0 alone can resolve all of the issues of efficiency and sustainability in the current agricultural model built upon the presuppositions of the Green Revolution is indeed a very ambitious projection [16, 19].

As mentioned at the start of this chapter, contemporary agriculture is not a single homogenous concept but made up of multiple actors and different manners of handling the earth. Through the framework of the Green Revolution, the farmer has come to be seen as a recipient of technology. However, the agricultural practice has always been a platform for producing new ideas, seeking solutions, and a permanent test field to meet adversity head-on in the search for results put forward by culture. Agriculture is a history of the permanent creation of knowledge and technology [14, 16].

Adapting and improving seeds to meet different cultural requirements is merely one expression of the knowledge network other forms of agriculture offer. The Green Revolution is also a network of expertise that has created epistemic privilege for a determined type of technology. However, this does not mean that it is the only network that exists or that other means of agriculture do not have their technologies.

A dialogue of knowledge presupposes that the members of these different networks be recognized as sources of knowledge and the production of technologies in the face of adversity. This generates two direct challenges to Agriculture 4.0: (1) How can it value, create dialog, and learn from other forms of knowledge? (2) How can it be a source for producing technologies and solutions for agricultural systems other than those based on the presuppositions of the Green Revolution?

Turning back to the issue of agrobiodiversity, Agriculture 4.0 needs to include informal seed systems as a permanent, stable source for offering germplasms. Seed banks are important for conservation, but the farmer, through on-farm conservation, can ensure that plant varieties are selected and permanently adapted to face innumerable adversities (climatic, biological, etc.). Unlike seed banks, which carry out the static conservation of seed, and in which genetic material is immutable, farmers are able to carry out dynamic conservation.

Current seed laws, epistemic prejudice, and the technologies that have emerged to meet the demands for homogenous agriculture created by the Green Revolution have put informal seed systems at risk. Another related challenge for Agriculture 4.0 is how to produce technology for an agricultural system other than agribusiness, recognizing and favoring diversity of actors, needs, and types of cultivation.

Compromising informal systems also involves many risks for formal seed systems. Respect for intellectual property cannot mean overlooking the idea of seed being a common good, thereby disregarding systems for trade and commercialization that are not governed by legal guarantees to plant breeders. Agriculture 4.0 is faced with the challenge of recognizing the plurality of existing agricultural systems, instilling efforts with reference to a dialog of knowledge, as well as supporting the development of technologies and strategies for strengthening these non-formal systems, which, even today, continue to be indispensable purveyors of agrobiodiversity.

References

1. Carvalho, H.M., de. (2003). O oligopólio na produção de sementes e a tendência à padronização da dieta alimentar mundial. *In*: Sementes: Patrimônio do povo a serviço da humanidade, 95–112. São Paulo, SP: Editora Expressão Popular, Brazil.
2. Fuck, M.P, Bonacelli, M.B. and Carvalho, S.P. (2008). Propriedade intelectual em melhoramento vegetal: Brasil e Argentina frente às possibilidades de mudanças institucionais. *Informações Econômicas*, 38(9), Brazil.
3. Gonçalves, C.W.P. (2004). Geografia da riqueza, fome e meio ambiente: Pequena contribuição crítica ao atual modelo agrário/agrícola de uso dos recursos naturais. INTER thesis. *Revista Internacional Interdisciplinar*, 1(1), Brazil.

4. Kloppenburg, J.R. (1990). *First the Seed: The Political Economy of Plant Biotechnology*. Univ. of Wisconsin Press, Madison, Wisconsin, USA.
5. Lei nº 20.247 de 30 de março de 1973. (1973). Ley de Semillas y Creaciones Fitogeneticas. Argentina.
6. Lei nº 9.456 de 25 de abril de 1997. (1997). Institui a Lei de Proteção de Cultivares e dá outras providências. Brasília, DF. Recovered from https://www.planalto.gov.br/ccivil_03/Leis/L9456.htm (accessed on 15 January 2023).
7. Lei nº 10.711 de 5 de agosto de 2003. (2003). Dispõe sobre o Sistema Nacional de Sementes e Mudas e dá outras providências. Brasília, DF. Recovered from https://www.planalto.gov.br/ccivil_03/LEIS/2003/L10.711.htm (accessed on 12 May 2023).
8. Massruchá, S.M.F.S. and Leite, M.A.A. (2017). Agro 4.0 – Rumo à agricultura digital. Revista JC na Escola Ciência, Tecnologia e Sociedade: Mobilizar o Conhecimento para Alimentar o Brasil, 28–35, Brazil.
9. Mcmichael, P. (2016). *Regimes alimentares e questões agrárias. Estudos Camponeses e Mudança Agrária.* (1ª Edn.) São Paulo: Editora Unesp e UFRGS Editora, Brazil.
10. Mooney, P.R. (1987). *O Escândalo das sementes: O domínio na produção de alimentos*. Brazil: Nobel.
11. Pereira, M.L. (2013). Valorização da agrobiodiversidade como estratégia de desenvolvimento territorial sustentável: Estudo exploratório na zona costeira centro-sul de Santa Catarina. (Tese de Mestrado). Universidade Federal de Santa Catarina, Florianópolis, SC, Brazil.
12. Perelman, M. (1978). *Farming for Profit in a Hungry World: Capital and the Crisis in Agriculture*. Maryland, USA: Rowman & Littlefield.
13. Perelmuter, T. (2017). *El rol de la propiedad intelectual en los actuales procesos de cercamientos. El caso de las semillas en la Argentina (1973–2015).* Buenos Aires: Facultad de Ciencias Sociales – Universidad de Buenos Aires, Argentina.
14. Ploeg, J.D. (2008). *Camponeses e impérios alimentares: Lutas por autonomia e sustentabilidade na era da globalização.* Porto Alegre: Editora da UFRGS.
15. Poth, C.M. (2019). Biotecnología, Ciencia y Poder. Un análisis crítico sobre la regulación en torno a las semillas genéticamente modificadas. *Revista Administración Pública y Sociedad* (*APyS*), *7*, 77–102. Retrieved from https://revistas.unc.edu.ar/index.php/APyS/article/view/24303 (accessed on 15 May 2023).
16. Reis, M.R. (2012). *Tecnologia social de produção de sementes e agrobiodiversidade.* Master thesis, University of Brasília, Brasília, DF, Brazil. Available from https://repositorio.unb.br/handle/10482/11240 (accessed on 15 May 2023).
17. Ribeiro, J.G., Marinho, D.Y. and Espinhosa, J.W.M. (2018, August 28–30). Agricultura 4.0: Desafios à produção de alimentos e inovações tecnológicas. Proceedings of the II Simpósio de Engenharia de Produção (SIENPRO), Federal University of Goiás – Catalão Regional, Catalão, GO, Brazil.
18. Santilli, J.F.R. (2009). Agrobiodiversidade e direito dos agricultores. Doctoral dissertation, Pontifical Catholic University of Paraná, Curitiba, PR, Brazil. Available from http://www.biblioteca.pucpr.br/tede/tde_busca/arquivo.php?codArquivo=1457 (accessed on 15 May 2023).
19. Santos, B.S. (Ed.). (2005). *Produzir para viver: Os caminhos da produção não capitalista* (2nd Edn.). Rio de Janeiro, RJ: Civilização Brasileira, Brazil.
20. Shiva, V. (2001). *Biopirataria: A pilhagem da natureza e do conhecimento* (*L.C.B. de Oliveira, Trad.*). Petrópolis, RJ: Vozes. (Original work published in 1997).
21. Shiva, V. (2015). *A violência da revolução verde: Agricultura, ecologia e política do terceiro mundo* (L.H. Teixeira, Trad.). Edições Mahatma. (Original work published in 1991), Brazil.

22. The Second Report on the State of the World. Plant Genetic Resources for Food and Agriculture Synthetic Account. (2010). *In*: Food and Agriculture Organization of the United Nations. Recovered from http://www.fao.org/3/i1500e/i1500e_brief.pdf (accessed on 15 May 2023).
23. Grupo Semillas. (2010). Las leyes de semillas aniquilan. La soberanía y autonomía alimentaria de los pueblos. Publicaciones web GRAIN. doi:/e/4098.
24. Machado C.D.C. (2020). Plantas alimentícias na Reserva de Desenvolvimento Sustentável Piagaçu-Purus. Amazônia Central. *Rodriguésia*, 71, e02332018. Available at doi: 10.1590/2175-7860202071076 (accessed on 15 May 2023).
25. Soleri, D. (2017). Civic Seeds: New Institutions for Seed Systems and Communities. A 2016 Survey of California Seed Libraries.
26. Boef, W.S., de, Dempewolf, H., Byakweli, J.M. and Engels, J.M.M. (2010). Integrating genetic resource conservation and sustainable development into strategies to increase the robustness of seed systems. *Journal of Sustainable Agriculture*, 34(5), 504–531.
27. Gill, T., Bates, R.M., Bicksler, A., Burnette, R., Ricciardi, V. and Yoder, L.S. (2013). Strengthening informal seed systems to enhance food security in Southeast Asia. *J. Agriculture, Food Systems, and Community Development*, 3, 139–153.
28. Almekinders, C.J., Louwaars, N.P. and Bruijn, G.D. (2004). Local seed systems and their importance for an improved seed supply in developing countries. *Euphytica*, 78, 207–216.
29. Fredenburg, P., Bishaw, Z., Niane, A.A. and Devlin, M. (2015). *Strengthening National Seed Systems for Household Food Security in Developing Countries*. International Center for Agricultural Research in the Dry Areas (ICARDA): Beirut, Lebanon, 2015. Available online: https://repo.mel.cgiar.org/handle/20.500.11766/7125 (accessed on 15 November 2022).
30. Coomes, O.T., McGuire, S., Garine, E., Caillon, S., McKey, D., Demeulenaere, E. et al. (2015). Farmer seed networks make a limited contribution to agriculture: Four common misconceptions. *Food Policy*, 56, 41–50.
31. Wattnem, T. (2016). Seed laws, certification and standardization: Outlawing informal seed systems in the Global South. *The Journal of Peasant Studies*, 43(4), 850–867. doi: 10.1080/03066150.2015.1130702 (accessed on 11 maio 2023).
32. Union for the Protection of New Varieties of Plants. (2021, February). Members of the International Union for the Protection of New Varieties of Plants. Retrieved from https://www.upov.int/edocs/pubdocs/en/upov_pub_423.pdf (accessed on 11 maio 2023).
33. United Nations Development Programme. (2020). Global Biodiversity Outlook 5. Retrieved from https://www.cbd.int/gbo/gbo5/publication/gbo-5-en.pdf (accessed on 11 maio 2023).

CHAPTER

7

Prospecting for Agro 4.0 Solutions to Control the Fruit Fly (*Anastrepha fraterculus*) in the Brazilian Context

Shana Sabbado Flores[1*][0000-0003-0414-6878], **Leonardo C. da Silva**[1][0000-0003-1526-2444], **Ronaldo Serpa da Rosa**[1][0000-0002-5202-3349], **Willian Bergonsi**[1][0000-0002-7333-1746], **João E. Gomes**[1][0000-0003-1418-0658], **Allana C. Decarli**[1][0000-0002-0312-1567], **Camila A. Mugnol**[1][0000-0003-4255-0919], **Teresa Enderle**[1][0000-0003-4257-7660], **Rafael Lauxen**[2][0000-0002-9596-1933] **and Alan Fantin**[1][0000-0002-3245-1805]

[1] Federal Institute of Education, Science et Technology (IFRS), Campus Bento Gonçalves, Av. Osvaldo Aranha, 540, Juventude da Enologia, 95700-206, Bento Gonçalves, Brasil

[2] Federal Institute of Education, Science et Technology (IFRS), Campus Ibirubá

This study aims to identify agro 4.0 solutions that can control the fruit fly (*Anastrepha fraterculus*) in Brazilian fruticulture. Fruit flies are considered one of the main pests of cultivated fruit around the world; it is also considered a phytosanitary problem as one of the main trade barriers in the commercialization of fresh fruit. The study is structured in prospecting and feasibility analysis. The prospecting is based on a systematic literature review and patent databases. The results point out some solutions that are not for the fruit fly nor have not been tested in Brazil, such as an automatic pest monitoring system using machine vision-based rice light traps. Also, some opportunities for further studies or product development are prospected.

1. Introduction

Fruit flies (*Anastrepha fraterculus*) are considered one of the main cultivated fruit pests worldwide and cause losses of 120 million dollars a year in Brazilian

* Corresponding author: shana.flores@bento.ifrs.edu.br

fruit production and more than two billion dollars in world fruticulture [1]. In addition, the fruit fly is also considered a phytosanitary problem as one of the main trade barriers in the commercialization of fresh fruit [2]. The incidence of fruit flies causes economic problems in Brazilian fruticulture, both due to the losses caused directly in the production as well as the increase in the cost of treatments for control [3, 4]. In citrus, for example, the fruit fly is the second pest that causes the most economic losses to producers, only behind greening. Thus, the treatment of the topic is directly related to the sustainability of fruticulture, whether from an economic point of view, with the reported impact and also in terms of environmental sustainability, since the control and combat are carried out with the application of agrochemicals, causing secondary impacts on biodiversity.

This study targets to identify agro 4.0 solutions that can control the fruit fly (*Anastrepha fraterculus*) in Brazilian fruticulture. The control management of this insect vs. pest dilemma is crucial to reduce losses and avoid the intensive use of agrochemicals. The control is hampered by the behavior of the insect, with the migration of adults from commercial orchards to native plants, and the fact that the fruit fly is not seasonal in Brazil, which causes infestations in orchards during all seasons of the year, and it requires ongoing follow-up and monitoring. To have a complete picture, a delay in the management actions to control the fruit fly can lead to production losses of between 20–30% for early varieties and between 70–80% for late varieties, reaching 100% of production.

Brazil has initiatives to monitor the fruit fly to reduce the negative impact on fruticulture. An example is the project "Alert System for Monitoring the Fruit Fly" developed by the "Brazilian Agricultural Research Corporation" (Embrapa) and coordinated by the "Technical Assistance and Rural Extension Corporation" (EMATER/RS), which groups together a set of strategies for managing the insect pest in peach orchards. Despite the initiative's relevance, some factors hinder its effectiveness, as most of the process is carried out manually, which generates a large operating cost. Another obstacle is that in some situations, the quality and comprehensiveness of the data can be compromised, as the data is collected from a reduced number of traps, and it can take up to seven days for an occurrence report to be issued. It is estimated that technological tools and concepts from Agriculture 4.0 can be applied to solve the critical points observed in current control systems. Technologies can be identified in the literature and prospective studies, and this work aims to contribute to the identification of solutions considering their technical and economic feasibility of application in the field, as well as their scalability.

The study is structured in two phases, involving the prospection of technologies in the literature and registration of patents, and feasibility analysis, with the participation of a multidisciplinary group of experts. This chapter addresses producers, technical experts, researchers, and stakeholders interested in improving performance and sustainability in Brazilian fruticulture.

2. Background

2.1 Fruit Fly and Its Control

The term fruit fly covers a series of fly species that cause significant economic losses for world fruticulture. In Brazil, the species *Ceratitis capitata* (Diptera: Tephritidae) can be found throughout the country, and the main species found in the southern region of the country is the South American fly *Anastrepha Fraterculorus* (Diptera: Tephritidae).

Adults *A. fraterculus* are 6–7 mm long, with yellowish bodies and transparent wings with dark shaded bands, one similar to an 'S' and another shaped like an inverted 'V'. The *C. capitata* is a bit smaller, 4–5 mm long. Their bodies are, in greater part, yellow, with the top of their thorax black and transparent wings with shaded yellow stripes [1]. On both species, the female possesses the final part of the abdomen tapered, giving rise to the ovipositor used to deposit the fly eggs in the host fruits.

The fly's life cycle depends on climatic conditions (usually longer in colder places) and the host fruit. Since flies cover a large amount of affected cultures, adults are detected virtually all year round, which becomes evident during the fruit maturation period [1]. Its life cycle is divided into four phases: egg, larva, pupa, and adult. It can last 22–24 days to *C. capitata* and 26–28 days to *A. fraterculus* [2].

The female deposits its eggs in the fruit through a break in the skin made by the ovipositor. The egg develops, turning into a larva, then migrating to the ground. The third stage is the pupal stage, which occurs on the soil until the complete development of the fly. After that, the fly becomes an adult insect, being fit for reproduction and restarting the cycle.

The fruit fly causes a series of damage to the host fruits, from the holes caused by the ovipositor to the galleries formed by the larvae. The action of the insect can cause the deformation of fruits, alteration in taste, early ripening, rotting, and even the fall of fruits in some cases [1]. These losses can be added to other direct damage, in which the opening made by the insertion of the ovipositor in the fruits causes the break of the skin, which allows the entry of phytopathogenic microorganisms [3].

It is known that *A. fraterculus* affects 67 species of host cultures while *C. Capitata* affects 374. Among the affected cultures are families with significant production loss: *Myrtaceae* (Guava, Pitanga, Jabuticaba, and Jambo); *Rosaceae* (apple, pear, plum, nectarine, and peach), and *Rutaceae* (orange-sour, orange-sweet, lemon-clove, gossip, and tangerine) [3].

The bigger damage to the orchards occurs at the beginning of the fruit maturation and when the fruits are almost ready for harvest [4, 5]. A large number of hosts for the insect brings negative consequences for producers when the cycle of culture ends. The fly goes to another culture at the ideal oviposition stage.

The comparative study of different capture peaks of *A. fraterculus*, for the production cycles of *gabiroba*, *cereja-do-rio-grande,* and *guava-serrana* cultures, can illustrate the problem. *Fraterculus* for the production cycles of *gabiroba*, *cereja-do-rio-grande,* and *guava-serrana* cultures.

The problem can be illustrated by the comparative study of different capture peaks of *A. fraterculus* for the production cycles of *gabiroba*, *cereja-do-rio-grande,* and *guava-serrana* cultures. These capture peaks showing the population's fluctuation and permanence [5].

Another negative point for fruit producers is the great diversity of hosts in the native forest surrounding the orchards because there are several succession of hosts for the fly. Considering this point, at the time of the monitoring of the plague, the boundaries of the orchards that are in contact with the native forest are of paramount importance to avoid the entrance of the plague into the cultures [10].

Control and monitoring of fruit flies is a practice in the "Integrated Pest Management" (IPM) context. For species such as *A. fraterculus* and *C. capitata*, the usage of these techniques is essential because "since it is a polyphagous pest (attacks several fruit species) and with high adaptive capacity, only the integrated use of practices of control allows positive and effective responses in the population reduction of these insects" [6]. These different methods effectively protect a large area, creating a barrier against insects coming from outside.

Mass capture is an alternative to control the fruit fly in small areas, in which traps made with a PET bottle containing hydrolyzed protein are used and distributed so that the flies are attracted and killed in the trap. Efficiency was tested in table grape vines and confirmed by the effectiveness of the results, demonstrating how the use of traps is an alternative method to the use of chemical control, especially the use of insecticides applied by spraying, which is extremely harmful to the environment [1, 7].

Traps are used to control the fruit fly as they make it possible to observe the pest density in each monitored part, as well as identify which types of fruit fly are present, thus being able to plan the appropriate control method only in the affected areas and at the exact time of population growth [7, 8]. The variety of insecticides for controlling fruit flies has been increasingly limited, and those available are limited to the adult stage of the fly. That is why traps are so important in the orchard, as it monitors the adult pest, the stage when insecticides can be used to control it [9]. Thus, for efficient management and punctual control of infestations of these species, it is necessary to monitor, that is, identify and quantify the presence of the insect, and for this, traps are an economical, environmentally, and technically viable alternative, which makes their use common [6].

The IPM includes several phases and stages, with monitoring being the most important to stop the pest's advance. For this monitoring, traps are used, which can be the Jackson type that employs pheromones to attract flies (used for *Ceratitis capitata*) or the McPhail type, which uses liquid attractants for the attraction (it is used mainly for *Anastrepha fraterculus*, but it is also effective for *Ceratitis capitata*) [10].

McPhail traps, for example, are affordable and can also be made with recyclable products such as PET bottles or other containers. Its attractant can be purchased or made with fruit juices, sugar, or torula yeast [8]. A factor that directly influences the efficiency of this monitoring and the attraction of insects (which does not have a simpler and cheaper way than using traps) is the bait used in the trap. In the case of McPhail traps, CeraTrap (hydrolyzed protein) was an advantage over other baits.

As a general rule, the most used trap for catching fruit flies is the McPhail trap. A considerable part of its attractiveness and efficiency come from the bait used, but it is not limited to that. According to studies, the yellow color is more efficient in attracting both species of flies when compared to transparent, blue, or red traps, with a presence index of insects very high compared to others [11].

In addition to the efficiency of the yellow color in the McPhail trap, another point to be taken into account is its shape and construction. It has a good sealing between the top and bottom parts so that no losses occur when attractants that release their odor are used through the volatilization of compounds such as CeraTrap [12]. The transparency of the upper part of the McPhail trap is of great importance to aid in capturing fruit flies efficiently. Once the fly is attracted to the trap, it will try to get out and fly to the top, not realizing that there is a transparent barrier. This attempt will repeat until the fly dies in the attractant. Thus, the trap's choice of color and shape works together with technologies and new attractive lures [12].

In addition to mass capture using traps, there are a number of control methods, usually more laborious, expensive, or biologically harmful to the environment. One example is the cultural control approach, in which fruits infested by fly larvae are gathered and dealt with so that they do not end their cycle. This can be done by discarding the fruits, burying them in a hole lined with quicklime, feeding animals (since the fruits are still in good condition), composting (liquid or solid with coverage), or eliminating alternative fruit hosts/multipliers near the orchard.

Chemical control can also be used, utilizing toxic bait (applied with a manual pump sprayer to attract insects) or applying insecticides by spraying. These two methods, especially the application of insecticide, attack the entire ecosystem of the orchard, altering the biological system present in the area.

Biological control is becoming an increasingly present alternative, in which fruit fly parasitoid insects (such as *Diachasmimorpha Longicaudata* and *Fopius Arisanus*) are used. Another alternative is the sterile insect method, which proposes inserting sterile males in the environment for population control. However, for both techniques, there is still a cost and time needed to produce the insects in the laboratory, making using traps cheaper and easier [6].

Another practice that can be adopted to control the fruit fly is the bagging of fruits, which prevents, through physical means, the oviposition or test bite of the female. Covering materials can range from waxed paper to non-woven fabric. As a control method requiring more labor, it is commonly used in small and medium-sized orchards [1].

Following the principle of physical protection against the fruit fly, plastics or screens cover the entire orchard, making the pest unable to access the plants and not only the fruits, as in the method above. If the producer wants screens to perform this protection, they must be at most 2 mm thick to block the insect's entry into the orchard. As with the fruit bagging control method, this is also not commonly used in very large orchards [1].

2.2 Agro 4.0

The technical term Agriculture 4.0 originated as an analogy to the term Industry 4.0, which was used by the "German Ministry of Education and Research" in a final report of the officially instituted working group to develop guidelines for German industry in the context of integrating it in a "smart networked world". The manufacturing environment includes a vertical network, end-to-end engineering, and horizontal integration across increasingly intelligent products and systems. These are set to inaugurate the fourth stage of industrialization, Industry 4.0 [13].

Using the Industry 4.0 concept, Agriculture 4.0 can also be divided into revolutions. Efrom & Bottom [14] describe the First Agricultural Revolution in 10,000 BC as the man using animals and simple tools. After over 82 centuries, the Second Agricultural Revolution emerged in the 19th century, with the adoption of the mechanical strength of tractors powered by combustion engines and agrochemicals. Until the late 1970s, during the era of mechanization, the contribution of Design for Agriculture was limited to the design of material artifacts, tools, and means to facilitate work in the field [15].

Computers became essential tools in the last quarter of the 20th century and the beginning of the 21st century. The development of agricultural robots began, in addition to hegemonic precision agriculture, leading to the Third Agricultural Revolution. In the 1990s, the first forms of technology and process innovation in agriculture emerged from the development of Precision Agriculture [16]. Such evolution paved the way for the Fourth Agricultural Revolution using the "Internet of Things" (IoT), "cloud computing" (CC), "artificial intelligence" (AI), generation and use of "big data" (BD), and the development of "cyber-physical systems" [14].

The digital revolution and the ICT innovations developed over the last 30 years with science, robotics, and AI were first applied to agriculture in terms of precision agriculture [16]. In a second moment, the Industry 4.0 paradigm has transposed to the agricultural sector, bringing the trend towards automation and interconnection of farming activities and all supply chain processes [22]. This paradigm defines Agriculture 4.0 or Digital Agriculture today and instituted new problems and new areas of experimentation within Agro 4.0.

Due to its results, Agriculture 4.0 extends from monitoring crops through the traceability of the entire production chain, reaching the whole supply chain. Agriculture 4.0 is essentially the digitalization of sociotechnical phenomena associated with agroecosystems [16], a phenomenon analogous to what occurred in urban and industrial systems.

The insertion of "information communication technologies" (ICT) in the psychosocial structure of people, cities, and industries enabled by smartphones, IoT, and AI occurred gradually and has great potential to be identified and explored due to the ease of operation and cost reduction [17]. Therefore, as systems constantly improve, their use by people and industries has become widespread, resulting in the dependence of increasing functionalities on these systems. Humanity currently exhibits an almost symbiotic synergy with digital technology [17].

The AI advance integrated into IoT in the field tends to accelerate in the coming years, using increasingly robust and precise algorithms that will directly benefit agribusiness in terms of cost reduction, increased productivity, and environmental sustainability [18]. It is important to consider that the growing globalization of agrifood production has been accompanied, in recent decades, by the importance of product sustainability and management practices in their production [26]. The sustainability of agricultural systems, which is fundamental for the survival and well-being of humans, is greatly aided by the adoption of IoT technology globally, contributing to better use of inputs based on more effective decision-making supported by data [19].

3. Materials and Methods

The study is structured in two stages: prospecting and feasibility analysis. The prospecting is based predominantly on secondary data, using systematic literature review, prospecting in patent databases, and gray literature. The bibliographic breakdown seeks to use scientific methods that objectively and reliably avoid human error. Thus, data processing, considering the same parameters and conditions, can always be replicated to obtain the same result. The study used a systematic literature review as it allows for an "entire quantitative assessment of knowledge structures and research trends in the field, without having to select or dismiss any title for the selection and mapping or representation" [20].

For the systematic review analysis, bibliometric techniques were used, as it is a systematic, reproducible process based on the statistical measurement of scientific activity. The data are imported into the RStudio platform to obtain the science mapping, which performs statistical and graphic calculations and uses the Bibliometrix package, consisting of bibliometric analysis tools [21, 22]. The search in the Scopus database used three keywords as input, chosen due to their close relation to the research topic. The terms were:

- 'Automatic': This term was chosen to focus the bibliographic research on texts dealing with processes without human interference since this is one of the characteristics of Industry 4.0.
- 'Trap': The word choice explored different types of traps and analyzed the

results obtained in the articles to provide a broader view of the other models that can be used for more automated detection of insects.

- 'Insect': The term was chosen to limit the scope and bring results related to pests (since there may be different types of pests such as fungi, insects, rodents, etc.), at the same time that left it wide enough to obtain texts from insects with different habits and characteristics, such as target crop, size, reproduction, habitats, etc.

With the keywords defined, the parameters were applied to the Scopus database. This tool was chosen because it is the most widely used database for evaluating scientific publications. The search in Scopus resulted in a total of 119 articles. As filters for prioritizing articles, publications between 2006 and 2021 written in English and Portuguese were considered. The final result was 93 articles (Fig. 1).

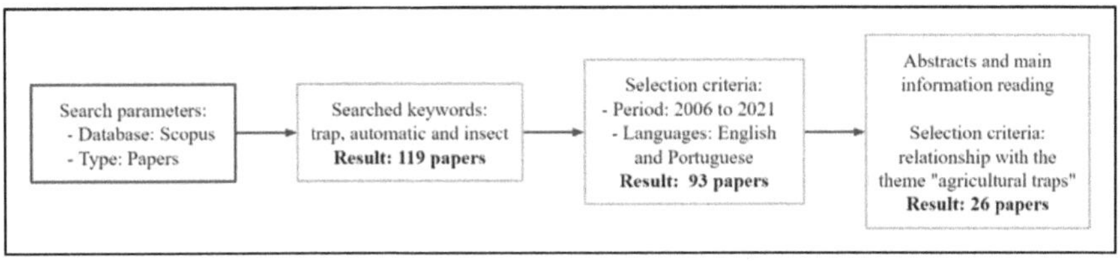

Fig. 1. Selection of articles for bibliometric analysis and systematic literature review.

"The existence of substantial, effective statistical algorithms, access to high-quality numerical routines, and integrated data visualization tools are perhaps the strongest qualities to prefer R to other languages for scientific computation." [21]. In the analysis stage, through the Biblioshiny web interface, graphics and data were generated, presenting the bibliographic information in a friendly way. Finally, with the help of the information obtained, titles and abstracts were analyzed to classify the most relevant publications for the study, highlighting the texts with a bias related to traps in agriculture. The result of this step is shown in items 4.1 and 4.2.

The search for patents was carried out in the Espacenet and Lens databases, using as input: ("fruit fly" OR "fruit flies") AND "control" AND "trap". The entry terms were chosen for their adherence to the theme, namely:

- Fruit fly/Fruit flies: These words were used so that the research showed traps for controlling fruit flies, particularly being more focused and avoiding searching for patents for other insects.
- Control: This word was used because it is the process in which the trap is included in the IPM. This step is the most important in the management of the fruit fly.
- Trap: Keyword used to direct all research on traps, considering that they are one of the main subjects of all writing and the greatest ally in controlling the insect pest.

In the Espacenet database, the only filter used for this search was publications from 2015 onward: 9 patents were found. As a result, 1 output was discarded because it was not a trap. In the Lens database, patents from 2015 and onward were also selected, but the filter patents with active statuses were added and returned 12 results. By crossing the two bases, 3 common patents and 1 duplicate patent were identified, which resulted in 16 patents in total. The results are shown in item 4.3.

In the second stage, the systematized solutions were submitted to a feasibility analysis that featured a panel of experts with technical and market representatives. The opinion of experts in the field was used to validate the selection; semi-structured interviews consulted them. Identifying participants' names and characteristics is reserved for preserving confidentiality commitments.

4. Results and Discussion

The presentation of the results begins with the bibliometric analysis, which will point out the main areas of concentration and trends in research on the topic comprehensively (4.1). Next, the systematic review scrutinizes the main studies and proposals involving monitoring and Agro 4.0 technologies, linking research and devices (4.2). On the other hand, patents show the registered products that can be produced or licensed (4.3). Finally, the feasibility analysis of the prospected technologies is carried out considering technical factors and the context of Brazil (4.4).

4.1 Bibliometric Analysis

The bibliometric analysis considered the data generated by the Scopus database (119 papers). From the resulting illustrations, it is possible to understand and evaluate the topic of study more extensively, analyzing the most cited countries and authors, the journals referenced on the subject, and the co-occurrence network of the keywords used.

4.1.1 Bibliometric Analysis: Broad View of Automatic Traps for Insects Publications

The bibliometric analysis highlights the great presence of Brazil in research related to insects (Fig. 2). Among the facts that help explain this great occurrence of the country among the articles found, we can mention: the geographic characteristics of the country, with large areas with tropical climates favorable to the presence of insects; and the consequent existence of a large number of cases of diseases transmitted by insects, such as dengue, leishmaniasis, and yellow fever [23]. This high incidence of epidemiological cases also explains the existence of relevant epidemiological research institutions, especially the Oswaldo Cruz Foundation. It serves as an incentive for much research related to the habits of these vector insects.

Regarding the number of citations, the United States and China appear right below Brazil but still with a significant number of sources (Fig. 3). The large

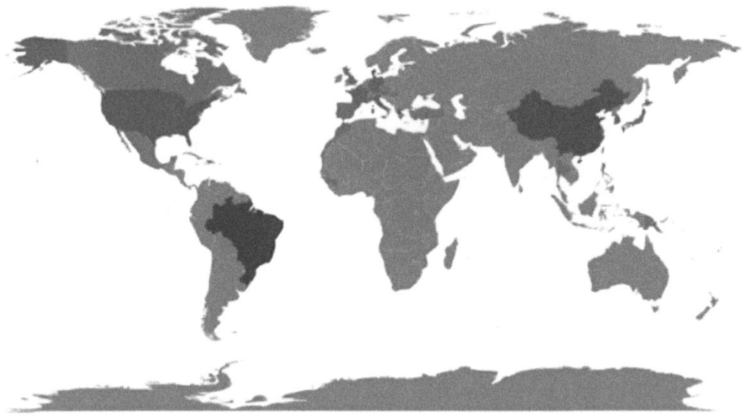

Fig. 2. Country scientific production.

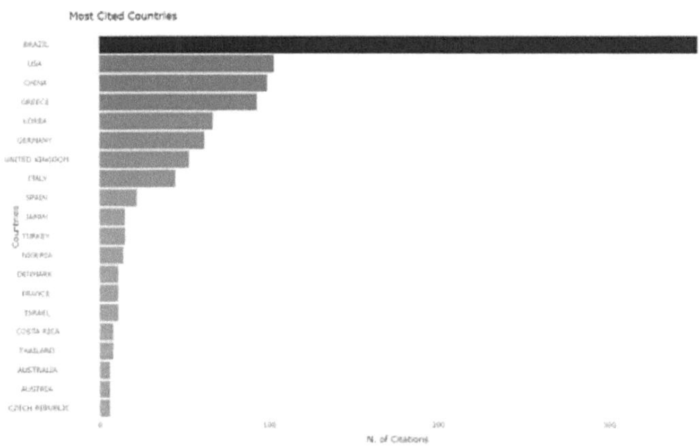

Fig. 3. Most cited countries in the articles.

technical production of both countries in recent years, as well as their continental dimensions and a large number of research institutions, justify many citations of both countries. Then there are Greece, Italy, and Spain (European countries with agricultural production highly affected by pests), the United Kingdom and Germany (countries with historical techno-scientific production), and South Korea (a country with relevant academic output). The other countries have a small number of citations and are close to each other in numbers, and rambling about the motives for their smaller number of citations is beyond the scope of this study [29, 30].

It is worth noting that several countries with climatic characteristics similar to Brazil and also plagued by diseases transmitted by insects were not mentioned. No research containing the chosen keywords may have been developed in these

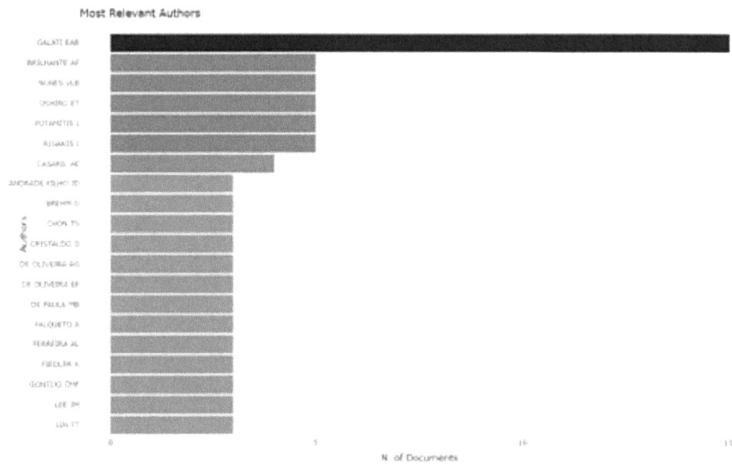

Fig. 4a. Most relevant authors.

countries. Still, considering that the keywords used can be regarded as loose, Brazilian research institutions' importance and prominence about the topic on the international scene is evident [31].

It is also possible to identify the most relevant authors (Fig. 4). Brazil also has the most relevant authors in the assessed set, similar to the indication of countries. With the largest number of documents, we present the author Eunice Aparecida Bianchi Galati, identified in the chart as Galati E.A.B., a professor of epidemiology and entomology at the University of São Paulo, Brazil. A great part of her research is focused on the areas of epidemiology, entomology, and parasitology [32]. From the second to the sixth position, five documents per author are exposed, classifying them in the second position referring to the number of publications. In addition, below the most cited author, Brilhante AF is mentioned, refers to Andreia Fernandes Brilhante, who is also a Brazilian Ph.D. in Science from the Faculty of Public Health of the University of São Paulo, Brazil, with experience in topics involving parasites and vectors of importance in public health, sandfly fauna, leishmaniasis, and sanitation education [33].

4.1.2 Reference Articles for the Topic

The bibliometric analysis allows us to point out the most relevant sources on the topic (Fig. 4). "Memories of the Oswaldo Cruz Institute" stands out, with a significant number of articles. The international journal is focused on biological and biomedical research, and contributions from researchers and scientists from around the world are accepted [34].

In the second place, as a reference journal, is the *Journal of the Brazilian Society of Tropical Medicine*, whose mission is to publish scientific papers on infectious and parasitic diseases, preventive medicine, public health, and related issues [35]. Therefore, because some insects are vectors of diseases in the community environment, this journal received a prominent position in the chart.

Another noticeable fact is the importance of Brazil for the research in the area compared to other countries, as both Mem. Instituto Oswaldo Cruz and the *Journal of the Brazilian Society of Tropical Medicine* have Brazilian origins. Again, it reiterates the importance of a country in Latin America in the search for technological solutions for insects.

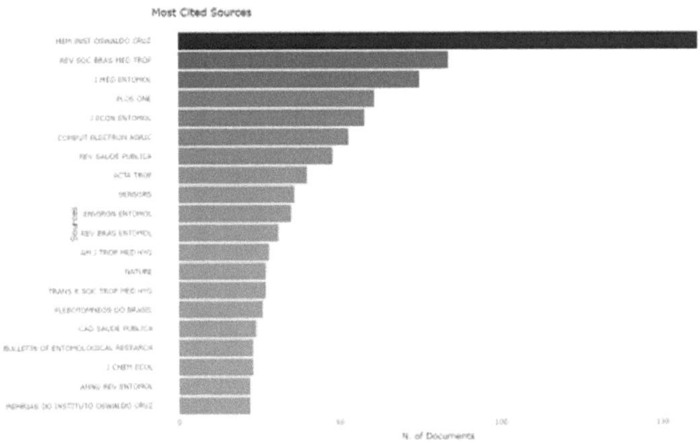

Fig. 4b. Most cited sources.

4.1.3 Co-occurrence Network

Co-occurrence assesses the frequency of occurrence and proximity of keywords in documents [36], enabling a greater understanding of how the keywords used to delimit the research complement each other. The font size used in each word written is related to how often the word appears in the articles found. The larger the font size, the more frequently the term was used.

The present co-occurrence network (Fig. 5) can be divided into three major clouds, displayed in the figure below, with red, blue, and green colors. The words highlighted in red are associated with health research, addressing the correlation of vector insects in transmitting diseases to humans. In addition, the term 'season' appears in singular and plural forms, indicating the influence of seasons on insects. In this cloud, Brazil is mentioned too, which, as mentioned above, has great representation in health research involving insect vectors of diseases.

The green cloud has several interconnected words and is lean towards insects and their influence in the agrarian area. The main correlations relate to how animals contribute to species diversity as well as to biodiversity. And according to the characteristics and habits of each insect, they can contribute both beneficially and negatively.

The last cloud is the blue one. It contains the smallest number of words connected with the medical part and the diversity and biodiversity of the fauna parts. The words 'article', 'fly', and "controlled study" also direct the study of insects to the articles found in bibliometrics.

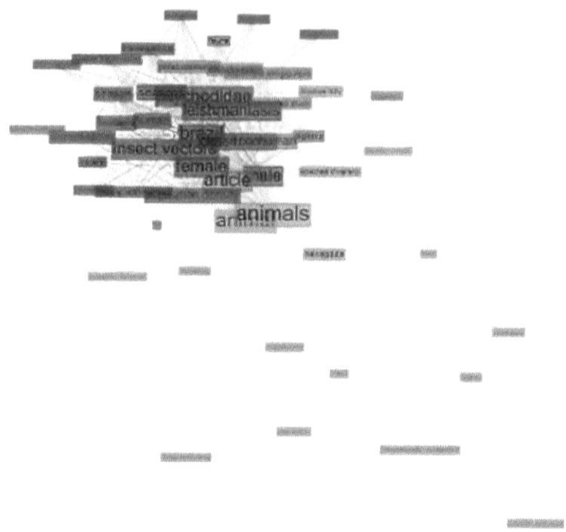

Fig. 5. Co-occurrence network.

4.2 Systematic Revision: Studies and Proposals for 4.0 Monitoring

Insects are a very important part of the global ecosystem and include pollinators, disease vectors, and agricultural pests, all of which fundamentally influence society. In cases where the aim is to determine their presence, and to monitor and control such insects with high priority, the use of traps with an automatic system has great relevance [14]. In addition, more accurate information allows a farmer to make a more precise decision, knowing if it is really necessary to seek some form of tackling the insects. As mentioned by Pereira et al. [25], insect traps also favor practical applications in integrated pest management (IPM). However, the manual counting of insects caught by traps is a task that requires time and logistics and results in slow results. Even if it has sufficient precision, the delay in issuing the results facilitates the target pests to cause advanced damage to the property [19]. Thus, this systematic review aims to expose different situations in which automatic insect traps were used worldwide, focusing on agriculture, based on the articles selected in the bibliographic analysis.

The selected studies for analysis can be classified into two approaches: biological control and automated control. In terms of biological control, an example is the Portuguese and Israeli study, which focused on pheromones produced by two species of cochineal, seeking to detect the moment when the production and detection of pheromones is maximum to disrupt the mating process of insects [15]. A study carried out in Metapa, Chiapas, Mexico, and Niayes, Senegal, which focused on the biological control of the fruit fly, used the dispersal of sterile individuals in productive zones. The study dealt with new equipment used for better dispersal of sterile insects and the best location for their spreading, maximizing the results of this biological management technique [29].

Studies involving automated control use capture techniques combined with an insect recognition system in an IoT approach. In most cases, the capture comes from devices adapted to existing traps, especially McPhail [16–18]. Most investigations employ cameras and image recognition algorithms to identify insects, but sensors, infrared, or probe can also make the recognition.

The use of an autonomous robot was studied for the collection and identification of insects in a pepper greenhouse. The robot could move around the greenhouse and had a compressed air jet capable of shaking the leaves of the plants, a surface with adhesive, and a camera, with which insects were captured and identified. The results presented indicate that the robot is efficient in detecting small quantities and being useful in providing warnings at the beginning of an infestation [19].

Another study [20] analyzed the efficiency of different trap models for the Oriental Fruit Moth (*Grapholita Molesta*). In the end, there was no difference between colors, but the shape and aperture showed different results for different periods of exposure. The results can serve as a basis for standardizing the traps.

A demonstrative study on automated traps tested two prototype traps with insect detection sensors for remote population monitoring [21]. The project's tests were handled in orchidariums located in Pennsylvania and Washington, USA, in 2009. The traps use light sensors (infrared) that detect the entrance but not the type of insect in the trap. The information is then forwarded to a center where it is processed, registered, and reported to technicians. The focus of the work was only to demonstrate the system's potential and not to present refined results or applications on a larger scale.

With a more applied perspective, the study [22] presents a monitoring station that has a trap (for capture), a camera (for recording), and a Zigbee module (for wireless transmission). Such a station is connected to a network and allows real-time monitoring of the level of infestation in a given area. The wireless network results were compared with manual checks of the traps, proving that the system is viable. The stations have low power consumption, being maintained by simple AA batteries. However, it was found that this choice of batteries has a short lifespan (about 34 hours), and another source of energy is suggested.

Along the same line, the study [23] presents a traditional computer vision technique for identifying moths in a rice field. The technique uses two cameras and a transparent surface to capture and process images. Despite carrying out the identification and counting (including different types of moths) automatically and accurately, it requires mechanical filtration (sieving) and manual collection of captured samples. Although the studies are not focused on the fruit fly, they point out solutions that may be viable for application to the insect in the Brazilian context.

Based on the McPhail trap, a study developed in Greece [17] focused on detecting the *Bactrocera Oleae* fly, a pest that affects olive trees. The method developed consists of light sensors installed at the trap entrance and a continuous light source (infrared LED). The interruption in the captured light caused by the flapping of the wings of insects entering the trap is then recorded, analyzed, and compared with known frequencies of insects. Estimated values for the cost of each set of sensors are provided, but they are likely to be outdated already. The study proposes integration with an alert system, allowing complete automation of detecting and sending information to farmers' smartphones.

In another application proposal by the same group of researchers [18], sensors were installed at the entrance of conventional traps (McPhail) that capture the flapping of fruit flies' wings. Again, the data are used to map the population of this pest, generating alerts and pest coverage zones for farmers. Although they obtained a high accuracy rate, they could not successfully distinguish between fruit fly species as their wing-beat frequency is very similar. For this case, it is interesting to note that, even though the publication focused on the proposal, the project's objective was to create a product.

4.3 Technological Patents' Prospecting

The result of technological prospecting in the databases found 16 patents, with a strong presence in China, with 12 patents (Table 1). Most patents are generally intended for fruticulture, except 2 for olive trees and 1for mango. The traps differ in format and technology, but all are of the "attract and kill" type, using volatile compounds, fermented fruits, or solutions as attractive lures.

The shape of the traps is also a visually attractive aspect for insects, and the patents depict solutions that differ in greater efficiency, durability, or lower cost. Although the main focus is on capture, traps can also be used for monitoring and integrated with other control systems, analog or digital. In this sense, it is highlighted that the patents do not present 4.0 solutions by themselves but can be integrated into a control system and combined with other IoT technologies for monitoring and alarm.

Another point that can be seen as limiting or an opportunity is that, despite having a consolidated presence in terms of academic production, Brazil does not have patents on the topic. The only patent in Latin America is from Mexico. The data do not provide further information on the development of research and validation. Still, the absence of patents may also indicate that such devices have not been tested and validated in the country. This issue can be an opportunity for new studies or even for technology transfer projects, in the sense of systematizing studies into improvements or as new products for Latin America.

4.4 Feasibility of Solutions in the Brazilian Context

Some selected proposals are analyzed for this topic, considering their potential application in the Brazilian context.

The proposal entitled "Automatic greenhouse insect pest detection and recognition based on a cascaded deep learning classification method" [26] relied on sticky paper with a colored attractive lure. The trap solves specific phytosanitary problems caused by insects in a greenhouse. The process attracts

Table 1. Survey of work done

Patent	Name	Country	Basis
CN104938460A	Fruit fly trap provided with fast-trapping type fermented fruit trapping agents	China	Espacenet
CN212279555U	Tubbiness trypetid trapper	China	Espacenet and Lens
CN212279554U	Cup-shaped fruit fly trapper	China	Espacenet and Lens
CN206165568U	Mango fruit fly trapper	China	Espacenet
EP3772278A1	Device and process for olive pick paste catching: horizontal tubular model	Portugal	Espacenet
CN205695184U	A kind of device utilizing olfactory sensation and vision trapping citrus fruit fly	China	Espacenet and Lens
EP3772277A1	Device and process for massive capture of olive flies - electrocutor trap	Portugal	Espacenet
MX2014001788A	Prolonged action device for monitoring and control of fruit flies (Diptera: Tephritidae)	México	Espacenet
CN104304337A	Plant source biotype fruit fly bisexual trapping and killing agent	China	Lens
CN105494411A	Citrus fruit fly adult trapping and killing agent and using method thereof	China	Lens
CN211458601U	Fruit fly trapper		
EP2849569B1	Insect attractant formulations and insect control	EscritórioEuropeu de Patentes	
CN207692798U	Gyplure suction-type traps and kills drosophila device	China	Lens
CN210184275U	Bactroceradorsalis trap	China	Lens
AU2017200488B2	Fruit fly olfactory attractant composition and use thereof	China	Lens
CN205756706U	A kind of modified model trapper	China	Lens

and kills by bonding, solving the problem of a few flying dipterous insects, excluding the fruit fly, whose attractiveness is given by the attractive lure's aroma and the trap's yellow color.

In another proposal for a monitoring system that uses sticky paper [24] with an attractive yellow lure, the trap solves specific insect problems, especially in greenhouses, which cause specific phytosanitary problems in specific crops. By bonding, the process attracts and kills, solving the problem of a few flying dipterous insects, which exclude the fruit fly whose attractiveness is given by the attractant's aroma, in addition to the trap's yellow color. Although the results are relevant, the commercial application becomes difficult because the tools make the process very manual and lose efficiency and agility.

The trap, "camera trap for perching dragonflies" [25], solves the specific problems of dragon flies. When identifying the insects, the apparatus performs the monitoring. However, the insects identified by the trap differ from the target insect-pest of this study. The proposal to use energy-saving systems is very important for solutions that will be used in the field, as they are normally difficult to access places where it is necessary to use photovoltaic systems and batteries. In these cases, the more energy efficient, the safer and more reliable the solution.

The pressure of insect pests in greenhouses is extremely high and problematic for crops that use this productive management. The "pre-programmed autonomous pest control robot" solves the specific problems of insects common in greenhouses. The process attracts and kills by bonding and solves the problem of a few flying dipterous insects. However, they exclude the fruit fly, whose attractiveness is given by the attractant's aroma, in addition to the trap's yellow color. Sensors for humidity, temperature, and light intensity levels are interesting to correlate with the incidence of insects, making it possible to develop infestation prediction models.

Also, by using a visual attractant with the emission of light as attractant and detection, this trap solves the adversities resulting from the attack of pests whose attractiveness is given by the light [26]. By attracting, counting, and releasing insects again to the production system, the apparatus performs monitoring that can be interesting for the fruit fly. The system proves to be interesting and flexible regarding installation in other traps, but it has difficulty identifying the pest insect. It proves to be a good option to use with other solutions.

The trap, "Automated Surveillance of Fruit Flies" [18], monitors the same insect pest in crops where the fruit fly causes significant damage. The proposed model is also an adaptation of the McPhail trap. Nevertheless, the way the insect is identified is different. The proposed model attracts, kills, and identifies the insect without requiring other human agents to interfere in the process through the neural network analysis and identification of images. The presented trap performs monitoring through the analysis of models referring to the flapping of the wings of fruit flies. As demonstrated in the study, the solution presents difficulties in identifying different fruit flies since the frequency of wing beats between the different types is similar. In this way, it is understood that using this solution with the presented ones is possible.

The solution, "digital image-processing method and fitted logistic model to the control of oriental fruit moths" [27], monitors a distinct insect-pest, from another Order, the Lepidoptera, whose attraction occurs with pheromones. The trap uses image recognition and machine learning, in which modeling was performed to do the alert process as population control. The process attracts and kills, by bonding through the attraction of pheromone, solves the problem of the flying lepidopteran insect, which excludes the fruit fly, whose attractiveness is due to the attractant's aroma, in addition to the yellow color of the trap. Although the results are relevant, the commercial application becomes difficult because the tools make the process very manual and lose efficiency and agility. Yet, using mathematical models is quite interesting to predict future infestations.

The trap, "Insect pest monitoring with camera-equipped traps" [28], monitors pest-insects attracted by pheromones. The trap also uses image recognition and machine learning. Still, the process is not recommended for the fruit fly. It is necessary to improve image capture quality to use the fruit fly as a target insect.

Regarding technological structure, 65.38% of the solutions use a digital camera as a sensor, that is, for image capture and subsequent processing, but only 26.92% use some artificial intelligence technique in image processing. AI can contribute in terms of efficiency in identifying pest insects and aldo scalability, expanding the target insect types. It is also possible to locate in the studies that 34.62% use the cloud for data storage or image processing, thus enabling real-time monitoring of data collected in the field.

Although the vast majority of studies have shown good results, there are still gaps in application and technology transfer. Only 11.54% aimed to generate a product and go to market. The rest are studies with academic objectives. The transposition of such technologies to the field will imply solutions in terms of connectivity and scalability.

5. Conclusion

Innovation is a process completed when technology is transferred, making the results available as products or services. In this sense, as important as investing in research is working on the accessibility of the results to the public involved, which in this case are producers, cooperatives, institutions, and improvement in learning processes.

Agro 4.0 is a new concept that brings new areas of knowledge to solve known problems, with enormous potential to increase efficiency and, if well worked, generate inclusion and competitiveness for smaller producers, contributing to sustainability in a broad scope. From the studies, it was possible to identify that Agro 4.0 technologies will contribute to reducing operating costs and improving pest monitoring.

It was possible to observe that Brazil has an enormous potential to develop solutions in the area, taking advantage of its experience combating other types of insects. The country is a reference in studies, but there is a gap in the fruit fly, especially in the monitoring systems proposition, even if the insect has economic importance. In this way, the article aims to contribute to state-of-the-art updating in the area and generate hypotheses for technological development and innovation projects, aiming to be another step towards fostering Agro 4.0 solutions and sustainability in Brazilian fruticulture.

Acknowledgments

We would like to thank the structure and the environment provided by IFRS.

References

1. Moscamed, Linhas. (2012). de ação, *Moscamed*, Jun. http://moscamed.org.br/linhas-de-acao. Accessed Sep. 29, 2022.
2. Silveria, N., Michereff, S. and Caatinga, S.O. (2005). Doenças fúngicas pós-colheita em frutas tropicais: patogênese e controle. *Revista Caatinga*, 18, 283–299. Accessed: Sep. 28, 2022. [Online]. Available: https://www.redalyc.org/pdf/2371/237121130015.pdf.
3. Nora, I., Hickel, E.R. and Prando, H.F. (2000). Moscas-das-frutas nos estados brasileiros: Santa Catarina. *In*: Malavasi, A., Zucchi, R.A. (Eds), *Moscas-das-frutas de importância econômica no Brasil: conhecimento básico e aplicado,* 271–276. Holos, SP, Brazil: Ribeirão Preto. .
4. de S. Filho, M., Costa, V. and Pazini, W. (2004). Manejo integrado de pragas na cultura da manga. *Entomologia*, 339–376. Accessed: Sep. 28, 2022. [Online]. Available: https://www.researchgate.net/profile/Wilson-Pazini-2/publication/266493214_MANEJO_INTEGRADO_DE_PRAGAS_NA_CULTURA_DA_MANGA/links/58ac2865aca27206d9bf92e2/MANEJO-INTEGRADO-DE-PRAGAS-NA-CULTURA-DA-MANGA.pdf.
5. Menezes-Netto, A.C., Arioli, C.J., Nava, D.E., dos Santos, J.P., da Rosa, J.M. and Botton, M. (2016). Combate às moscas-das-frutas em pomares domésticos. *Boletim Didático*, 16–16. Accessed: Sep. 27, 2022. [Online]. Available: https://publicacoes.epagri.sc.gov.br/BD/article/view/401.
6. Raga, A. and de Souza Filho, M.F. (2021). Manual de moscas-das-frutas: medidas para o controle sustentável. *Fundo de Defesa da Citricultura – Fundecitrus.* Accessed: Sep. 27, 2022. [Online]. Available: http://repositoriobiologico.com.br/jspui/handle/123456789/733.
7. Nava, D.E. and Botton, M. (2010). Bioecologia e Controle de Anastrepha fraterculus e Ceratitis capitata em Pessegueiro. *Empresa Brasileira de Pesquisa Agropecuária – Embrapa Clima Temperado.* Accessed: Sep. 27, 2022. [Online]. Available: www.cpact.embrapa.br.
8. Carvalho, R.D.S. (2005). Metodologia para monitoramento populacional de moscas-das-frutas em pomares comerciais. *EMBRAPA Mandioca e Fruticultura Tropical*, 17. Accessed: Sep. 27, 2022. [Online]. Available: https://www.embrapa.br/busca-de-solucoes-tecnologicas/-/produto-servico/1506/metodologia-para-monitoramento-populacional-de-moscas-das-frutas-em-pomares-comerciais
9. de Souza, R.B. and Kovaleski, A. (2012). Incidência e flutuação populacional de Anastrepha fraterculus (Diptera: Tephritidae) em diferentes frutíferas nativas no

município de Vacaria/RS., *XXII Congresso Brasileiro de Fruticultura*, 2248–2251. Accessed: Sep. 27, 2022. [Online]. Available: https://www.alice.cnptia.embrapa.br/bitstream/doc/945015/1/Souza22CPF2012p22482251.pdf.

10. da Silva, J.G., de Brito, C.H., de Oliveira, R., Araújo, E.L., Gonzaga, K.S. and Lopes, G.N. (2022). Biodiversity and population distribution of fruit Fly (Diptera: Tephritidae) in DOMESTIC orchards OF PARAÍBA State, Brazil, *Revista Caatinga*. Universidade Federal Rural do Semi-Árido, Apr. [Online]. Available: https://www.scielo.br/j/rcaat/a/tkL6drjHzS8Cwpnthk976pG/?lang=en#
11. Coleção, S. (2016). Fruticultura: moscas-das-frutas (biologia e manejo), *Serviço Nacional de Aprendizagem Rural.*
12. Botton, M., Junior, M.R., Nava, D.E. and Aroli, C.J. (2012). Novas alternativas para o monitoramento e controle de anastrepha fraterculus (wied., 1830) (Diptera: Tephritidae) na fruticultura de clima temperado., *Embrapa Clima Temperado-Artigo em Anais de Congresso.* Accessed: Sep. 27, 2022. [Online]. Available: https://www.alice.cnptia.embrapa.br/bitstream/doc/939420/1/DoriEdsonNava.pdf
13. Paranhos, B. and Barbosa, F. (2006). Monitoramento de moscas-das-frutas e o seu manejo na fruticultura irrigada do Submédio São Francisco. *Empresa Brasileira de Pesquisa Agropecuária - EMBRAPA – Semiárido.* Accessed: Sep. 27, 2022. [Online]. Available: http://www.cpatsa.embrapa.br/public_eletronica/downloads/OPB702.pdf.
14. Efrom, C. and Botton, M. (2012). Bioecologia e controle da moscas-das-frutas em citros. *Empresa de Assistência Técnica e Extensão Rural - EMATER/RS.* Accessed: Sep. 27, 2022. [Online]. Available: https://www.infoteca.cnptia.embrapa.br/infoteca/handle/doc/925110.
15. da Silva, A., da Silva, A., Marques, F. and da Silva, J. (2016). Uso de armadilhas alternativas para o monitoramento de adultos de moscas-das-frutas (anastrepha spp.) em pomar de acerola, *I Congresso Internacional das Ciências Agrárias – COINTER.* Accessed: Sep. 27, 2022. [Online]. Available: https://cointer-pdvagro.com.br/wp-content/uploads/2016/12/USO-DE-ARMADILHAS-ALTERNATIVAS-PARA-O-MONITORAMENTO-DE-ADULTOS-DE-MOSCAS-DAS-FRUTAS-Anastrepha-spp.-EM-POMAR-DE-ACEROLA.pdf.
16. Esposti, M., Caetano, L. and Caiado, L. (2016). Influência da cor da armadilha modelo PET na 1 captura de mosca-das-frutas em pomar de laranjeiras' Bahia na região Sul do Estado do Espírito Santo. *XXII Congresso Brasileiro de Fruticultura.* Accessed: Sep. 27, 2022. [Online]. Available: http://biblioteca.incaper.es.gov.br/digital/handle/item/1535.
17. Isca Tecnologias, Ficha técnica: Armadilha Bola. http://www.isca.com.br/pt/produtos/p/58b45aa2-393b-48bc-9076-d34bb784a6b8/armadilha-bola. Accessed Sep. 27, 2022.
18. Kagermann, H., Wahlster, W. and Helbig, J. (2013). Recommendations for Implementing the Strategic Initiative INDUSTRIE 4.0 – Securing the Future of German Manufacturing Industry, München, Sep. [Online]. Available: http://forschungsunion.de/pdf/industrie_4_0_final_report.pdf
19. Zhai, Z., Martínez, J.F., Beltran, V. and Martínez, N.L. (2020). Decision support systems for agriculture 4.0: Survey and challenges. *Computers and Electronics in Agriculture*, 170, 105256. doi: 10.1016/j.compag.2020.105256.
20. Bollini Letizzia, C.A. and Martino, C. (2019). Interfaces of the Agriculture 4.0. *Proceedings of the 15th International Conference on Web Information Systems and Technologies* (WEBIST), 273–280. doi: 10.5220/0008164802730280.
21. Auernhammer, H. (2001). Precision farming the environmental challenge. *Computers and Electronics in Agriculture*, 30, 31–43.

22. MIT Media Lab, Group overview open agriculture, *MIT Media Lab*. https://www.media.mit.edu/groups/open-agriculture-openag/overview/} Accessed Sep. 29, 2022.
23. Albiero, D. (2019). Agricultural robotics: A promising challenge. *Current Agriculture Research Journal*, 7, 1–3, Sep. doi: 10.12944/CARJ.7.1.01.
24. Albiero, D., Paulo, R.L., Junior, J.C., Santos, J.D. and Melo, R.P. (2020). Agriculture 4.0: A terminological introduction. *Revista Ciência Agronômica*, 51. Special Agriculture 4.0 Ceara, Brazil, ISSN 1806-6690 (online) 0045-6888 (printed).
25. Pereira, I.M., de Senna Carneiro, T.G. and Figueiredo, E. (2021). Understanding the context of IoT software systems in DevOps, *CoRR*, *abs/2104*, 10147. [Online]. Available: https://arxiv.org/abs/2104.10147.
26. Coelho, F.C., Coelho, E.M. and Egerer, M. (2018). Local food: Benefits and failings due to modern agriculture. *Scientia Agricola*. São Paulo - Escola Superior de Agricultura Luiz de Queiroz. [Online]. Available: https://www.scielo.br/j/sa/a/DNPPhm8L8RD9zGQtzrjCwhf/?lang=en#
27. Lima, G.C., Figueiredo, F.L., Barbieri, A.E. and Seki, J. (2020). Agro 4.0: Enabling agriculture digital transformation through IoT. *Revista Ciência Agronômica* [online].
28. Camarasa, C., Nägeli, C., Ostermeyer, Y., Klippel, M. and Botzler, S. (2019). Diffusion of energy efficiency technologies in European residential buildings: A bibliometric analysis. *Energy and Buildings*, 202, 109339. doi: https://doi.org/10.1016/j.enbuild.2019.109339.
29. Aria, M. and Cuccurullo, C. (2017). Bibliometrix: An R-tool for comprehensive science mapping analysis. *Journal of Informetrics*, 11(4), 959–975. doi: https://doi.org/10.1016/j.joi.2017.08.007.
30. Domingues, M. and Bianchini, I. (2018). Mapeamento da Ciência com o Pacote R Bibliometrix: Uma aplicação no estudo de Empreendedorismo Acadêmico. *9th International Symposium on Technological Innovation*, 9(1), 287–294. doi: 10.7198/S2318-3403201800010033.
31. Pignatti, M.G. (2004). Saúde e ambiente: As doenças emergentes no Brasil., *Ambiente & Sociedade* [online], 7(1), 133–147. doi: https://doi.org/10.1590/S1414-753X2004000100008.
32. Lattes, Currículo Eunice Aparecida Bianchi Galati, *Currículo Lattes*. http://lattes.cnpq.br/1359035734723864. Accessed Sep. 29, 2021.
33. Lattes, Currículo Andréia Fernandes Brilhante, *Curriculo Lattes*. http://lattes.cnpq.br/0029213795464192. Accessed Sep. 29, 2021.
34. SciELO, Open access Memórias do Instituto Oswaldo Cruz, *SciELO*. https://www.scielo.br/journal/mioc/about/ Accessed Sep. 29, 2021.
35. SciELO, Revista da Sociedade Brasileira de Medicina Tropical, *SciELO*. https://www.scielo.br/j/rsbmt/}. Accessed Sep. 29, 2021.
36. de Moraes, L.L. and Kafure, I. (2020). Bibliometria e ciência de dados: um exemplo de busca e análise de dados da Web of Science (WoS), *RDBCI: Revista Digital de Biblioteconomia e Ciência da Informação*, 18(00), 20016, Jun. doi: 10.20396/rdbci.v19i0.8658521.

CHAPTER

8

The Smart Agricultural Revolution Across Global Nations – A Technological Perspective

R. Jenice Aroma[1*][0000-0003-4022-1898], Emmanuel Joy[1][0000-0002-4070-8594], Kumudha Raimond[1][0000-0001-8680-8390], Nikolaos Andreopoulos[2][0000-0001-9975-7043], Vania V. Estrela[3][0000-0002-4465-7691] and Andrey Terziev[4][0000-0002-7069-367X]

[1] Department of CSE, Karunya Institute of Technology & Sciences, Karunya University, Karunya Nagar, Coimbatore, 641114 - India

[2] Computer Science Department, Technological Institute of Iceland Sæmundargata 2, 102, Reykjavik, Ssn. 600169-2039, Iceland

[3] Telecommunications Department, Universidade Federal Fluminense (UFF), Brazil

[4] TerziA, Bul Vazkresenie 103, 1504, Sofia, Bulgaria

Agriculture advanced in multiple locations, and it also includes divergent species from all over the Earth. As the planet's population augments, so does the food production demand, resulting in several innovations, equipment, and tools to magnify the yield. Agriculture 4.0 refers to the digitalization of food and agronomy structures through the usage of "artificial intelligence" (AI), the "Internet of Things" (IoT), automation/robotics, and other technologies. This text serves as a proof-of-concept for various strategies employed in the smart agricultural revolution. It aims to highlight the knowledge gap between subsistence and sustainable farming with key solutions from established technologies to bridge it. The state of affairs, from the green to this age's smart revolution, is studied to observe the transition effects. The challenges and implications of adopting smart technologies in farming for individual farmers and governments are discussed. The current status of smart applications worldwide, besides the progress of smart transition in developing countries, is also discussed. This manuscript tackles comparatively cumulative approaches to Agriculture 4.0 that would (i) greatly help governments and research organizations to make decisions, and (ii)

* Corresponding author: jenicearoma@gmail.com

function as a cautionary report for planning proper measures to empower food, and eco-friendly and social sustainability.

1. Introduction

Food is indispensable to life on Earth; neither evolution nor revolution can deny this. Every individual on the planet earns to feed their dependent family; even world nations trade to safeguard the well-being of their citizens. Our ecosystem is an interconnected mode of existence, with species dependent on one another in a hierarchical fashion to craft a far-reaching food chain. Humans require animals and plants for food, so light and water are essential nutrients for food production. It also confirms the agricultural prominence as a line of work since trades, live stocks, and other assets/services rely directly or indirectly on it.

Farming is extremely manual labor-intensive, and deeply counts on the farmer's proficiency, causing concerns about farm management. The low yield and lower returns force most farmers to regret the quotidian hard labor. As a result, many folks left occupational farming, not wanting to continue practicing for their generations, thus slowly terminating agriculture. However, numerous modern tools and methods can resolve this situation. These modern aids must be produced cheaper to improvise this worse situation [1]. There should be a shift from farmers physically working with machines to supervising autonomous gears. To adopt revolutionary schemes, the farmers need effective training and assistance.

1.1 Objectives

This study's main objective is to revise merging "information and communication technologies" (ICT) applications into agricultural practices among Western and developing countries and detail the contributions of two major technologies, "Remote Sensing" (RS) and "Augmented Reality" (AR), among them to harness smart farming practices. RS is a leading upfront technology in enriching the quality of agriculture, while AR is the upcoming technology that aids both in field sustenance and yield marketing [2]. This study deliberates on the distinct roles of unarmed vehicles, satellite imaging, sensor instruments, and other remote sensing tools for the beneficial progress toward achieving Agriculture 4.0 [3]. The highlight of this study is presenting an indigenous report on food customs and the current status of Agriculture 4.0 across the global nations with suitable recommendations. In brief, this chapter presents an abridged narrative of the technological perspectives of an ingenious agricultural revolution.

This chapter is systematized as follows: The background information is outlined in Section 2. The review of literature, challenges, and best practices for smart agriculture are enlisted in Section 3. The role of RS in smart agriculture is portrayed in Section 4. Adopting AR in Agriculture 4.0 and its current trends are listed in Section 5. Discussions and inferences on the role of RS and AR in smart farming are highlighted in Section 6. This study's limitations, future directions, and conclusion are summarized in Section 7.

2. The Background

2.1 Brief History

Agriculture 1.0 embraces only manual labor and is rooted in traditional methods. Agriculture 2.0 introduced machines, fertilizers, and better seeds with agronomic management practices and synthetic pesticides, creating innovative agro-businesses resulting in industrial agriculture. Food wastes were dumped into landfills, generating approximately 23 times deadlier methane emissions than carbon-di-oxide [4]. Hence, specialists must cogitate the climatic changes vs. the food scarcity conundrum. Technology should be tapped more efficiently to address the pitfalls. Agriculture 4.0 revolutionized the concept of connected farms. For example, a company known as *Sundrop* in South Australia used hydroponics, a type of hydroculture that involves growing vegetables in seawater without soil using water-based mineral nutrients [5]. Algal blooms in water bodies can also be used as an effective feedstock for biofuel synthesis and food for fish [6]. The desert farming practice also makes a case in the smart agriculture revolution [57].

2.2 Key Concepts

Agriculture 4.0 is the most anticipated revolution on the globe. Numerous cutting-edge robotic-age technologies modernize traditional agricultural schemes and values, from seed selection to food supply. Throughout the literature, innumerable descriptions have been developed. Rose et al. [7] viewed Agriculture 4.0 as adopting emerging technologies, such as sensors integrated into the cloud, robotics, and artificial intelligence. Zambon et al. [8] compared Agriculture 4.0 to Industry 4.0 and defined it as the combined internal and external interaction of farming operations, offering digital information in all farm sectors and processes. On generalizing the views and definitions from the literature, we describe Agriculture 4.0 as a collective effort to adopt emerging technologies to overcome current barriers and challenges impeding the advancement of farming practices in improving food production while maintaining ecological balance and sustainability. The importance of adopting sustainable approaches in farming is connected to the protection of human health, as sustained individual crop health reflects in the health of livestock and human being. Conventional farming methods are transitioned with modern tools and scientific approaches to bring this custom into practice. On adopting these approaches, the loss of cultivation due to pests, weeds, or any crop diseases is minimal, as the continuous monitoring of fields could render many benefits.

Subsistence farming is the practice of independent small farms for one's needs and not much for trade. Sustainable farming stands for the agribusiness practices that impart control over ecological balance and biological cycles. *Sustainable farming* is defined as farming that is environmentally responsible by promoting methods and practices that are economically viable, ecologically sound, and protect public health. It focuses on the economic aspects of farming and the effective use of non-renewable resources.

While satellite image-based agricultural monitoring applications effectively track the health of vegetation and natural ecosystem on a large scale. *Precision farming* is termed for adopting highly technological practices to cater to the needs of small-scale farms and even individual crops [9]. It assures the increase in crop yield and checks crop failure rate by using cutting-edge technologies, such as RS, IoT, objects "unmanned aerial vehicles" (UAVs), AR, and "Machine Learning" (ML). It adopts new technologies to improve crop productivity while lowering the traditional inputs like fertilizers, water, etc. Precision agriculture is needed in developing countries like India to meet the growing population's food demands. It also helps to resolve the challenges of biotic and abiotic stress faced by the crops. Precision agriculture is also known as satellite farming. It offers site-specific crop management (SSCM) [10]. In the 1930s, the first revolution mechanized agriculture. The green revolution in the 1990s rendered genetically modified new sets of crops that are pest resistant and need less water. Now, precision agriculture with technologies assimilates together in resolving and uplifting the food journey.

2.3 Phases in Agriculture 4.0

The technological action regimen for Agriculture 4.0 includes four different phases (Fig. 1): (1) Pre-farming, (2) Farming, (3) Tracking, and (4) Supply chain.

1. *Pre-farming* includes the role of scientific methods such as genomics and nanotechnology for gene-editing, protein synthesis, high-yield breeds, seed development, and cellular agriculture to improve seed varieties and production. It aids farming by augmenting the yield for extended trade. It also checks food security amidst the huge rising population.
2. *Farming* comprises modern tools and methods for seeding under dry conditions, drip irrigation, measures for increasing surface water, and soil health assessment sensors. RS and IoT-based tools could support farming easily and without any environmental defects.
3. *Tracking* includes health tracking of crops, "greenhouse gas" (GHG) mitigation due to livestock holding, IoT sensors, and UAV monitoring [62]. AR-based analytics are applied during the entire growth period of crops to ensure quality. IoT-based data analytics and cloud-based models could assist decision-making over consistently tracking crop health.
4. The *supply chain* represents the actions from food supply to food security that nourishes the end-user. The quality of food and the authenticity of the suppliers need to be tracked and maintained for a complete modernized chain of processes, and it can be made possible by using Blockchain, cybersecurity, and cloud-AI models [59–61].

Implementing these four phases in individual farms need investments and technical assistance.

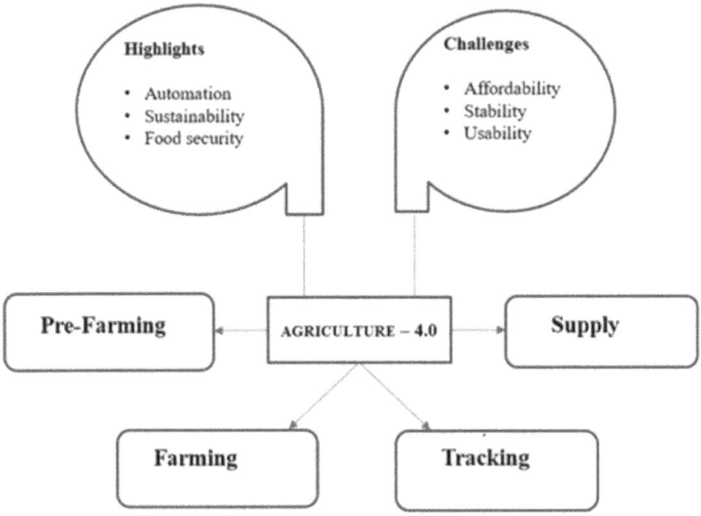

Fig. 1. Agriculture 4.0 – roles and outcomes.

The traditional approaches for improving income in subsistence farming include livestock holding, inter-cropping, and cash crop cultivation. Recent studies have mentioned farm methane emissions due to cattle rearing. Hence, global governing bodies have formulated many GHG mitigation strategies to overcome this issue [11].

3. Review of Literature

Agriculture 4.0 primarily originated from precision farming and upgrading towards attaining complete automation in "Agriculture 5.0". Accurate in-field positioning of agricultural robots (agribots) is a prerequisite for precision farming applications [12]. A few of the agriculture methods boosted with the integration of "computer vision" (CV) and AR methods are as follows:

Xue et al. [13] developed a robot that uses a camera and image processing technique for supervision in the field. An SVM-based classifier model detected parasites in strawberry plants in [14]. Li et al. adopted a multifractal analysis for segmenting whitefly images [15]. Therefore, promising CV algorithms established on high-resolution cameras will greatly help precision farming. Image processing and CV techniques play critical roles in pest management in agriculture. In addition to these technologies, RS, AR, IoT, and AI play a huge role in the smart agricultural revolution.

The study in [16] presented an AR architecture for supervising multi-robot systems in agricultural field operations. For monitoring autonomous machines, a farmer's job is fleet management and taking care of safe operations with spatial cognizance [29]. This method uses AR, a powerful tool for visualizing information in real-time to help the farmer supervise the operation of two autonomous agricultural machines. A previous study by the same authors (in 2018) on precision farming proposed a method for soil sampling with drones and AR. The study highlights are: prototyping AR application as wearable technology in agriculture, drone-based aerial imaging of bare soil fields, automatic determination of management zones for precision farming, software for AR smart glasses, and a complete toolchain from aerial imaging to identifying the soil characteristics [17]. A prototype application called FarmAR for cultivation optimization using AR and IoT blended with the semantic web to support the development of knowledge bases and intelligent applications regarding crops emerges in [18]. The application exploits AR to provide useful information on the health status of plants to farmers through a mobile device. Maulana, H., & Kanai, H. proposed a method for land suitability evaluation by incorporating multi-criteria decision-making, "geographic information system" (GIS), and AR. The intention is to make it easier for farmers to read the land suitability map, which can provide better visualization [19]. With AR-GIS, the farmers' understanding of land suitability can be improved, which could reduce the risk of crop failure.

Comparative studies and investigations [2, 19–22] show that employing AR in agricultural practices has increased production manifold. Therefore, AR can make crucial influences on the prime management of agriculture farms and improve future production.

3.1 Challenges

The challenges in implementing Agriculture 4.0 comes in three ways: (a) affordability, (b) usability, and (c) stability. Technologically advanced and developed countries such as the United States, Europe, and Israel are well equipped and can afford to implement upgrading technologies to increase food production. However, developing countries are still heavily reliant on subsistence farming. Due to a lack of modern telecommunications in rural areas, developing countries cannot afford the most recent technologies for instantaneous crop health monitoring. The technologies must be user-friendly, with real-time data collection facilities for quick decision-making [59]. Even inexperienced handlers should be able to use them without human assistance. However, expert help is required to install and monitor *in-situ* sensors. The Government policies and strategies towards the modernization of agriculture must be carried out overtime to attain the target increase in food production. Suitable campaigns and regional execution groups must be planned to bring awareness to every independent farmer in the region. The policy formulators must collaborate with the incubation centers at universities or R&D firms to design innovative solutions for solving Agriculture 4.0 barriers [24]. The progressive development of technologies, supporting policy formulation by governing bodies, and awareness of technological aids among the farmers are the major focus areas that could fix the above-listed challenges.

3.2 Best Practices for Smart Farming

The best practices for smart agriculture and systematic approaches implemented across the globe are reviewed, and a few most promising practices have been briefed in this section. Farmers most common productivity-boosting practices include intercropping, crop rotation, and shifting cultivation. Inter-cropping is seeding two or more crops together for increased production in a small patch of land. Crop rotation is a common technique to sow crops alternatively to retain soil nutrients. Shifting cultivation involves abandoning a piece of farmland for a few years and moving to a new location until the soil fertility has been naturally restored. It is otherwise termed "slash-and-burn agriculture". These are the most common traditional practices, but modernized techniques could help achieve much better results [25].

Israel is a pioneer in the adoption of such modern agricultural techniques. It does not all fall into place overnight; it requires the collective effort and hard work of generations. Since the country's independence in 1948, the area of agricultural lands has grown from 165,000 to 433,000 hectares. Though the grounds are arid and water scarcity is a major societal issue, revolutionized agrarian practices and the intentional efforts of the Government aid a tremendous impact on agricultural welfare. Various institutions in Israel have researched and devised novel methods for cultivating under dry conditions, with less or no water, and with reduced groundwater. Yatir Forest was planted by the Jewish National Fund in 1950 and has since grown into a full forest [26]. It is regarded as one of the most promising research areas for new ideas. Israel scientists have even uncovered the mechanisms that resist water loss and CO2 overload, enabling plants or trees to survive during hot weather [27]. Most Western countries like North America, Europe, and other developed countries have afforded the cost of implementing innovative agricultural solutions. Many R&D institutions, including universities, collaborate with such governments to test solutions for increased yield and groundwater levels. But still, many African, South Asian, and other developing countries fell apart to afford these novel practices.

India is a diverse country, and agriculture has been the backbone of the Indian economy since its inception. Traditional agricultural methods, however, which are labor-intensive and heavily reliant on rain, are more prevalent in these countries. Crop production is jeopardized if the monsoons fail. *The adoption of drip irrigation over traditional irrigation methods accelerates agricultural productivity growth. Drip irrigation is a micro-irrigation system of valves, emitters, and tubes that allow water to reach the roots slowly and directly. China and India are the two major drip irrigation countries* [28]. However, the application of modern technologies benefits China more than India. As per the report by Piesse and Thirtle [63], due to the adoption of smart agricultural practices, China's annual crop yield is 415 million tonnes, whereas India, which has more agrarian lands and a wider variety of crops than China, could only achieve 218 million tons per year. The statistics may differ, but the importance of innovative technology adoption in developing countries is evident [29].

4. The Role of Remote Sensing in Smart Agriculture

The act of imaging our planet began around 150 years ago with the invention of the camera. The primary purpose of imaging was topographical assessment, as cameras were attached to air balloons or pigeon fleets to capture the view of the land. This mode of aerial photography transitioned to aircraft during the First World War period, and from then, it was the major source of imaging until the early 1960s. The term "Remote Sensing" was originally used by Evelyn Pruitt, U.S. Office of naval research, in 1950 [30]. It is the science of acquiring spatial information about objects through spectral reflection and without direct physical contact. RS can have either a passive or active mode. In passive mode, the naturally available energy of an object due to sunlight and other thermal radiations of the electromagnetic spectrum reflected from the Earth's surface will be detected by the sensors as in Fig. 2a. These objects are RS-monitored using artificially emitted radiation as in Fig. 2b. These observed signals are then digitized as satellite images [31].

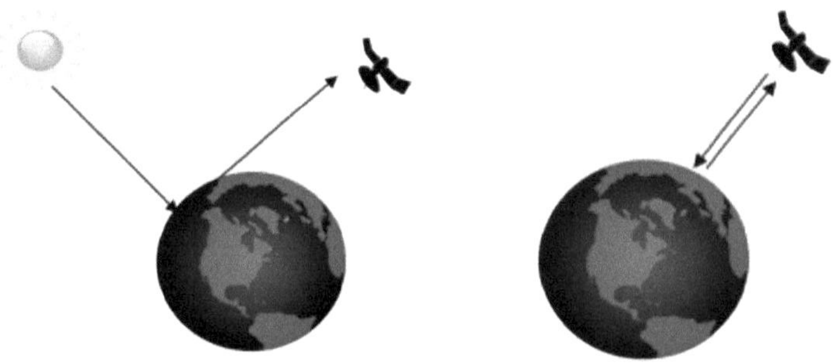

Fig. 2. (a) Passive remote sensing. (b) Active remote sensing.

The satellite images acquired are differentiated based on the spatial resolution, either moderate or high-resolution imageries. This spatial resolution is determined by the level of sensor instruments used for imaging [35]. In 1960, the "Television and Infrared Observation Satellite" (TIROS-1) was launched as the first weather satellite for monitoring. Today, numerous satellites revolve around the planet for instantaneous surveillance, and even space stations are deliberately installed for interplanetary missions [32]. In satellite image analysis, spectral reflectance observed from the spatial objects could be differentiated for categorizing objects' classes in any land cover. However, persistent monitoring of any spatial object through individual sensors can offer live tracking of such things, which can be possible by UAV-assisted monitoring. Figure 3 clearly illustrates the spectral reflection mechanism in plants.

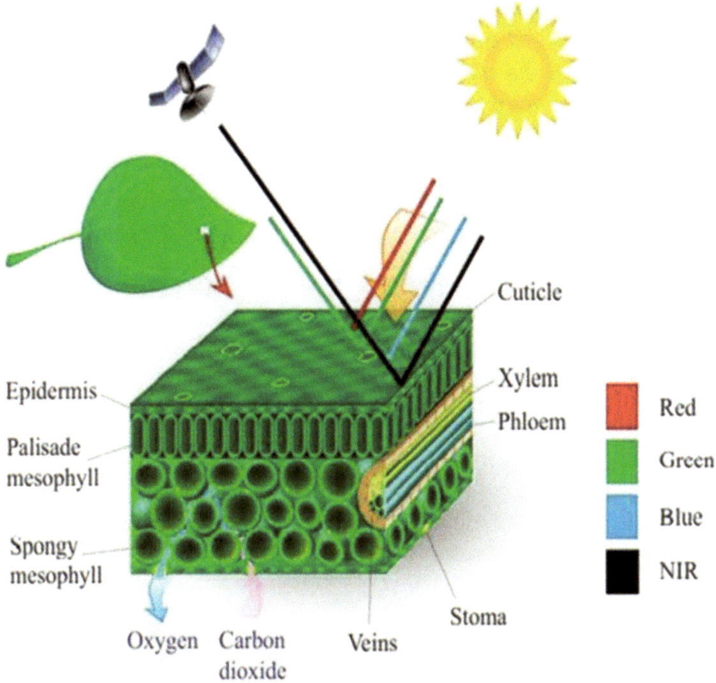

Fig. 3. Spectral influence-based vegetation mapping [31].

The sunlight is the major energy source for photosynthesis, and the green and "near infrared" (NIR) bands from the incident light are reflected, whereas blue and red bands are absorbed. Photosynthesis is a phenomenon through which plants intake carbon dioxide and water to release oxygen. Mesophyll is an inner plant tissue consisting of chloroplasts responsible for producing chlorophyll, which imparts the green color of plants and is also responsible for converting light energy into chemical energy during photosynthesis. If the plant is healthy, green light is highly reflected, and the NIR bands are highly reflected on any stress. Thus, plant diseases can be monitored by tracking the higher reflections of NIR. The "Normalized Difference Vegetation Index" (NDVI) is the traditional spectral index with a combination of green and red bands for spectral band rationing. There are numerous such vegetation indices in practice [23].

The spectral indices are the mathematical combination of different spectral bands which could contrast or enhance the underlying characteristics of the spatial land cover classes in common. It is a vegetation index if such traits depend on the vegetation class. The NDVI is a notable vegetation index from 1970. However, the advancements in sensors and other technologies have led to numerous vegetation indices. Apart from devising spectral indices, ready-to-use vegetation indices using different sensors are available in the data portal of the "United States Geological

Survey" (USGS) [33]. The datasets are free, as is a sample NDVI product from the National Oceanic and Atmospheric Administration (NOAA). These readily available NDVI products can assist large-scale vegetation health monitoring across states and long-term drought analysis. It reduces the computational task of generating vegetation products and accounts for regular health monitoring. A sample NDVI image for South America is in Fig. 4. Apart from NDVI, numerous vegetation indices are available globally (Table 1) to monitor high vegetation cover, soil characteristics, and atmospheric influences on the health of vegetation.

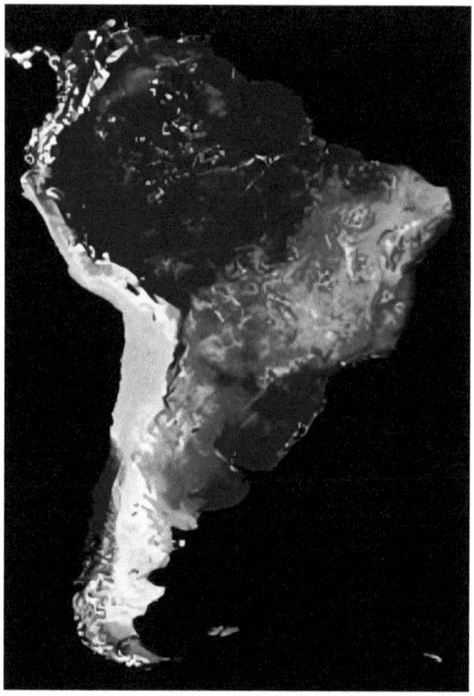

Fig. 4. Sample NOAA CDR – NDVI of South America during September 2021.

Next to vegetation indices, UAV-assisted monitoring of croplands assists farmers in all sorts of agricultural activities, from seeding to harvest. It also predicts the yield even before harvest through regular crop health assessment. Even irrigation, soil moisture, soil nutrients, application of pesticides, and fertilizers rate can also be checked. Further, crop damages can be shown to insurance providers for reimbursement of claims [34]. On the combination of both spectral indices and UAV monitoring along with a blend of IoT sensors, data analytics using AI, and cloud-based processing, many innovative applications for agriculture are in practice. One such GIS-based application is CropIn, a SaaS-based smart farming application that provides solutions to the worldwide agriculture industry, targeting maximizing productivity [11]. Similarly, Arable, Mothive, Precision Hawk,

Table 1. Highlights of a few important vegetation indices

Spectral indices for vegetation	**Authors and year**	**Band ratio**	**Highlights**
RVI - Ratio Vegetation Index	Birth and McVey, 1968 [39]	$\frac{NIR}{RED}$	A much earlier index of more than 5 decades old, otherwise termed as simple ratio index. But it is still good at monitoring vegetation cover.
SAVI - Soil Adjusted Vegetation Index	Huete, 1988 [40]	$\left[\frac{(NIR - red)}{(NIR + red + L)}\right] * (1 + L)$ where, L = 0.5 which is a correction factor that ranges from 0 to 1 for very high to low vegetation cover, respectively.	Intensely minimizes the soil noise effects in NDVI results. Suitable for monitoring sparse vegetation.
ARVI – Atmospherically Resistant Vegetation Index	Y.J. Kaufman & D. Tanre, 1992 [41]	$\frac{(NIR - RB)}{(NIR + RB)}$ where, 'RB' stands for the combination of *Red* and *Blue* bands.	Highly suitable for monitoring the vegetation over tropical mountainous regions highly influenced by atmospheric aerosol.
GNDVI - Green Normalized Difference Vegetation Index	Buschmann and Nagel, 1993 [42]	$\frac{(NIR - GREEN)}{(NIR + GREEN)}$	• Nitrogen discrimination is higher than in NDVI. • To assess different concentrations of chlorophyll that are highly correlated at nitrogen
EVI – Enhanced Vegetation Index	Liu and Huete, 1995 [43]	$G * \left[\frac{(NIR - red)}{(NIR + C1 * red - C2 * blue + L)}\right]$ where, $G = 2.5$, $C1 = 6$, $C2 = 7.5$, $L = 1$	Enhanced vegetation monitoring under high biomass with reduced atmospheric noise.

Contd...

Contd...

OSAVI -Optimized Soil Adjusted Vegetation Index	Rondeaux et al., 1996 [44]	$\left[\frac{(\text{NIR} - \text{red})}{(\text{NIR} + \text{red} + L)}\right] * (1 + L)$ where L is a correction factor that equals 0.16 (replaced correction factor of 0.16 instead of 0.5 in SAVI)	1. Soil backgrounds can be well differentiated from vegetation. 2. The soil brightness effects are highly minimized.
GARI - Green Atmospherically Resilient Index	Gitelson et al., 2002 [45]	$\frac{(\text{NIR} - [\text{GREEN} - (\text{BLUE} - \text{RED})]}{(\text{NIR} * [\text{GREEN} - (\text{BLUE} - \text{RED})])}$	1. Can assess different concentrations of chlorophyll than in NDVI. 2. Less affected by atmospheric disturbances.
GRVI - Green Ratio Vegetation Index	Sripada et al., 2006 [46]	$\frac{\text{NIR}}{\text{GREEN}}$	1. A modified RVI using a Green band with higher reflectance for vegetation. 2. Can monitor changes amidst different phenological stages of plant growth. However, less sensitive in differentiating various changes in high vegetative cover.

and CropX are smart agricultural technologies that offer smart assistance for agriculture [35]. The subsequent section discusses adopting AR for Agriculture 5.0.

5. Adopting Augmented Reality for Agriculture 5.0

AR technology overlays virtual content on top of a real-world experience in real-time. In recent years, this technology has sparked significant interest in a wide range of research fields. Clinical diagnosis, retail and logistics, design, modeling, maintenance, tourism, entertainment, education, and agriculture are just a few examples [36]. Agriculture and livestock are two of the most important industries where AR-based methods can effectively increase production and quality in the supply chain. Due to the complexity of soil changes and weather conditions, the large size of agricultural lands makes efficient monitoring an extremely time-consuming process. Extensive research has been conducted to investigate the potential use of AR tools to assist with on-site farming tasks as depicted in Fig. 5. The agrarian AR research and the availability of commercial products show that these facets are strategic and promising for future goals.

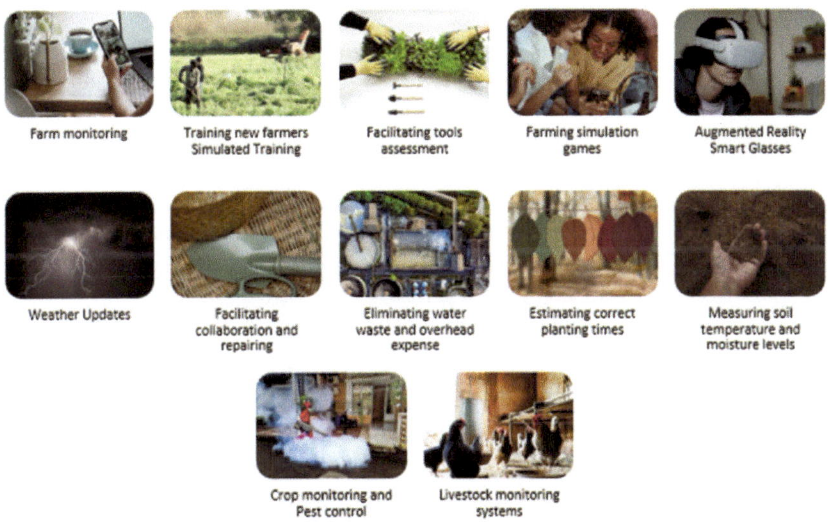

Fig. 5. Application of AR in agriculture
(*Source:* www.pexels.com)

Because of these products' improved usability, AR, in conjunction with other technologies such as CV and the IoT, aids the seamless transition from Agriculture 4.0 to Agriculture 5.0. Table 2 illustrates a few of the most popular smart farming applications globally.

Table 2. Smart farming applications and services

S. No	Smart farming applications	Services	Customers
1	Cropin [47]	• Satellite based monitoring • Weather analysis • Crop growth tracking • Crop failure forecast	• Crop Insurance companies • Banks • Governments • Commodity Traders
2	AgEagle [48]	• Monitoring using drones • A complete mapping and surveying drone solution • Measure Ground Control	• Aerial intelligent companies • Crop Registration, Oversight & Compliance agencies
3	Yeppar [49]	• Real-time satellite observation data of the farmers' lands. • Create AR content • Drone API	• Farmers -provide useful information regarding the plant through a mobile device. • Government and private entities.
4	FarmVR [50]	• 360 videos for agriculture • Virtual farm tours • Virtual reality farming experiences • Think digital coach	• Teachers • Educational institutions • Agricultural event organizers and businesses.
5	FarmApp [51]	• Scouting app - iOS/Android-based mobile app for crop monitoring • Greenhouse automation - integrate with insights from pests and diseases • Spray diary - understand where and how the agrochemicals are applied.	• Smart technologies for more productive and efficient agricultural firms • Agriculture managers • End-users - farmers
6	Growlink [52]	• Smart motor controller • Smart water wireless gateway • Smart IO wireless gateway • Fertigation skid–feeding made simple	• Designed to simplify and perfect nutrient dosing for Greenhouse workers and horticulturists. • Advanced data analytics and AI for agriculture professionals.
7	Väderstad AR [53]	• The Väderstad ARapp gives a true AR experience through a smartphone camera to a Väderstad ARtrigger.	• AR businesses • Farming entities • Technology enabled farmers

Contd...

Contd...

S. No	Smart farming applications	Services	Customers
		• Tasks like tillage, drilling, planting, and effective farming methods can be visualized and monitored.	• Large-scale production farming units.
8	Allmeteo [54]	• Monitoring of climate conditions • Weather-map • Meteoshield • Meteowind • Sensor protection testing • Road air temperature sensor • Meteorain	• Forester. • Landscape architect • Greenhouse worker • Smart Farming organizations
9	GreenIQ [55]	• Smart irrigation system • Control irrigation saves water • Connect to leading home automation platforms • Add accessories to a smart garden hub	• Lawn irrigation groundskeeper • Florist • Vineyard worker • Landscaper. • Media and event adornment agencies
10	CropX [56]	• Precision farming • Crop-specific management: Solving crop variability and risk • Optimized fertilizer management to reduce crop loss	• Large-scale agriculture companies • Top-tier business agencies • Modern agriculturists

5.1 Current Trends of AR in Smart Agriculture

The answer to the looming question of whether AR can help agriculture flourish should be a resounding yes. Farmers are moving towards employing augmented farming practices, such as AR, for agricultural field checking, simulated training, and tool assessment. In the agrarian and livestock fields, demonstrating novel equipment does not always result in immediate on-farm projects. Exploration of smart glasses, which can assist farmers in operations such as rapid field visualization, obtaining diverse information related to cattle, plants, feed, and machines, or receiving remote assistance through AR video calls, is one of the most valuable and integrative tools in precision agriculture and livestock farming. Studies have shown that smart glasses and head-mounted displays (HMDs) contribute to assisting and educating farmers in a significant way.

A smart infrastructure framework for smarter husbandry concerning food, energy, soil moisture, and water supplies is another key add-on for sustainable farming. Soil sampling is a paramount information acquisition tool to make timely judgments about field manuring. Precision farming necessitates constant sample collection and information retrieval for data analysis, which CV-based AR solutions can help. Examples of such data include: (i) How far does urban land stretch? (ii) What percentage of the land is planted for crops?(iii) How big will the wheat crop be? (iv) How much rainforest remains? (v) What are the chances of crop failure in extreme temperatures? among others (Forsyth and Ponce [37]). When combined with wearable devices, these technologies can make operations more convenient. Phupattanasilp, P., and Tong, S.R. discuss AR integration with the IoT (AR-IoT) with an application for precision farming to improve the usability of adopted modern strategies [38]. The proposed system has a high potential for integrating IoT data with AR resolution, which will significantly contribute to modernizing precision agricultural practices in an environmentally sustainable manner. Another example is aquaculture ponds, where AR may help with well-organized farm administration activities like providing food for aquatic species in coastal and inland areas, water quality monitoring for their survival, remote collaboration, and AR-assisted expert conferences.

6. Discussions and Inferences

6.1 Inferences on the Role of RS and AR in Smart Farming

Agriculture should be a key priority for every country because it is a life source. As a result, Governments and other private enterprises invest huge funds in agricultural research to obtain higher yields in the future. Smart agriculture, which answers agronomic intelligence and computerization demands, has arrived in the agriculture 4.0 era [58]. However, possible dangers to information security should not be overlooked in the process, necessitating proactive study into ways to prevent unauthorized data access and privacy control. Despite developments in AR that can improve agro-productivity, there are a few major issues and concerns with the platform's implementation and target users. AR presents a number of additional hurdles in addition to data privacy and security concerns. High-priced hardware devices are one of the AR adoption barriers.

Traditional farmers cannot afford expensive farm technologies like AR devices and smart gadgets. Another problem is persuading farmers who have been using conventional methods for years that these technologies will add value to old ways rather than replace them. They must assimilate how these technologies will reduce their workload while increasing total production. While this revolution may enhance productivity in the short term, the long-term health consequences, such as food safety problems, higher pricing, and less transparency for end customers, are unclear and outside the scope of this study. Few investment programs that rely on monetary mechanisms, such as private shareholding, can increase corporates' involvement in financing Agricultural 4.0 technological adaptations. On a more positive note, funding scientific methods in globalization may help to expand the agricultural sector and speed up the supply chain process. It can potentially restructure the agriculture business and make it far more productive in the direction of long-term change. Ecological and physical frameworks for self-managing and sustainable farming must be established to maintain worldwide enthusiasm and long-term food security.

7. Conclusion

RS and IoT technology in agriculture 4.0 offers enormous potential for reaping benefits quickly. AR and CV are technological add-ons that provide significant benefits at various stages of cultivation. These innovations will increase agricultural production if carefully implemented into farming practices. It has the potential to alleviate the food crisis among refugees, those regions where the population is swarming, and those suffering from malnutrition due to multiple poverty-related factors. The research on AR in agriculture and the availability of commercial products indicate that using AR in agriculture is beneficial. As a result, additional research in this area is precious. This smart agriculture revolution will be incomplete until all the aforementioned emerging technologies are fully integrated. In the future, the robotics revolution may even result in less manual labor due to replacing human intervention.

References

1. Bonke, V., Fecke, W., Michels, M. and Musshoff, O. (2018). Willingness to pay for smartphone apps facilitating sustainable crop protection. *Agronomy for Sustainable Development*, 38, 51.
2. Klerkx, L., Jakku, E. and Labarthe, P. (2019). A review of social science on digital agriculture, smart farming and agriculture 4.0: New contributions and a future research agenda. *NJAS-Wageningen J. Life Sciences*, 90, 100315.
3. Aroma, R.J., Raimond, K., Razmjooy, N., Estrela, V.V. and Hemanth, J. (2020). Multispectral vs. hyperspectral imaging for unmanned aerial vehicles: Current and prospective state of affairs. *In*: Imaging and Sensing for Unmanned Aircraft Systems, Vol. Chapter 2, 133–155. IET.
4. Smith, P., Reay, D. and Smith, J. (2021). Agricultural methane emissions and the potential for mitigation. *Phil. Trans. R. Soc. A.*, 379.
5. Sundrop Farms - [Online]: https://www.sundropfarms.com/
6. Yanglei, Yi. (2021). Tiny bugs play big role: Microorganisms' contribution to biofuel production. *Advances in 2nd Generation of Bioethanol Production*. Woodhead Publishing Series in Energy, 113–136.
7. Rose, D.C. and Chilvers, J. (2018). Agriculture 4.0: Broadening responsible innovation in an era of smart farming. *Frontiers in Sustainable Food Systems*, 2, 87.
8. Zambon, I., Cecchini, M., Egid, G., Saporito, M.G. and Colantoni, A. (2019). Revolution 4.0: Industry vs. agriculture in a future development for SMEs. *Processes*, 7(1), 36.

9. Awad, M.M. (2019). Toward precision in crop yield estimation using remote sensing and optimization techniques. *Agriculture,* 9(3), 54.
10. Plant, R.E. (2001). Site-specific management: The application of information technology to crop production. *Computers and Electronics in Agriculture* (*CEA*), 30(1–3), 9–29.
11. Leahy, S., Clark, H. and Reisinger, A. (2020). Challenges and prospects for agricultural greenhouse gas mitigation pathways consistent with the Paris agreement. *Front. Sustain. Food Syst*., 22.
12. Ponnusamy, V. and Natarajan, S. (2021). Precision agriculture using advanced technology of IoT, unmanned aerial vehicle, augmented reality, and machine learning. *In*: Smart Sensors for Industrial Internet of Things, 207–229. Cham.: Springer.
13. Xue, J., Zhang, L. and Grift, T.E. (2012). Variable field-of-view machine vision based row guidance of an agricultural robot. *CEA*, 84, 85–91.
14. Ebrahimi, M.A., Khoshtaghaza, M.H., Minaei, S. and Jamshidi, B. (2017). Vision-based pest detection based on SVM classification method. *CEA*, 137, 52–58.
15. Li, Y., Xia, C. and Lee, J. (2015). Detection of small-sized insect pest in greenhouses based on multifractal analysis. *Optik-International Journal for Light and Electron Optics*, 126(19), 2138–2143.
16. Huuskonen, J. and Oksanen, T. (2019). Augmented reality for supervising multi-robot system in agricultural field operation. *IFAC-PapersOnLine*, 52(30), 367–372.
17. Huuskonen, J. and Oksanen, T. (2018). Soil sampling with drones and augmented reality in precision agriculture. *CEA*, 154, 25–35.
18. Katsaros, A. and Keramopoulos, E. (2017). FarmAR, a farmer's augmented reality application based on semantic web. *Proc. 2017 South Eastern European Design Automation, Computer Engineering, Computer Networks and Social Media Conference* (SEEDA-CECNSM), 1–6. IEEE.
19. Maulana, H. and Kanai, H. (2021). Land suitability evaluation by integrating multi-criteria decision-making (MCDM), geographic information system (GIS) method, and augmented reality-GIS. *In*: Intelligent Decision Technologies, pp. 309–320. Singapore: Springer.
20. Yang, X., Shu, L., Chen, J., Ferrag, M.A., Wu, J. et al. (2020). A survey on smart agriculture: Development modes, technologies, and security and privacy challenges. *IEEE/CAA J. Automatica Sinica*, 8(2), 273–302.
21. Caria, M., Todde, G., Sara, G., Piras, M. and Pazzona, A. (2020). Performance and usability of smart glasses for augmented reality in precision livestock farming operations. *Applied Sciences*, 10(7), 2318.
22. Xi, M., Adcock, M. and McCulloch, J. (2018, March). Future agriculture farm management using augmented reality. *Proc. 2018 IEEE Workshop on Augmented and Virtual Realities for Good* (*VAR4Good*), pp. 1–3. IEEE.
23. Aroma, R.J., Raimond, K., Estrela, V.V. and de Jesus, M.A. (2023). A coastal band spectral combination for water body extraction using Landsat 8 images. *Int. J. Environ. Sci. Technol.* https://doi.org/10.1007/s13762-023-05027-z.
24. Agri 4.0 – Challenges [online]: https://en.yeeply.com/blog/trends-challenges-agriculture-4-0/ (accessed on 15-10-21).
25. Shifting Cultivation – Stefan Hauser and Lindsey Norgrove. (2013). Slash-and-burn agriculture, effects of, reference module in life sciences. *Encyclopedia of Biodiversity* (2nd Edn.), 551–562.
26. Yatir Forest – [online]: www.theecologist.org/2014/jun/23/israels-forest-yatir-expand-over-bedouin-village (accessed on 15-10-21).

27. Best practices Israel – [online]: Israeli Agro-Technology (jewishvirtuallibrary.org).
28. Kumar, M., Sun, Y., Rathour, R., Pandey, A., Thakur, I.S. and Tsang, D.C. (2020). Algae as potential feedstock for the production of biofuels and value-added products: Opportunities and challenges. *Science of the Total Environment*, 716, 137116.
29. Katsaros, A., Keramopoulos, E. and Salampasis, M. (2017). A prototype application for cultivation optimization using augmented reality. *Proc. 8th International Conf. Information and Communication Technologies in Agriculture*, pp. 805–811.
30. Estrela, V.V., Hemanth, J., Saotome, O., Nikolakopoulos, G. and Sabatini, R. (2020). Introduction to advances in UAV avionics for imaging and sensing. *In*: Imaging and Sensing for Unmanned Aircraft Systems (Vol. 1): Control and Performance, 1–21. https://doi.org/10.1049/PBCE120F_ch1.
31. Aroma, R.J. and Raimond, K. (2015). Multispectral vs. hyperspectral imaging for unmanned aerial vehicles: Current and prospective state of affairs. *Proc. 2015 IEEE Technological Innovation in ICT for Agriculture and Rural Development* (*TIAR*), 150–155.
32. Space stations – interplanetary [online]: www.history.nasa.gov/conghand/stations.htm. (accessed on 29-12-21).
33. Saporito, M.G. and Colantoni, A. (2019). Revolution 4.0: Industry vs. agriculture in a future development for SMEs. *Processes*, 7(1), 36.
34. GIS in agriculture – [online]: https://smallfarms.cornell.edu/2017/04/use-of-gis/
35. GIS in agriculture 4.0 – [online]: Digital Farming for Sustainable Agriculture | CropIn (accessed on 29-12-22).
36. Deshpande, A., Estrela, V.V., Patavardhan, P. and Kallimani, G. (2023). Super-resolution in a world of scarce resources for medical imaging applications. *In*: Vyas, S. et al. (Eds.), Chapter 13, *Edge-AI in Healthcare*: *Trends and Future Perspectives*, London, UK: CRC Press.
37. Forsyth, D.A. and Ponce, J. (2015). *Computer Vision: A Modern Approach.* Pearson Higher Ed.
38. Phupattanasilp, P. and Tong, S.R. (2019). Augmented reality in the integrative internet of things (AR-IoT): Application for precision farming. *Sustainability*, 11(9), 2658.
39. Birth, McVey. (1968). Measuring the color of growing turf with a reflectance spectrophotometer 1. *Agronomy Journal*, 60, 640–643.
40. Huete. (1988). A soil-adjusted vegetation index (SAVI). *Remote Sensing of Environment*, 25(3), 295–309.
41. Kaufman, Y.J. and Tanre, D. (1992). Atmospherically resistant vegetation index (ARVI) for EOS-MODIS. *IEEE Trans. Geoscience and Remote Sensing*, 30(2), 261–270.
42. Spectral Indices for vegetation – [online]: http://static1.1.sqspcdn.com/static/f/891472/12276139/1305728074980/Vegetation+spectral+indices.pdf?token=ibMFPrn4CLSdO813AX3CcPU7xkE%3D, (accessed on 17-12-21)
43. Matsushita, B., Yang, W., Chen, J., Onda, Y. and Qiu, G. (2007). Sensitivity of the enhanced vegetation index (EVI) and normalized difference vegetation index (NDVI) to topographic effects: A case study in high-density cypress forest. *Sensors (Basel)*, 7(11), 2636–2651.
44. Rondeaux, G., Steven, M. and Baret, F. (1996). Optimization of soil-adjusted vegetation indices. *Remote Sensing of Environment*, 55(2), 95–107.
45. Gitelson, A.A., Vina, A., Ciganda, V., Rundquist, D.C. and Arkebauer, T.J. (2002). Remote estimation of canopy chlorophyll content in crops. *Geophysical Research Letters*, 32.

46. Mulla, D.J. (2013). Twenty-five years of remote sensing in precision agriculture: Key advances and remaining knowledge gaps. *Biosystems Engineering*, 114(4), April, 358–371.
47. Cropin – [online] : www.cropin.com (accessed on 17-12-22).
48. AgEagle – [online]: www.ageagle.com (accessed on 17-12-22).
49. Yeppar – [online]: www.yeppar.com (accessed on 17-12-22).
50. FarmVR – [online]: www.farmvr.com (accessed on 17-12-22).
51. FarmApp – [online]: www.farmapp.io (accessed on 17-12-22).
52. Growlink – [online]: www.growlink.com (accessed on 17-12-22).
53. Väderstad AR –[online]: www.vaderstad.com/ (accessed on 17-12-22).
54. Allmeteo – [online]: www. allmeteo.com (accessed on 17-12-22).
55. GreenIQ – [online]: www.greeniq.eu (accessed on 17-12-22).
56. CropX – [online]: www.cropx.com (accessed on 17-12-22).
57. Zhang, B., Wu, D., Zhang, L., Jiao, Q. and Li, Q. (2012). Application of hyperspectral remote sensing for environment monitoring in mining areas. *Environmental Earth Sciences*, 65(3), 649–658.
58. Lezoche, M., Hernandez, J., Alemany-Diaz, M.M., Panetto, H. and Kacprzyk, J. (2020). Agri-food 4.0: A survey of the supply chains and technologies for the future agriculture. *Computers in Industry*, Elsevier.
59. Tantalaki, N., Souravlas, S. and Roumeliotis, M. (2019). Data-driven decision making in precision agriculture: The rise of big data in agricultural systems. *J. Agricultural & Food Information*, 20(4), 344–380.
60. Deshpande, A., Patavardhan, P., Estrela, V.V. and Razmjooy, N. (2020). Deep learning as an alternative to super-resolution imaging in UAV systems. *In*: Estrela, V.V., Hemanth, J., Saotome, O., Nikolakopoulos, G., Sabatini, R. (Eds.), *Imaging and Sensing for Unmanned Aircraft Systems*, 2(9), 177–212. London, UK: IET. doi: 10.1049/PBCE120G_ch9.
61. Laghari, A.A., Zhang, X., Shaikh, Z.A., Khan, A., Estrela, V.V. and Izadi, S. (2023). A review on quality of experience (QoE) in cloud computing. *J. Reliable Intell. Environ.* https://doi.org/10.1007/s40860-023-00210-y.
62. Monteiro, A.C.B., Franca, R.P., Estrela, V.V., Fernandes, S.R., Khelassi, A., Aroma, R.J. et al. (2020). UAV-CPSs as a test bed for new technologies and a primer to Industry 5.0. *In*: Estrela, V.V., Hemanth, J., Saotome, O., Nikolakopoulos, G., Sabatini, R. (Eds.), *Imaging and Sensing for Unmanned Aircraft Systems*, 2(1), 1–22. London, UK: IET. doi: 10.1049/PBCE120G_ch1.
63. Piesse, Jenifer and Thirtle Colin, G. (2010). Agricultural R&D, Technology and Productivity (September 1, 2009). *Philosophical Transactions of the Royal Society, Series B*, 364, 429-437, Available at SSRN: https://ssrn.com/abstract=2399474.

CHAPTER
9

Coastal Agriculture in Latin America: Typical Abiotic Stresses and the Use of Microbial Inoculants

Jussara Tamires de Souza Silva Goulart [1[0000-0002-3632-3851]], Gabriel Quintanilha-Peixoto[2**[0000-0002-7476-5845]], Clarisse Maximo Arpini[3[0000-0001-5814-6393]], Alessandro Ramos Coutinho[3[0000-0002-4082-0076]] and Aline Chaves Intorne[1,4*[0000-0001-8015-6926]]**

[1] Laboratory of Physiology and Biochemistry of Microorganisms, Universidade Estadual do Norte Fluminense Darcy Ribeiro (UENF), Campos dos Goytacazes, RJ, 28013-602, Brazil

[2] Laboratory of Molecular and Computational Biology of Fungi, Universidade Federal de Minas Gerais (UFMG), Belo Horizonte, MG, 31270-901, Brazil

[3] Laboratory of Biotechnology and Environmental Microbiology, Vila Velha University (UVV), Vila Velha, ES, 29102-623, Brazil

[4] Laboratory of Chemistry and Biology, Instituto Federal de Educação, Ciência e Tecnologia do Rio de Janeiro, Volta Redonda, RJ, 27213-100, Brazil

Coastal areas are employed for agriculture worldwide. Nonetheless, compared with other arable areas, these sites present typical challenges, especially abiotic stresses such as high salinity and drought. Even though cultivation types vary across different countries, beneficial microorganisms play a crucial role in helping plants tolerate these stresses. The microbes are naturally present in the soil. They might be enriched in such environments by using inoculants designed to improve plant growth while decreasing or eliminating the need for chemical fertilizers, which might be detrimental to soil and water quality. This represents a sustainable alternative to traditional methods in agriculture, which have, over the years, polluted water bodies and shifted soil quality through spikes in nutrient availability. In this chapter, we review the challenges of Latin American coastal agriculture and the perspectives on using microbial inoculants as a sustainable and effective technology in these areas.

* Corresponding author: aline.intorne@ifrj.edu.br

** These authors contributed equally to this work.

1. Introduction

Coastal zones are the most populated and economically active regions worldwide. Despite places with lower population density and areas of environmental conservation (mangroves, estuaries, and restingas, for example), one can also find regions with great urbanization accompanied by industries, ports, tourist activities, and large cultivated areas. Therefore, it is common to see problems of anthropogenic origin in these coastal areas, such as soil and water contamination [1–3].

In Latin America (LA), some coastal ecosystems such as xeric shrublands (such as the *matorrales* in Mexico and the *Caatinga* biome in Brazil) have aridity as their hallmark [4, 5]. In some cases, such as in the Chilean and Peruvian deserts and Patagonia to the south of the continent, this is due to the influence of the Andes Mountains on the Pacific coast of South America, which is reflected in stressful environmental conditions for agriculture due to low rainfall and availability of nutrients in the soil [6]. Microorganisms residing in such ecosystems have specific mechanisms for acquiring nutrients (carbon, nitrogen, and phosphorus), which can help plant growth, playing a fundamental role in biogeochemical cycles [7].

Temperature stress, especially high temperatures, is capable of causing irreversible damage to plant growth. The increase in temperature will inevitably lead to increased evaporation and more salinization. In the case of coastal areas, the influence of salinity on crop growth is critical [8]. Other major challenges for advancing agriculture in these coastal environments are high sunlight intensity, soil pH (commonly acidic in tropical areas) [9], and different soil textures (sandy soils have difficulty retaining water) [10–12]. Water scarcity and increased temperature can negatively impact crops, limiting plant development. Increased salinity in soils associated with low nutrient content and long periods of drought also cause a decrease in yield [12, 13].

The environmental conditions in a given space shape the associated living community. This dynamic poses one of the greatest ongoing agrarian challenges since microorganisms improve plant growth, and when their survival is threatened, yielding is equally at risk [12–14]. Thus, research on the association between the adverse conditions found in arid and desert regions and the native soil microorganisms is of great importance, as they are better adapted to these extreme conditions [15, 16]. This work is dedicated to reviewing agricultural activities in LA's coastal areas based on distribution maps made available by the United States Department of Agriculture (USDA – available on <https://ipad.fas.usda.gov/ogamaps/cropproductionmaps.aspx>, accessed on April 12, 2022), reviewing the advantages of microbial inoculants for crops under specific abiotic stresses such as salinity and drought, and the biotechnological potential of the application of inoculants composed of fungi and bacteria as a sustainable alternative to recurrent techniques with chemical fertilizers. Mainly, studies on so-called 'commodity' crops are explored: wheat, sunflower, soybean, sorghum, rye, rice, rapeseed, peanuts, oats, corn, cotton, barley, and oil palm. It is important to note that the crop distribution maps by USDA do not include several countries in LA, especially the Caribbean islands, Honduras, El Salvador, Nicaragua, and Venezuela (Fig. 1). For this reason, the scientific literature relating to these regions may be underrepresented in this material.

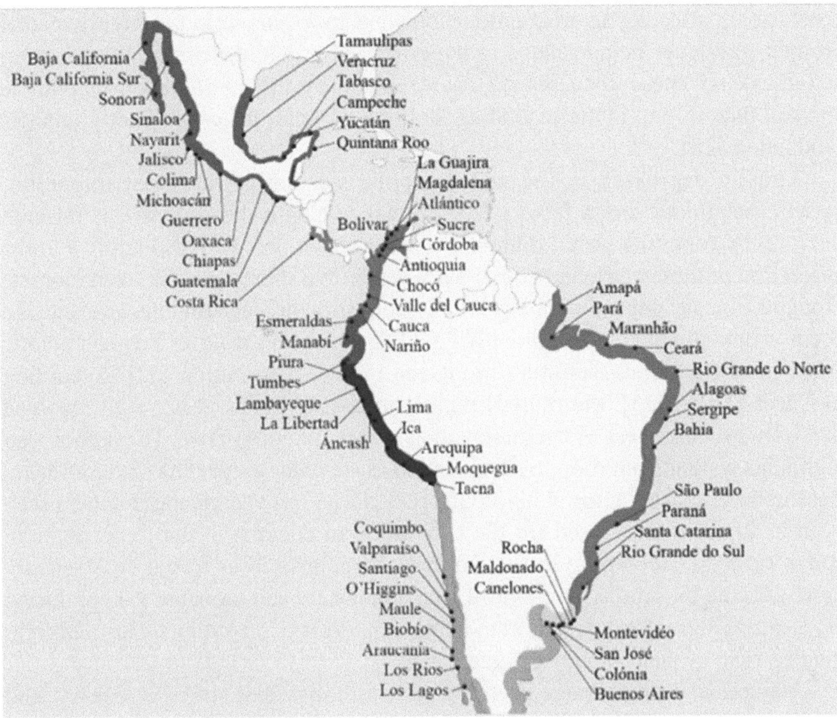

Fig. 1. Coastal regions of LA are considered in this review. Colors vary in each country. In gray: Countries not included in USDA maps and continental regions. Coastal lines are for illustration purposes only and are not to be scaled. Regions south of Los Lagos (Chile) were not represented in USDA maps and are thus not shown.

2. Abiotic Stresses in Coastal Agriculture

Abiotic stresses strongly affect agricultural production [17]. In the case of coastal regions, the main stress factors for crops are drought, salinity, and high temperatures, which may be accentuated due to climate change [18]. For example, studies show that sea-level rise increased salinity in coastal areas of India from 1% to 33% over 25 years, directly impacting agricultural areas in the region [19].

In the 2017 book, *Handbook of Drought and Water Scarcity*, Saeid Eslamian & Faezeh Eslamian deeply review the concepts and characteristics of drought. According to the authors, a consensus definition would be: "an extended period—a season, a year, or several years—of poor rainfall relative to the statistical multiyear mean for a region." According to the authors, there are four types of drought: meteorological (dry weather that lasts long enough to unbalance the availability of water in a given region), hydrological (when the amount of water in reservoirs and water bodies is below average), agricultural (rainfall deficit capable of considerably affecting agrarian production), and socioeconomic and environmental drought (the latter being related to the effects of previous definitions on human activities). All these concepts are characterized by the combination of climate and soil factors, and in the case of agricultural drought, the crop directly impacts production [20].

Drastic decreases in the water supply are one of the most impacting agricultural abiotic stress factors [20, 21]. Since plants absorb nutrients through the roots by capturing water, if there is a water deficit, this will also imply a lower absorption of these nutrients. Therefore, whether in a short period or even months, droughts directly impact plant physiology, root structure, and architecture and can occur in any climate regime globally [20, 22]. Drought is another frequent abiotic stress in coastal areas, resulting in uneven rainfall distribution [22]. According to Curtis (2019) [23], the rainfall regime in coastal zones is not well resolved precisely because these are common areas where rain originates. The author also highlights a significant drop in precipitation as we enter deeper into the continent. Research by Sousa Santos et al. (2020) [24] shows relevant nuances can be seen even in adjacent areas used for the cultivation of coconut in the Ceará State in Brazil (Fig. 1). According to the authors, the palms closer to the coastline are more affected by salinity stress. Still, one area where soil moisture is kept during dry seasons was also present, alleviating salt stress. The authors also highlight the decrease in yield in non-irrigated areas.

Natural and anthropogenic forms can establish salt stress in coastal and continental soils [25]. Naturally, it is caused by weather with the release of different types of salts (e.g., chlorides, sulfates, and bicarbonates of Na, Ca, and Mg) carried by water in the soil [26] and by marine spray from the ocean transported by winds and rainfall [17]. Anthropogenic forms include the deposition of saline water used for irrigation [25] and inadequate irrigation management, which leads to the elevation of salts present in underground water to the soil surface. Two relevant examples can be found in Peru; the aquifer La Yarada, in the Tacna region (Fig. 1), one of the few water resources in that area, has been overexploited for agriculture and other activities, including the expansion of the Tacna city. Such overexploitation now causes seawater intrusion into the soil and undergrounds, intensifying the water salination and scarcity in that region [27]. Unusual salinity levels can also be found in the Nepeña Valley region of Áncash (Fig. 1), which according to [28], happens due to intensive agriculture, even though it is confusing if this salination is recent or from pre-Hispanic times.

More than 830 million hectares of arable land are estimated to be affected by salinization worldwide [12]. This process is relatively common in arid and semi-arid coastal and continental regions, being related to the geological formation, low rainfall, soil type, and, in some instances, due to the agricultural development of that given region [17]. For example, in Argentina alone, approximately 600,000 ha of irrigated soil is affected by salinity. In Brazil, 30% of the 500,000 ha of irrigated areas in the Northeast are salinized. In Mexico, 20% of agricultural land is also undergoing salinization. In Ecuador and Cuba, the main cause of salinization is the use of poor-quality or saline water and excessive use of chemical fertilizers [29].

Increased salinity due to maritime influence and groundwater overexploitation is typical in coastal regions, which decreases the enzymatic activities in the soil, and microbial biomass and, consequently, reduces the microbial growth rate [12]. As a result, salinity directly affects the concentration of soil organic matter in these regions and its availability since microorganisms are responsible for a large part of the processing of nutrients in biogeochemical cycles [30].

The predictive study by Gupta et al. (2020) [22] warns that by 2050 we will have a decrease in the water supply for agriculture, while, in the same period, there will be a growing demand for arable land. This study describes that drought has historically been more influenced by low precipitation. Still, the authors also predict the influence, mainly of the increase in temperature associated with climate change alternating with drought cycles. Such cycles are common, for example, in the Maipo River and the Santiago region of Chile (Fig. 1). Nonetheless, research by Peña-Guerrero et al. (2020) [31] highlights that the ever-growing demand for agricultural products in that region will likely be affected by the water availability and quality in the Maipo River basin. Loitzenbauer & Bulhões Mendes (2012) [32] describe a mathematical model to understand how much of the water in the Tramandaí River in the State of Rio Grande do Sul, Brazil (Fig. 1) can be used without harm to salinity dynamics, suggesting the application of licenses to regulate water use.

Biotechnology is also increasingly used in agriculture, ranging from classical plant improvement through crossing wild and cultivated plant varieties to genetic engineering, using genetically modified, transgenic, and strass-adapted ecotypes of plants [33]. For example, in Chile, different ecotypes of quinoa, native to the Andes region, have different tolerance to salinity. Ruiz et al. (2016) [34] describe that the Salares ecotype, native to the salt flats of Bolivia, Chile, and Argentina, tolerates up to 500 mM (millimolar) of NaCl, while ecotypes native to the Andean lowlands are more sensitive to salinity. As another example, using beneficial microorganisms in plants contributes to the sustainable development of crops and simultaneously influences economic growth, as it is a low-cost tool. Paul & Nair (2008) [35] describe the potential of *Pseudomonas fluorescens* MSP-393 for use as a microbial inoculant in coastal areas, considering its high salt tolerance. The authors have applied sophisticated techniques such as HPLC and MALDI-TOF for this research. Such experimental design raises the awareness that even though biotechnological tactics might be essential in achieving greater productivity and eco-efficiency, they require high technical skills for implementation and maintenance, which is a challenge for small farmers [36]. Thus, the application of PGPMs is a biotechnological tool with the potential to improve plant growth and mitigate the effects of abiotic stresses on crops [37–39]. The use of plant growth-promoting microorganisms (PGPMs) in agricultural production systems has been used in specific crops such as soybeans [40–42] and sugarcane [43–45]. Coastal environments are a source of prospection for these microorganisms. They can improve their agricultural production by formulating and applying microbial inoculants, which are biological preparations from cultures of beneficial microorganisms that can help the host plant control pathogens, acquire nutrients, and better resist environmental stresses [46], which the next topics review.

3. The Role of Microorganisms in Agriculture

Microorganisms (bacteria, fungi, unicellular eukaryotes, and some algae and metazoa) are the most diverse and abundant living beings in nature [47]. It is hypothesized that bacteria were the first organisms to appear on the planet, having colonized even the most inhospitable environments and playing essential roles in biogeochemical cycles (Fig. 2). Thus, microorganisms are required for the availability of macronutrients, such as nitrogen, through biological nitrogen fixation, nitrification, and denitrification [48]. Alongside fungi, bacteria also contribute to the solubilization of phosphate, potassium, and zinc [49]. Furthermore, microorganisms actively act in the metabolism of vital greenhouse gases (GHGs), e.g., carbon dioxide, methane, and nitric dioxide [48] (Fig. 2).

Bacteria and fungi that exhibit beneficial traits for plant development are called plant growth-promoting microorganisms (PGPM). In addition to nutrient assimilation, they can help plants directly by producing phytohormones (e.g., auxins and gibberellins); and siderophores to capture iron, thus covering the main limiting factors for plant growth [50]. Evermore studies are carried out on the prospection of these PGPMs and their potential to alleviate plant stress symptoms. Zahir et al. (2008) [51] demonstrated a direct action of PGPM in *Pisum sativum* (peas); the action of ACC Deaminase, an enzyme responsible for the hydrolysis of 1-carboxylic acid 1-aminocyclopropane (ACC) into α-ketobutyrate and ammonia, allows the plant to become more tolerant to the effects of ethylene, when under conditions of salt and water stress. Meanwhile, PGPMs can also indirectly help plants by inhibiting phytopathogen action, synthesizing antimicrobial metabolites as active biocontrol agents, competing for nutrients, and even stimulating systemic plant resistance [50] (Fig. 3). Such concepts are crucial to the development of microbial inoculants. Their use is an interesting strategy in Agriculture 4.0, and if the included microorganisms are adapted to existing stressors, the chances of success increase considerably [52–54].

Changes in land use for agriculture and urbanization negatively impact soil quality [55]. GHG emissions, typical of anthropic activities and one of the leading causes of climate change, also affect soil quality and harm the rainfall regime, whether due to drought or flooding, which washes and breaks down soil aggregates [56]. About 40% of the habitable land worldwide (excluding deserts, glaciers, and rocks) has been transformed into agricultural (cultivated

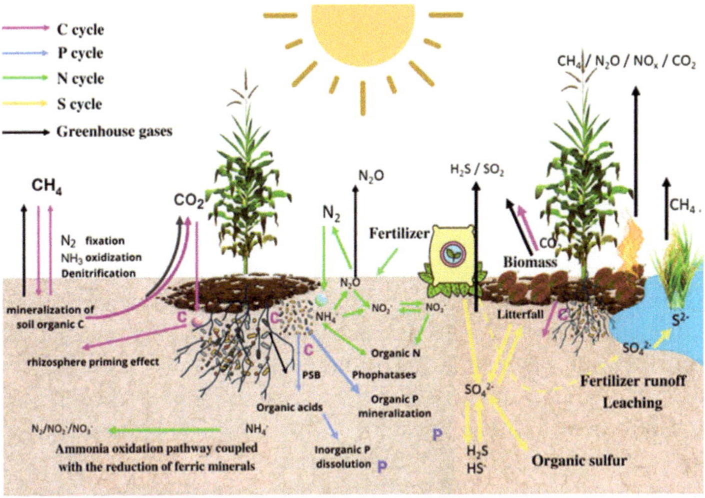

Fig. 2. Biogeochemical cycles in the soil and greenhouse gas emissions. Pink arrows correspond to the processes of carbon cycling (C), considering the mineralization of organic C in the soil, fixation of carbon dioxide (CO_2), and oxidation of methane (CH_4). The green arrows correspond to the nitrogen (N) cycling processes, highlighting the fixation of atmospheric nitrogen (N_2), nitrification, denitrification, organic N mineralization, and ammonia (NH_3) oxidation associated with the reduction of ferric minerals. Blue arrows represent the phosphorus (P) cycling processes, from its dissolution into inorganic phosphorus and mineralization of organic phosphorus, including the action of Phosphate Solubilizing Bacteria (PSB). Yellow arrows represent the Sulfur (S) cycling processes, and black arrows highlight the emission of greenhouse gases, including gases emitted due to the heating of compounds that can occur in the burning processes that are eventually carried out in some agricultural systems, such as Nitrogen Oxides (NOx).

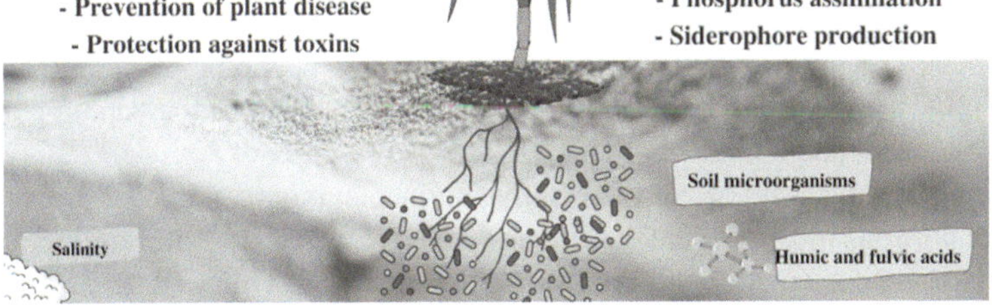

Fig. 3. Direct and indirect mechanisms of plant growth promoting microorganisms (PGPM) against abiotic stresses.

and/or pasture) areas. Nonetheless, currently cultivated areas correspond to approximately 25% of that total [57–59]. Land overexploitation causes a decrease in the soil organic matter content, directly impacting fertility and biodiversity, especially microbial diversity. Shifts in microbial diversity are indicators of soil degradation [60]. Land preparation processes in annual crops, for example, break down soil aggregates, causing an imbalance in the microbiota (e.g., bacteria and fungi involved in organic matter degradation processes) which increases CO_2 consumption, decreasing soil organic carbon stocks [61].

How organic contaminants act on microbial communities regarding carbon decomposition is still poorly understood [7]. However, it is known that plant-microbe interactions are directly affected by environmental changes, both positively and negatively [62]. Studies have shown that high levels of CO_2 favor root colonization by mycorrhizae and improve the rhizobia's and legumes' interaction [63]. The literature describes that the presence of symbiotic microorganisms can mitigate or even neutralize the effects of aridity or adverse temperatures on plants (native and cultivated), increasing their tolerance and allowing better development [64, 65]. Therefore, understanding the complex action of microorganisms in the soil is essential to ensure greater agricultural productivity. Throughout time, bacteria and fungi remain key elements for the availability of nutrients and the response of their plant hosts to the challenges presented by the environment.

4. Insights into the Role and Characteristics of the Soil in Coastal Agriculture

Cultivation forms and main crops change according to the region across LA's coast, primarily due to different climatic conditions, nutrient availability, and ecosystem type. In coastal zones, these factors are influenced by rivers and seas, which change their dynamics and course over time on Earth [66]. This influence can be verified through the soil, local ecosystem, and geopolitical characteristics [67].

LA's coast has a great diversity of landscapes, with the influence of the Andes mountain range on the Pacific coast, which contributes to the formation of commonly arid regions by blocking winds and rain in the region known as Diagonal Arid [6]. On the other hand, the Atlantic coast, mainly due to the tectonic movements of the South American plate, does not have comparable mountain ranges, being influenced by the deltas of large rivers, such as the Rio de la Plata, São Francisco, and Amazonas [68].

Among the coastal or predominantly continental biomes and ecosystems with coastal portions most affected coastal agriculture in LA (as seen in [69] and [70]) are: (i) savannas such as the pampas between Brazil, Uruguay, and Argentina and a portion the cerrado; (ii) tropical forests such as the Atlantic Forest, which covers most of the Brazilian coast, the Chocó-Darién Forest in Colombia, forests

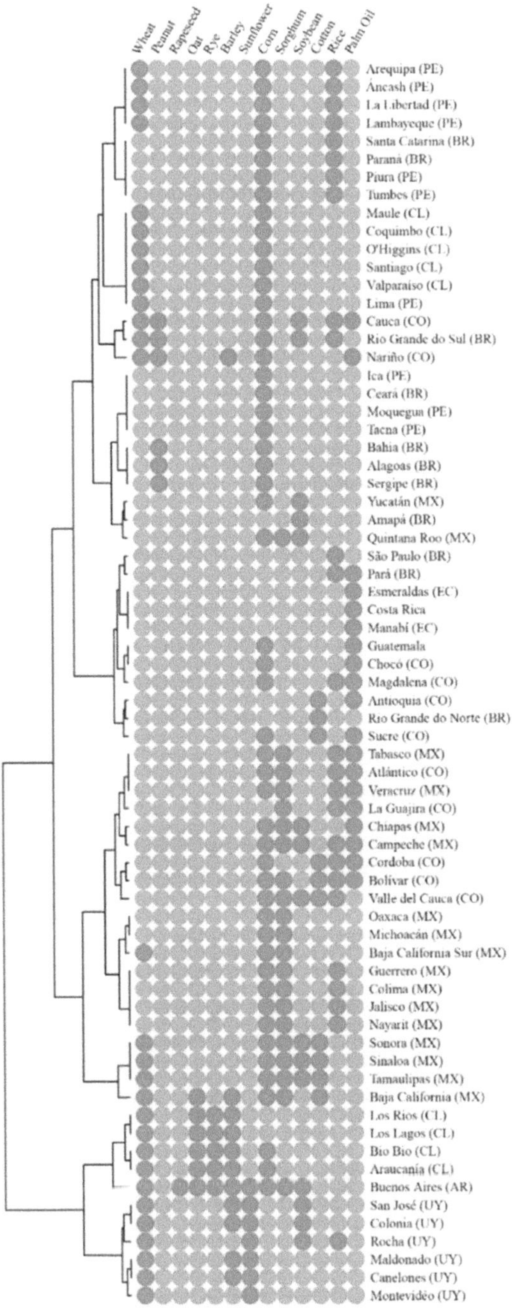

Fig. 4. Presence-absence map and adjacent clustering of regions based on cultivated crops according to USDA.

in southern Mexico and a small coastal portion of the Amazon forest, and (iii) arid and semi-arid coastal regions, in Andean biomes in Colombia, Peru, Chile, and Mexico, and a small portion of the Brazilian Caatinga (Fig. 1).

Historical maps from FAO [71] show different LA farming systems, some characteristic of coastal regions. In these more arid parts of the continent, irrigated farming systems predominate, including northern Mexico and the entire Andean chain; "Pampas and Campos" farms dedicated to the production of cereals and livestock, covering the Mar del Plata region and southern Brazil; mixed intensive agriculture, covering part of the Brazilian Atlantic Forest; and "Coastal plantation and mixed", in most of the non-arid LA's coast. Although such historical documents might be outdated, it is clear that a large part of this distribution of systems is maintained, considering the main areas of palm oil production [72], cocoa [73], and sugarcane [74], for example.

As seen in Fig. 1, when analyzing the USDA geographic distribution data [72], there is slight geographic segmentation among crops in LA. Simultaneously, wheat, corn, and sorghum are widely distributed, and crops such as oats, rye, barley, and sunflower are more restricted and co-occur (Fig. 1). However, according to the present data review, some regions and provinces are multifaceted in crop distribution. For example, the province of Buenos Aires, Argentina (not to be confused with the Autonomous City of Buenos Aires, the capital of the country) concentrates the planting of nine of the 15 included crops, missing only the cultivation of peanuts, rice, cotton, and palm oil (Agricultural data for Argentina is available under <https://datos.magyp.gob.ar/dataset>, accessed on April 12, 2022). Other geographically close regions, especially the provinces of Colonia and San José in Uruguay, and the three states that make up southern Brazil (Rio Grande do Sul, Santa Catarina, and Paraná), are united by the cultivation of wheat, corn, and soybeans, with exception to the state of Santa Catarina, where soybean planting in coastal regions was not detected [72]. These similarities can be partially explained by climatic conditions, soil formation in the Rio de la Plata basin, and the Pampa biome, common to Uruguay, parts of Rio Grande do Sul, and Argentina. Some planting differences include the presence of sunflower and barley in Uruguayan regions, absent in neighboring zones of Brazil, and corn planting in Brazilian states, absent in Uruguayan regions.

Other agricultural hubs are present on the Pacific coast of northern LA, including the departments of Nariño, Cauca, and Valle del Cauca in Colombia and the departments of Áncash, Arequipa, La Libertad, and Lambayeque in Peru, which are major producers of wheat, rice, and corn. In Cauca and Nariño, peanuts and oil palm are also planted [75, 76]. These regions are quite geographically distinct; the Colombian coast has more humid forests, where the soil can contain large amounts of organic matter in superficial layers deposited by litter and other decaying elements [77]. Even so, deeper layers are often nutrient-poor, making them unsuitable for agriculture [78]. Meanwhile, the Peruvian coast, like the Chilean coast, is arid and bounded by the Andean mountain range.

5. Microbial Inoculants as a Sustainable Alternative

The use of microbe-based agricultural inoculants is not a recent technology. According to Santos et al. (2019) [79], the first patent describing the use of microbial inoculants dates to 1896, indicating its informal use years prior. Microorganisms participating in these formulations generally improve plant growth through the increase in phosphorus and nitrogen uptake, macronutrients essential to all living beings, especially in nucleotide and amino acid synthesis processes. Thus, such elements are commonly limiting factors in plant growth, and their supplementation brings better yields [21]. Likewise, other factors can limit agricultural production, such as diseases of various types caused by all domains of life, especially bacteria, fungi, nematodes, and viruses. The monoculture scheme, still widely used despite its known risks, has greater susceptibility to pests since the cultivated plants are usually clonal or genetically similar. These factors can help to explain the need for fertilizers and pesticides. However, their wide use and difficulty of replacement bring significant field challenges because chemical fertilizers have a high environmental cost despite being profitable.

Estimated emissions of NO_2 (nitrous dioxide) indicate that agriculture is responsible for the emission of 86.6% of this gas, which means that the fertilizer industry is a considerable emitter of GHGs [80]. N_2O is released into the atmosphere during the application of nitrogen fertilizers. This greenhouse gas is 300 times more potent than carbon dioxide [81]. Phosphate fertilizers also have limitations; from rocks, phosphate reserves are currently running out [82]. In this scheme, it is estimated that only 15–25% of applied phosphorus is absorbed by plants [83]; that part of the fertilizers not used by the plants leaches into water bodies (rivers, lakes, estuaries, and finally, the sea), which then become eutrophic due to the increase in nutrients load (especially nitrogen and phosphorus), which profoundly unbalances the community of microorganisms, and therefore, all the fauna and flora of these environments [84]. Eutrophication also leads to the proliferation of cyanobacteria and dinoflagellates, causing the so-called algal blooms. In addition to preventing the entry of light by covering the surface of water bodies and consequently corrupting the oxygen cycle, these microorganisms commonly produce harmful toxins for human and environmental health [85]. In summary, microbial inoculants bring positive perspectives to the agricultural sector since they generate an increase in productivity and present a low risk of environmental contamination, in addition to affordable selling prices [86] or being a nature-based solution that optimizes the role already played by beneficial microorganisms in biogeochemical cycles.

Some studies propose complementing chemical fertilization with microbial inoculants; for example, offering rock phosphate together with phosphate-solubilizing microorganisms, ensuring the gradual release of that nutrient, as required by the plant, thus reducing costs and avoiding problems with leaching and eutrophication [87]. A study conducted by Ditta et al. (2018) [88] compared the effectiveness of organic fertilizers enriched with natural phosphate, associated with phosphate-solubilizing microorganisms, and the use of diammonium phosphate used in legumes, showing very similar results between the two treatments, especially when there is an association of the organic fertilizer with PSM and diammonium phosphate, with an increase in soil aggregate stability by up to 37%, an increase of up to 24% in the nodules and an increase in the aerial part by up to 19%.

Another interesting strategy is the formulation of microbial inoculants using microorganisms native to the given environment where they will be applied. In this case, a previous phase of bioprospecting for PGPM strains is necessary. In this way, the survival of microorganisms is not compromised, as they are already adapted to the environment. Two examples are available in the Northeastern region of Brazil; Fernandes & Fernandes (2000) [89] characterized rhizobia from coastal flatlands in the city of Aracaju, in the State of Sergipe (Fig. 1) with biological nitrogen fixation capacity when associated with pigeon pea (*Cajanus cajan*). The native rhizobia of coastal tablelands showed great potential for use as a pigeon pea inoculant, as they have a high capacity for biological nitrogen fixation and characteristics that favor survival in the soil. Souza & Martins (2010) [4] selected isolates of native rhizobia from agricultural soils in the municipality of Areia, State of Paraíba (Fig. 1), and found native nitrifying bacteria capable of nodulating and fixing nitrogen in *Vigna unguiculata*. Studies carried out in several countries with microorganisms from coastal regions with dunes have revealed potential microorganisms for use in agriculture, mainly due to the production of Indole Acetic Acid (IAA), increasing the yield of bean seedlings, production of Gibberellin used in the production of rice demonstrating root growth. Rhizobia isolated from legumes from dune regions on the west coast of India demonstrated the ability to increase nodulation and nitrogen fixation, in addition to increasing the development of cowpea roots and shoots.

The most severe difficulties in the microbial inoculants' implantation in world agriculture have been mainly the competition of traditional methods already widely used of chemical fertilization and, to a lesser extent, the need for more resilient protocols which serve different cultures in different regions worldwide [90–92]. However, we hypothesize that this technology will become progressively more widely used and well-established due to unsustainability and environmental damage caused by years of chemical fertilizer use. The excessive use of agrochemicals, mainly in coastal areas, causes harm because the high amount of these products make the soil even more salinized, causing losses in agricultural production [93]. New sustainable and non-toxic alternatives must be explored to minimize these losses and improve soil quality. Thus, the use of microorganisms (bacteria and fungi) that maintain, and at the same time, provide beneficial associations with plants and resistance to different abiotic stresses is a viable alternative [21, 94, 95]. Some products have been developed and commercialized as microbial inoculants. As an example, some plant growth-promoting bacteria were listed by [96]: *Agrobacterium radiobacter, Azospirillum brasilense, Azospirillum lipoferum, Azotobacter chroococcum, Bacillus fimus, Bacillus licheniformis, Bacillus megaterium, Bacillus mucilaginous, Bacillus pumilus, Bacillus* spp., *Bacillus subtilis, Bacillus subtilis* var. *amyloliquefaciens, Burkholderia cepacia, Delfitia acidovorans, Paenobacillus macerans, Pantoea agglomerans, Pseudomonas aureofaciens, Pseudomonas chlororaphis, Pseudomonas fluorescens, Pseudomonas solanacearum, Pseudomonas* spp., *Pseudomonas syringae, Serratia entomophilia, Streptomyces griseoviridis, Streptomyces* spp., *Streptomyces lydicus,* and various *Rhizobia* spp.

Microbial inoculants can be composed of single microorganisms or microbial consortia. They can be applied to plant seeds [97] or soil [98]. Generally, microbial inoculants used in agriculture are classified into three categories depending on the microorganisms involved: Bacterial inoculants involve nitrogen-fixing bacteria, phosphorus-solubilizing bacteria, and symbiotic (fixing with the host plant), non-symbiotic (free-living) or associative (free-living with the host plant) bacteria (i), Fungal inoculants: arbuscular mycorrhizae and phosphorus-solubilizing fungi; (ii) or compound inoculants and consortia: a combination of two or more microbial inoculants; (iii) [99]. Some LA countries were cited by Basu et al. (2021) [100] as manufacturers of microbial inoculants based on nitrogen-fixing bacteria (Table 1).

Table 1. An overview of PGPM-based biofertilizer products available in LA

Name	PGPR strains	Country	Crops
Zadspirillum	*Azospirillum brasilense*	Argentina	Corn
Rizo-Liq	*Bradyrhizobium* sp., *Mesorhizobium ciceri, Rhizobium* spp.	Argentina	Soybean
Rizo-Liq Top	*Bradyrhizobium japonicum*	Argentina	Soybean
BiAgro 10®	*Bradyrhizobium japonicum*	Argentina Brazil Bolivia	Soybean
Dimargon®	*Azotobacterchroococcum*	Colômbia	Cotton Soybean
Nitrasec	*Rhizobium* sp.	Uruguay	Soybean

Table 2 presents crops that utilize commercial microbial inoculants and investigation papers that showed microorganisms with potential for future application. The microbial inoculants presented are formulated mainly by bacteria, emphasizing the genus *Bradyrhizobium*. The main cultures that employ these microbial inoculants are soy and sugarcane, while most of the related research was conducted in Brazil, followed by Argentina. In sugarcane, we also highlight the isolation of PGPM with great biotechnological implications, such as *Gluconacetobacter diazotrophicus* [101, 102] and *Herbaspirillum seropedicae* [103]. Research works were found with other cultures, such as (*Zea mays, Oryza* sp.), related to microorganisms that promote plant growth and which are less

Table 2. Microbe-mediated benefits to plant growth and yield. 'Yield' means biomass, the aerial part of the plant, or grain size

Crops	Microorganism	Main impacts	Country	Reference
Soybean Cotton	*R. clarus*	+20% root colonization	Brazil	[104]
Soybean Corn	*R. clarus*	+16,3% yield (soybean) +20% yield (corn)	Brazil	[105]
Soybean.	*B. japonicum*	+6,39% yield	Argentina	[106]
Soybean.	*B. elkanii*	+40% beans yield	Uruguay	[42]
Corn	*Bacillus, Brevibacillus, Staphylococcus, Paenibacillus*	+56–86% yield	Brazil	[107]
Corn Wheat	*A. brasilense* *A. lipoferum*	+16–30% yield	Brazil	[108]
Rice	*A. brasilense* *P. fluorescens*	+11% yield	Argentina	[109]
Wheat	*Bacillus* sp.	+38% height in aerial parts	Mexico	[110]
Peanut	*O. intermedium, Pantoea* sp.	+92% dry biomass in aerial parts	Argentina	[111]
Sunflower	*A. xylosoxidans* *Alcaligenes* sp. *B. pumilus*	Increase in dry biomass of roots	Argentina	[112]

represented in research, even though these cultures are economically and culturally significant. More information regarding other cultures was also found, with fewer results obtained in the search (e.g., cotton, banana, peanut, sunflower, and wheat). This shows a lot of room for the growth of research related to these cultures and the formulation of microbial inoculants to improve cultivation, mainly to alleviate the abiotic stresses that affect LA coastal areas.

6. Conclusions

Agriculture is one of the oldest human activities, and it is always innovating to meet the population's food needs. As our growth has been exponential, it is an activity with high demand, which increasingly seeks to be productive, offering quality products to people. However, we are experiencing times of profound change, especially concerning climate change, which intensifies abiotic stresses such as salinity and water deficit, bringing challenges to the era of Agriculture 4.0. One perspective to circumvent such adverse effects is complementing non-renewable chemical fertilizers and pesticides with microbial inoculants. In this way, it will be possible to improve the quality of soils, avoid the loss of diversity, and conserve water bodies, promoting sustainable and safe agriculture.

Abiotic stresses, including environments with high salinity and water deficit, adversely affect crops and soil microorganisms' structure, composition, and physiological activities. In addition, these stresses disrupt crops' physiological and metabolic processes, leading to large crop losses worldwide. Habitats with these characteristics, such as coastal areas, are, however, a source of MPCVs that, in addition to providing nutrients to plants and soil, can improve the response to abiotic stresses of crops, maintaining agricultural productivity. The ability to facilitate plant growth by providing essential nutrients such as nitrogen and phosphorus, especially in soils with these deficiencies, is one of the functional properties of abiotic stress-tolerant MPCV. Microbial inoculants are biotechnological products used for several purposes, mainly complementing or replacing chemical fertilizers. Interest in employing microbial inoculants containing bacteria and fungi, which promote plant growth, is likely to increase in the coming years due to the higher fertilizers costs, mainly with concerns about pollution and emphasis on sustainable agriculture. So, to guarantee food security for the population, it is necessary to explore new ways, and microbial inoculants are a successful practice that can solve 21st-century agriculture's problems.

References

1. Scholvin, S. and Malamud, A. (2020). Is Brazil a Geoeconomic Node? Geography, Public Policy, and the Failure of Economic Integration in South America. *Bras. Pol. Sc. Rev.*, 14(2).
2. Scherer, M., Costa, M.F., Boski, T., Azeiteiro, U.M. and Dias, J.A. (2014). Integrated coastal management in Latin America: The ever new world. *Rev. Gestão Costeira Integr.*, 14(4), 663–668. doi: 10.5894/rgci575.
3. Barragán Muñoz, J.M. (2020). Progress of coastal management in Latin America and the Caribbean. *Ocean Coast. Manag.*, 184(October 2019). doi: 10.1016/j.ocecoaman.2019.105009.
4. da S. Souza, J. and Martins, A.F. (2010). Caracterização e seleção de rizóbios nativos de solos paraibanos associados à Vigna unguiculata (L.) Walp. *Agropecuária Técnica*, 40(3–4), 71–81.
5. Garcia-Franco, J.G. (1996). Distribución de epífitas vasculares en matorrales costeros de Veracruz, México. *Acta Bot. Mex.*, 37, 1. doi: 10.21829/abm37.1996.764.
6. de Porras, M.E. and Maldonado, A. (2018). Metodologías y avances del tardío a lo largo de la diagonal árida Sudamericana. *Asoc. Paleontológica Argentina*, 18(2), 18–38.
7. Vicuña, R. and González, B. (2021). The microbial world in a changing environment. *Rev. Chil. Hist. Nat.*, 94(1). doi: 10.1186/s40693-021-00099-y.
8. Corwin, D.L. (2021). Climate change impacts on soil salinity in agricultural areas. *Eur. J. Soil Sci.*, 72(2), 842–862. doi: 10.1111/ejss.13010.
9. Fageria, N.K. and Nascente, A.S. (2014). *Management of Soil Acidity of South American Soils for Sustainable Crop Production.* Elsevier. 128.

10. Tapia-Torres, Y., Elser, J.J., Souza, V. and García-Oliva, F. (2015). Ecoenzymatic stoichiometry at the extremes: How microbes cope in an ultra-oligotrophic desert soil. *Soil Biol. Biochem.*, 87, 34–42. doi: 10.1016/j.soilbio.2015.04.007.
11. Santoyo, G., Hernández-Pacheco, C., Hernández-Salmerón, J. and Hernández-León, R. (2017). The role of abiotic factors modulating the plant-microbe-soil interactions: Toward sustainable agriculture. A review. *Spanish J. Agric. Res.*, 15(1), 1–15. doi: 10.5424/sjar/2017151-9990.
12. Wen Zhang, W., Wang, C., Xue, R. and Jie Wang, L. (2019). Effects of salinity on the soil microbial community and soil fertility. *J. Integr. Agric.*, 18(6), 1360–1368. doi: 10.1016/S2095-3119(18)62077-5.
13. Balestrini, R., Chitarra, W. and Fotopoulos, V. (2017). Soil biological communities and ecosystem resilience. *Soil Biol. Communities Ecosyst. Resil.*, 191–207. doi: 10.1007/978-3-319-63336-7.
14. Suleiman, A.K., Harkes, P., van den Elsen, S., Holterman, M., Korthals, G.W., Helder, J. and Kuramae, E.E. (2019). Organic amendment strengthens interkingdom associations in the soil and rhizosphere of barley (Hordeum vulgare). *Sci. Total Environ.*, 695, 133885. doi: 10.1016/j.scitotenv.2019.133885.
15. Alsharif, W., Saad, M.M. and Hirt, H. (2020). Desert microbes for boosting sustainable agriculture in extreme environments. *Front. Microbiol.*, 11 July. doi: 10.3389/fmicb.2020.01666.
16. Meena, V.S., Mishra, P.K., Bisht, J.K. and Pattanayak, A. (2017). *Agriculturally Important Microbes for Sustainable Agriculture.* (Vol. 2). doi: 10.1007/971-981-105588-1.
17. Pedrotti, A., Chagas, R.M., Ramos, V.C., do N. Prata, A.P., Lucas, A.T.A. and dos Santos, P.B. (2015). Causas e consequências do processo de salinização dos solos. *Rev. Eletrônica em Gestão, Educ. e Tecnol. Ambient.*, 19(2), 1308–1324. [Online]. Available: http://periodicos.ufsm.br/reget/article/view/16544/pdf.
18. Ullah, A., Bano, A. and Khan, N. (2021). Climate change and salinity effects on crops and chemical communication between plants and plant growth-promoting microorganisms under stress. *Front. Sustain. Food Syst.*, 5 June, 1–16. doi: 10.3389/fsufs.2021.618092.
19. Rahman, A.K.M.M., Ahmed, K.M., Butler, A.P. and Hoque, M.A. (2018). Influence of surface geology and micro-scale land use on the shallow subsurface salinity in deltaic coastal areas: A case from southwest Bangladesh. *Environ. Earth Sci.*, 77(12), 1–8. doi: 10.1007/s12665-018-7594-0.
20. Eslamian, S. and Eslamian, F.A. (2017). *Handbook of Drought and Water Scarcity: Environmental Impacts and Analysis of Drought and Water. Scarcity*. New York: CRC Press, LCCN 2016030589 ISBN 9781498731089 (V.1: hardback) ISBN 9781315404226 (v.1: e-book).

Handbook of Drought and Water Scarcity: Environmental impacts and analysis of drought and water/ edited by Saeid Eslamian and Faezeh A. Eslamian.

New York: CRC Press, 2017- LCCN 2016030589 ISBN 9781498731089(V.1: hardback) ISBN9781315404226(v.1:e-book)

21. Kour, D., Rana, K.L., Yadav, A.N., Sheikh, I., Kumar, V., Dhaliwal, H.S. and Saxena, A.K. (2020). Amelioration of drought stress in Foxtail millet (*Setaria italica* L.) by P-solubilizing drought-tolerant microbes with multifarious plant growth promoting attributes. *Environ. Sustain.*, 3(1), 23–34. doi: 10.1007/s42398-020-00094-1.
22. Gupta, A., Rico-Medina, A. and Caño-Delgado, A.I. (2020). The physiology of plant

responses to drought. *Science,* (80-.), 368(6488), 266–269. doi: 10.1126/science. aaz7614.
23. Curtis, S. (2019). Means and long-term trends of global coastal zone precipitation. *Sci. Rep.*, 9(1), 1–9. doi: 10.1038/s41598-019-41878-8.
24. Sousa Santos, M.M. (2020). Ecophysiology of the tall coconut growing under different coastal areas of northeastern Brazil. *Agric. Water Manag.*, 232 June, 106047. doi: 10.1016/j.agwat.2020.106047.
25. Ribeiro, P.H.P., Gheyi, H.R., Uyeda, C.A., Teixeira, M.B., Soares, F.A.L. and Dias, N.D.S. (2016). Taxa de Crescimento e Produção de Girassol Irrigado com Água Salina sob Doses de Nitrogênio. *Irriga*, 233–247. 2016.
26. Várallyay, G. (1994). Climate change, soil salinity, and alkalinity. *Soil Responses to Clim. Chang.*, 39–54. doi: 10.1007/978-3-642-79218-2_4.
27. Pino, E., Ramos, L., Avalos, O., Tacora, P., Chávarri, E., Angulo, O. et al. (2019). Factors affecting depletion and pollution by marine intrusion in the La Yarada's coastal aquifer, *Tacna, Peru. Tecnología y ciencias del agua*, 10(5), 177–213. Epub 15 de febrero de 2020.https://doi.org/10.24850/j-tyca-2019-05-07 10(5).
28. Chicoine, D., Mcnabb, C. and Clement, B. (2020). Commensality, irrigation and agrarian entanglements in Early Horizon Nepeña, Coastal Peru. *Am. Eur. J. Am. Archaeology*, 5, 7–22.
29. F.A.O., WWC. (2015). *Towards a Water and Food Secure Future: Critical Perspectives for Policymakers*. Rome and Marseille.
30. Setia, R., Smith, P., Marschner, P., Gottschalk, P., Baldock, J., Verma, V. et al. (2012). Simulation of salinity effects on past, present, and future soil organic carbon stocks. *Environ. Sci. Technol.*, 46(3), 1624–1631. doi: 10.1021/es2027345.
31. Peña-Guerrero, M.D., Nauditt, A., Muñoz-Robles, C., Ribbe, L. and Meza, F. (2020). Drought impacts on water quality and potential implications for agricultural production in the Maipo River Basin, Central Chile. *Hydrol. Sci. J.*, 65(6), 1005–1021. doi: 10.1080/02626667.2020.1711911.
32. Loitzenbauer, E. and Bulhões Mendes, C.A. (2012). Salinity dynamics as a tool for water resources management in coastal zones: An application in the Tramandaí River basin, southern Brazil. *Ocean Coast. Manag.*, 55, 52–62. doi: 10.1016/j. ocecoaman.2011.10.011.
33. Gao, C. (2021). Genome engineering for crop improvement and future agriculture. *Cell*, 184(6), 1621–1635. doi: 10.1016/j.cell.2021.01.005.
34. Ruiz, K.B., Aloisi, I., Del Duca, S., Canelo, V., Torrigiani, P., Silva, H. and Biondi, S. (2016). Salares versus coastal ecotypes of quinoa: Salinity responses in Chilean landraces from contrasting habitats. *Plant Physiol. Biochem.*, 101, 1–13. doi: 10.1016/j.plaphy.2016.01.010.
35. Paul, D. and Nair, S. (2008). Stress adaptations in a Plant Growth Promoting Rhizobacterium (PGPR) with increasing salinity in the coastal agricultural soils. *J. Basic Microbiol.*, 48(5), 378–384. doi: 10.1002/jobm.200700365.
36. Rose, D.C. and Chilvers, J. (2018). Agriculture 4.0: Broadening Responsible Innovation in an Era of Smart Farming. *Front. Sustain. Food Syst.*, 2(December), 1–7. doi: 10.3389/fsufs.2018.00087.
37. de Morais, M.B., Azevedo, R.A., Camara, T.R., Ulisses, C., de Albuquerque, C.C. and Willadino, L. (2020). Antioxidative metabolism in sugarcane (*Poaceae*) varieties subjected to water and saline stress. *Rev. Bras. Eng. Agric. e Ambient.*, 24(11), 776–782. doi: 10.1590/1807-1929/agriambi.v24n11p776-782.

38. Siddikee, M.A., Chauhan, P.S., Anandham, R., Han, G.H. and Sa, T. (2010). Isolation, characterization, and use for plant growth promotion under salt stress, of ACC deaminase-producing halotolerant bacteria derived from coastal soil. *J. Microbiol. Biotechnol.*, 20(11), 1577–1584. doi: 10.4014/jmb.1007.07011.
39. Mayak, S., Tirosh, T. and Glick, B.R. (2004). Plant growth-promoting bacteria confer resistance in tomato plants to salt stress. *Plant Physiol. Biochem.*, 42(6), 565–572. doi: 10.1016/j.plaphy.2004.05.009.
40. De Paula, G.F., Demétrio, G.B. and Matsumoto, L.S. (2021). Biotechnological potential of soybean plant growth-promoting rhizobacterial. *Rev. Caatinga*, 34(2), 328–338. doi: 10.1590/1983-21252021v34n209rc.
41. Ambrosini, V.G., Fontoura, S.M.V., de Moraes, R.P., Tamagno, S., Ciampitti, I.A. and Bayer, C. (2019). Soybean yield response to Bradyrhizobium strains in fields with inoculation history in Southern Brazil. *J. Plant Nutr.*, 42(16), 1941–1951. doi: 10.1080/01904167.2019.1648680.
42. Riviezzi, B., Cagide, C., Pereira, A., Herrmann, C., Lombide, R., Lage, M. et al. (2020). Improved nodulation and seed yield of soybean (*Glycine max*) with a new isoflavone-based inoculant of Bradyrhizobium elkanii. *Rhizosphere*, 15, 100219. doi: 10.1016/j.rhisph.2020.100219.
43. Martins, D.S., Reis, V.M., Schultz, N., Alves, B.J., Urquiaga, S., Pereira, W. et al. (2020). Both the contribution of soil nitrogen and of biological N_2 fixation to sugarcane can increase with the inoculation of diazotrophic bacteria. *Plant Soil*, 454(1–2), 155–169. doi: 10.1007/s11104-020-04621-1.
44. Ferreira, N.S., Matos, G.F., Meneses, C.H., Reis, V.M., Rouws, J.R., Schwab, S. et al. (2020). Interaction of phytohormone-producing rhizobia with sugarcane mini-setts and their effect on plant development. *Plant Soil*, 451(1–2), 221–238. doi: 10.1007/s11104-019-04388-0.
45. Lima, C.O.D.C., Vaz, A.B., De Castro, G.M., Lobo, F., Solar, R., Rodrigues, C. et al. (2021). Integrating microbial metagenomics and physicochemical parameters and a new perspective on starter culture for fine cocoa fermentation. *Food Microbiol.*, 93 103608. doi: 10.1016/j.fm.2020.103608.
46. Wu, L., Wu, H.J., Qiao, J., Gao, X. and Borriss, R. (2015). Novel routes for improving biocontrol activity of Bacillus based bioinoculants. *Front. Microbiol.*, 6 Dec., 1–13. doi: 10.3389/fmicb.2015.01395.
47. Singh, N., Singh, J. and Singh, K. (2018). Small at size, big at impact: Microorganisms for sustainable development. *Microb. Bioprospecting Sustain. Dev.*, 3–28. doi: 10.1007/978-981-13-0053-0_1.
48. Madsen, E.L. (2011). Microorganisms and their roles in fundamental biogeochemical cycles. *Curr. Opin. Biotechnol.*, 22(3), 456–464. doi: 10.1016/j.copbio.2011.01.008.
49. Kumawat, N., Kumar, R., Kumar, S. and Meena, V.S. (2017). Nutrient Solubilizing Microbes (NSMs): Its role in sustainable crop production. *In:* V.S. Meena (Ed.), *Agriculturally Important Microbes for Sustainable Agriculture* (Vol. 2), January, 1–374.
50. Bargaz, A., Lyamlouli, K., Chtouki, M., Zeroual, Y. and Dhiba, D. (2018). Soil microbial resources for improving fertilizers efficiency in an integrated plant nutrient management system. *Front. Microbiol.*, 9 July. doi: 10.3389/fmicb.2018.01606.
51. Zahir, Z.A., Munir, A., Asghar, H.N., Shaharoona, B. and Arshad, M. (2008). Effectiveness of rhizobacteria containing ACC deaminase for growth promotion of peas (*Pisum sativum*) under drought conditions. *J. Microbiol. Biotechnol.*, 18(5), 958–963.

52. Rizvi, A., Ahmed, B., Khan, M.S., Umar, S. and Lee, J. (2021). Psychrophilic bacterial phosphate-biofertilizers: A novel extremophile for sustainable crop production under cold environment. *Microorganisms*, 9(12). doi: 10.3390/microorganisms9122451.
53. Kumar, S., Nand, S., Pratap, B., Dubey, D. and Dutta, V. (2021). Removal kinetics and treatment efficiency of heavy metals and other wastewater contaminants in a constructed wetland microcosm: Does mixed macrophytic combinations perform better? *J. Clean. Prod.*, 327 July, 129468. doi: 10.1016/j.jclepro.2021.129468.
54. Cornejo-Ríos, K., Osorno-Suárez, M.D.P., Hernández-León, S., Reyes-Santamaría, M.I., Juárez-Díaz, J.A., Pérez-España, V.H. et al. (2021). Impact of trichoderma asperellum on chilling and drought stress in tomato (*Solanum lycopersicum*) *Horticulturae*, 7(10), 385. doi: 10.3390/horticulturae7100385.
55. Parikh, S.J. and James, B.R. (2012). Soil: The foundation of agriculture. *Nature Education Knowledge,* 3(10), 2.
56. Malhi, G.S., Kaur, M. and Kaushik, P. (2021). Impact of climate change on agriculture and its mitigation strategies: A review. *Sustain.*, 13(3), 1–21. doi: 10.3390/su13031318.
57. Hooke, R.L.B., Martín-Duque, J.F. and Pedraza, J. (2012). Land transformation by humans: A review. *GSA Today*, 22(12), 4–10. doi: 10.1130/GSAT151A.1.
58. Kanianska, R. (2016). Agriculture and its impact on land-use, environment, and ecosystem services. *In*: Landscape Ecology: The Influences of Land Use and Anthropogenic Impacts of Landscape Creation land, 3–26.
59. Poore, J. and Nemecek, T. (2018). Reducing food's environmental impacts through producers and consumers. *Science* (80-.), 360(6392), 987–992. doi: 10.1126/science.aaq0216.
60. Teixeira, A.F., Silva, J.S., Vilela, L.A., Costa, P.F., Costa, E.M.D., Guimaraes, A. et al. (2019). Microbiological indicators of soil quality under native forests are influenced by topographic factors. *An. Acad. Bras. Cienc.*, 91(4), e20180696. doi: 10.1590/0001-3765201920189696.
61. Dhaliwal, J., Kukal, S.S. and Sharma, S. (2018). Soil organic carbon stock in relation to aggregate size and stability under tree-based cropping systems in Typic Ustochrepts. *Agrofor. Syst.*, 92(2), 275–284. doi: 10.1007/s10457-017-0103-8.
62. Cheng, Y.T., Zhang, L. and He, S.Y. (2019). Plant-microbe interactions facing environmental challenge. *Cell Host Microbe*, 26(2), 183–192. doi: 10.1016/j.chom.2019.07.009.
63. Alshaal, T., El-Ramady, H., Al-Saeedi, A.H., Shalaby, T., Elsakhawy, T., Omara, A.E.D. et al. (2017). The rhizosphere and plant nutrition under climate change. Essential Plant Nutrients: Uptake, Use Efficiency, and Management, 275–308. https://doi.org/10.1007/978-3-319-58841-4_11
64. De Zelicourt, A., Al-Yousif, M. and Hirt, H. (2013). Rhizosphere microbes as essential partners for plant stress tolerance. *Mol. Plant*, 6(2), 242–245. doi: 10.1093/mp/sst028.
65. Timmis, K. and Ramos, J.L. (2021). The soil crisis: The need to treat as a global health problem and the pivotal role of microbes in prophylaxis and therapy. *Microb. Biotechnol.*, 14(3), 769–797. doi: 10.1111/1751-7915.13771.
66. Brêda, J.P.L.F., de Paiva, R.C.D., Collischon, W., Bravo, J.M., Siqueira, V.A. and Steinke, E.B. (2020). Climate change impacts on South American water balance from a continental-scale hydrological model driven by CMIP5 projections. *Clim. Change*, 159(4), 503–522. doi: 10.1007/s10584-020-02667-9.
67. Yoon, J., Klassert, C., Selby, P., Lachaut, T., Knox, S., Avisse, N. et al. (2021). A

coupled human-natural system analysis of freshwater security under climate and population change. *Proc. Natl. Acad. Sci. U.S.A.*, 118(14), e2020431118. doi: 10.1073/pnas.2020431118.
68. Seeliger, U. and Kjerfve, B. (2001). *Coastal Marine Ecosystems of Latin America*. Springer.
69. Boit, A., Sakschewski, B., Boysen, L., Cano Crespo, A., Clement, J., Garcialaniz, N. et al. (2016). Large-scale impact of climate change vs. land-use change on future biome shifts in Latin America. *Glob. Chang. Biol.*, 22(11), 3689–3701. doi: 10.1111/gcb.13355.
70. Reis, C.R., Pacheco, F.S., Reed, S.C., Tejada, G., Nardoto, G.B., Forti, M.C. et al. (2020). Biological nitrogen fixation across major biomes in Latin America: Patterns and global change effects. *Sci. Total Environ.*, 746, 140998. doi: 10.1016/j.scitotenv.2020.140998.
71. F. and A.O. of T.U.N. FAO. (2001). *Major Farming Systems: Latin America and Caribbean*.
72. U.S.D. of A. USDA. (2021). *Global: Crop Production Maps*.
73. Beckett, S.T. (1994). *Industrial Chocolate Manufacture and Use*. Wiley Online.
74. Dias, H.B. and Sentelhas, P.C. (2018). Sugarcane yield gap analysis in Brazil: A multi-model approach for determining magnitudes and causes. *Sci. Total Environ.*, 637–638, 1127–1136. doi: 10.1016/j.scitotenv.2018.05.017.
75. Gutiérrez, L.-F., Rosada, L.-M. and Jiménez, A. (2011). Chemical composition of Sacha Inchi (*Plukenetia volubilis* L.) seeds and characteristics of their lipid fraction. *Grasas y aceites*, 62(1), 76–83.
76. Yusoff, M.N.A.M., Zulkifli, N.W.M., Sukiman, N.L., Chyuan, O.H., Hassan, M.H., Hasnul, M.H. et al. (2021). Sustainability of palm biodiesel in transportation: A review on biofuel standard, policy and international collaboration between Malaysia and colombia. *Bio. Energy Res.*, 14, 43–60. [Online]. Available: https://doi.org/10.1007/s12155-020-10165-0.
77. Zhou, Z., Wang, C. and Luo, Y. (2018). Effects of forest degradation on microbial communities and soil carbon cycling: A global meta-analysis. *Glob. Ecol. Biogeogr.*, 27(1), 110–124. doi: 10.1111/geb.12663.
78. Ranger, J. (2018). Forest soils: Characteristics and sustainability. *In*: Soils as a Key Component of the Critical Zone 1: Functions and Services, 163–186. Wiley (online).
79. Santos, M.S., Nogueira, M.A. and Hungria, M. (2019). Microbial inoculants: Reviewing the past, discussing the present and previewing an outstanding future for the use of beneficial bacteria in agriculture. *AMB Express*, 9(1). doi: 10.1186/s13568-019-0932-0.
80. Cunha-Zeri, G. and Ometto, J. (2021). Nitrogen emissions in Latin America: A conceptual framework of drivers, impacts, and policy responses. *Environ. Dev.*, 38, November, 2020,100605. doi: 10.1016/j.envdev.2020.100605.
81. Signor, D. and Cerri, C.E.P. (2013). Emissões de óxido nitroso em solos agrícolas: Uma revisão. *Pesqui. Agropecu. Trop.*, 43(3), 322–338. doi: 10.1590/S1983-40632013000300014.
82. Alewell, C., Ringeval, B., Ballabio, C., Robinson, D.A., Panagos, P. and Borrelli, P. (2020). Global phosphorus shortage will be aggravated by soil erosion. *Nat. Commun.*, 11(1). doi: 10.1038/s41467-020-18326-7.
83. Corrêa, J.C., Mauad, M. and Rosolem, C.A. (2004). Phosphorus in soil and soybean growth as affected by phosphate fertilization and cover crop residues. *Pesqui. Agropecu. Bras.*, 39(12), 1231–1237. doi: 10.1590/s0100-204x2004001200010.

84. Khan, M.N., Mobin, M., Abbas, Z.K. and Alamri, S.A. (2017). *Fertilizers and Their Contaminants in Soils, Surface and Groundwater* (Vols. 1–5). Elsevier Inc.
85. Glibert, P.M. and Burkholder, J.M. (2006). The complex relationships between increases in fertilization of the earth, coastal eutrophication and proliferation of harmful algal blooms. *Ecol. Harmful Algae*, January, 341–354. doi: 10.1007/978-3-540-32210-8_26.
86. Kour, D., Rana, K.L., Yadav, A.N., Yadav, N., Kumar, M., Kumar, V. et al. (2019). Microbial biofertilizers: Bioresources and eco-friendly technologies for agricultural and environmental sustainability. *Biocatal. Agric. Biotechnol.*, 23, 101487. doi: 10.1016/j.bcab.2019.101487.
87. Billah, M., Khan, M., Bano, A., Nisa, S., Hussain, A., Dawar, K.M. et al. (2020). Rock phosphate-enriched compost in combination with Rhizobacteria: A cost-effective source for better soil health and wheat (*Triticum aestivum*) productivity. *Agronomy*, 10(9), 1390. doi: 10.3390/agronomy10091390.
88. Ditta, A., Imtiaz, M., Mehmood, S., Rizwan, M.S., Mubeen, F., Aziz, O. et al. (2018). Rock phosphate-enriched organic fertilizer with phosphate-solubilizing microorganisms improves nodulation, growth, and yield of legumes. *Commun. Soil Sci. Plant Anal.*, 49(21), 2715–2725. doi: 10.1080/00103624.2018.1538374.
89. Fernandes, M.F. and Fernandes, R.P.M. (2000). Seleção inicial e caracterização parcial de rizóbios de tabuleiros costeiros quando associados ao guandu. *Rev. Bras. Ciência do Solo*, 24(2), 321–327. doi: 10.1590/s0100-06832000000200009.
90. Islam, M.N., Germida, J.J. and Walley, F.L. (2021). Survival of a commercial AM fungal inoculant and its impact on indigenous AM fungal communities in field soils. *Appl. Soil Ecol.*, 166, March, 103979. doi: 10.1016/j.apsoil.2021.103979.
91. Harris, B.A., Bauske, E.M. and Pennisi, S.V. (2021). Cultural practices and microbial inoculants have variable impact on a bedding plant (*Lantana camara* L.) performance in the landscape. *Sci. Hortic.* (Amsterdam), 282, February, 110059. doi: 10.1016/j.scienta.2021.110059.
92. Salomon, M.J., Demarmels, R., Watts-Williams, S. J., McLaughlin, M. J., Kafle, A., Ketelsen, C. et al. (2022). Global evaluation of commercial arbuscular mycorrhizal inoculants under greenhouse and field conditions. *Appl. Soil Ecol.*, 169, 104225. doi: 10.1016/j.apsoil.2021.104225.
93. Lopes, C.V.A. and de Albuquerque, G.S.C. (2018). Agrotóxicos e seus impactos na saúde humana e ambiental: Uma revisão sistemática. *Saúde em Debate*, 42(117), 518–534. doi: 10.1590/0103-1104201811714.
94. Hashem, A., Abd-Allah, E.F., Alqarawi, A.A., Al-Huqail, A.A., Wirth, S. and Egamberdieva, D. (2016). The interaction between arbuscular mycorrhizal fungi and endophytic bacteria enhances plant growth of Acacia gerrardii under salt stress. *Front. Microbiol.*, 7, Jul., 1–15. doi: 10.3389/fmicb.2016.01089.
95. Fu, L., Penton, C. R., Ruan, Y., Shen, Z., Xue, C., Li, R. et al. (2017). Inducing the rhizosphere microbiome by biofertilizer application to suppress banana Fusarium wilt disease. *Soil Biol. Biochem.*, 104, 39–48. doi: 10.1016/j.soilbio.2016.10.008.
96. Glick, B.R. (2012). Plant growth-promoting bacteria: Mechanisms and applications. *Scientifica* (*Cairo*), 2012, 1–15.
97. Fipke, G.M., Martin, T.N., Muller, T.M., Cunha, V.D.S., Munareto, J.D., Schonell, A.T. et al. (2020). Osmoprotectant in soybean seeds can increase the inoculation and co-inoculation time in pre-sowing. *Aust. J. Crop Sci.*, 14(6), 905–912. doi: 10.21475/ajcs.20.14.06.p1696.

98. Scagliola, M., Valentinuzzi, F., Mimmo, T., Cesco, S., Crecchio, C. and Pii, Y. (2021). Bioinoculants as promising complement of chemical fertilizers for a more sustainable agricultural practice. *Front. Sustain. Food Syst.*, 4, January, 1–12. doi: 10.3389/fsufs.2020.622169.
99. Gosal, S.K. and Kaur, J. (2017). Microbial inoculants: A novel approach for better plant microbiome interactions. *In*: Probiotics in Agroecosystem, 1–537. ResearchGate.
100. Basu, A., Prasad, P., Das, S.N., Kalam, S., Sayyed, R.Z., Reddy, M.S. et al. (2021). Plant growth promoting rhizobacteria (PGPR) as green bioinoculants: Recent developments, constraints, and prospects. *Sustain.*, 13(3), 1140. doi: 10.3390/su13031140.
101. De Oliveira, M.V.V., Intorne, A.C., Vespoli, L.D.S., Madureira, H.C., Leandro, M.R., Pereira, T.N. et al. (2016). Differential effects of salinity and osmotic stress on the plant growth-promoting bacterium Gluconacetobacter diazotrophicus PAL5. *Arch. Microbiol.*, 198, 287–294. doi: 10.1007/s00203-015-1176-2.
102. Intorne, A.C., De Oliveira, M.V.V., Lima, M.L., Da Silva, J.F., Olivares, F.L. and De Souza Filho, G.A. (2009). Identification and characterization of Gluconacetobacter diazotrophicus mutants defective in the solubilization of phosphorus and zinc. *Arch. Microbiol.*, 191(5), 477–483. doi: 10.1007/s00203-009-0472-0.
103. Pessoa, D.D.V., Dos-Santos, C.M., Vidal, M.S., Baldani, J.I., Tadra-Sfeir, M.Z., de Souza, E.M. et al. (2021). Herbaspirillum seropedicae strain HRC54 expression profile in response to sugarcane apoplastic fluid, *3 Biotech*, 11(6), 292. doi: 10.1007/s13205-021-02848-y.
104. Cely, M.V.T., De Oliveira, A.G., De Freitas, V.F., De Luca, M.B., Barazetti, A.R., Dos Santos, I.M. et al. (2016). Inoculant of arbuscular mycorrhizal fungi (*Rhizophagus clarus*) increase yield of soybean and cotton under field conditions. *Front. Microbiol.*, 7, 720. doi: 10.3389/fmicb.2016.00720.
105. Barazetti, A.R., Simionato, A.S., Navarro, M.O.P., dos Santos, I.M.O., Modolon, F., de Lima Andreata, M.F. et al. (2019). Formulations of arbuscular mycorrhizal fungi inoculum applied to soybean and corn plants under controlled and field conditions. *Appl. Soil Ecol.*, 142, 25–33. doi: 10.1016/j.apsoil.2019.05.015.
106. Leggett, M., Diaz-Zorita, M., Koivunen, M., Bowman, R., Pesek, R., Stevenson, C. et al. (2017). Soybean response to inoculation with Bradyrhizobium japonicum in the United States and Argentina. *Agron. J.*, 109(3), 1031–1038. doi: 10.2134/agronj2016.04.0214.
107. Nascimento, R.D.C., Cavalcanti, M.I.P., Correia, A.D.J., Escobar, I.E.C., de Freitas, A.D.S., Nóbrega, R.S.A. et al. (2021). Maize-associated bacteria from the Brazilian semi-arid region boost plant growth and grain yield. *Symbiosis*, 83(3), 347–359. doi: 10.1007/s13199-021-00755-7.
108. Hungria, M., Campo, R.J., Souza, E.M. and Pedrosa, F.O. (2010). Inoculation with selected strains of Azospirillum brasilense and A. lipoferum improves yields of maize and wheat in Brazil. *Plant Soil*, 331(1), 413–425. doi: 10.1007/s11104-009-0262-0.
109. de Salamone, I.E.G., Funes, J.M., Di Salvo, L.P., Escobar-Ortega, J.S., D'Auria, F., Ferrando, L. et al. (2012). Inoculation of paddy rice with Azospirillum brasilense and Pseudomonas fluorescens: Impact of plant genotypes on rhizosphere microbial communities and field crop production. *Appl. Soil Ecol.*, 61, 196–204. doi: 10.1016/j.apsoil.2011.12.012.
110. Ibarra-Villarreal, A.L., Gándara-Ledezma, A., Godoy-Flores, A.D., Herrera-Sepúlveda, A., Díaz-Rodríguez, A.M., Parra-Cota, F.I. et al. (2021). Salt-tolerant

Bacillus species as a promising strategy to mitigate the salinity stress in wheat (*Triticum turgidum* subsp. *durum*). *J. Arid Environ.*, 186. doi: 10.1016/j.jaridenv.2020.104399.
111. Taurian, T., Anzuay, M.S., Angelini, J.G., Tonelli, M.L., Ludueña, L., Pena, D. et al. (2010). Phosphate-solubilizing peanut associated bacteria: Screening for plant growth-promoting activities. *Plant Soil*, 329, 421–431. doi: 10.1007/s11104-009-0168-x.
112. Forchetti, G., Masciarelli, O., Izaguirre, M.J., Alemano, S., Alvarez, D. and Abdala, G. (2010). Endophytic bacteria improve seedling growth of sunflower under water stress, produce salicylic acid, and inhibit growth of pathogenic fungi. *Curr. Microbiol.*, 61(6), 485–493. doi: 10.1007/s00284-010-9642-1.

CHAPTER
10

The Use of Sensors for Soil Fertility Mapping in the Context of Agriculture 4.0

Sandra Meinen da Cruz[1*][0000-0003-0497-2387], Daniel Uhry[1][0000-0002-3480-8652], Ronaldo Serpa da Rosa[2][0000-0002-5202-3349], Juliano Dalcin Martins[3][0000-0003-2501-1067], Nidgia Maria Nicolodi[1][0000-0002-4006-1313], Letícia Guadagnin Vogel[1][0000-0002-9794-1135], Natália Sornberger Adam[1][0000-0003-2292-5904], Alexandre Schafer[4][0000-0002-9264-2427], Guilherme Silva de Oliveira[4][0000-0001-6603-0980], Rafael Eduardo da Costa[4][0000-0002-1241-568X] and Leonardo Seibel Sander[4][0000-0003-2690-2451]

[1] Federal Institute of Education, Science and Technology IFRS, Ibirubá Campus, 111, Nelsi Ribas Fritsch Street, Esperança, 98200-000, Ibirubá, RS, Brasil

[2] Federal Institute of Education, Science and Technology IFRS, Bento Campus, 540, Osvaldo Aranha Avenue, Juventude da Enologia, 95700-206, Bento Gonçalves, RS, Brazil

[3] Department of Rural Engineering, Federal University of Santa Maria - UFSM, 1000, Roraima Avenue, 97195-000, Santa Maria, RS, Brazil

[4] Stara S/A. Agricultural Implement Industry, Stara Avenue, 519. 99470-000, Não-Me-Toque, RS, Brazil

The aim of this study was to evaluate a sensor for measuring soil pH composed of two Sb electrodes, comparing it with the standard laboratory method in order to validate it for using in Brazilian soils. Initially, sensor evaluations were carried out at bench scale and later in the field. On bench, consistent values were obtained with the pH standard method when the sensor's measurements were carried out in friable soil. The field-level evaluation indicates that the readings performed with the Sb electrodes indicated a consistent result with the laboratory analysis. The results obtained with the Sb ISE sensor indicated adequate accuracy and precision in the soil pH determination, under the evaluated humidity conditions. The pH data collection with the Sb ISE provides a greater sampling density,

* Corresponding author: sandra.cruz@ibiruba.ifrs.edu.br

allowing the creation of more detailed maps in a more sustainable and faster way compared to the traditional method.

1. Introduction

The global food demand and the current system of agricultural production lead to high levels of productivity, intensifying the use of inputs. Given this, it is necessary to develop systems that allow rationalizing the use of these inputs and, consequently, the production costs, the environmental impact and, at the same time, increasing the production level [1].

In this context, Precision Agriculture (AP) has been contributing to handlings and inputs application in a rational way, considering the soil's heterogeneity and the nutrients availability variation [2]. The use of sensors in agriculture optimizes the collection of data by area, allowing the acquisition of a greater density of sampling points [3]. Thus, the accuracy of fertility maps is high, enabling the rational use of chemical reagents and a quick obtainment of results, bringing economic and environmental benefits.

Technical analysis reports are commonly used to recommend liming and fertilization [4]. Most impactful soil parameters are organic matter (OM), hydrogenionic potential (pH), SMP index, clay and macronutrients (Ca, Mg, K, P, S). The determination of these parameters in the laboratory requires soil samples' collection, drying, grinding and preparation with generally toxic and bioaccumulative solutions.

The PA is characterized by the Ministry of Agriculture, Livestock and Supply (MAPA), as "an agricultural management system based on the spatial and temporal variability of the productive unit and committed to increasing economic return, sustainability and minimization of environmental impacts". PA demands created the ideal scenario for machinery industries to seek the creation of the first generation of agricultural machines with embedded electronics [5].

Based on PA principles, Agriculture 4.0 was developed, a term derived from Industry 4.0, also known as Digital Agriculture. Agriculture 4.0 aims to innovate and improve software dedicated to agricultural use [6]. To this, new technologies were added, which go beyond the idea of simple field mechanization, proposing the generation of greater connectivity, allowing instantly field data capture and generating useful information for rural property management [5, 7].

The insertion of equipment embedded sensors for agricultural purposes improves applications in agriculture, and it is fundamental in Agriculture 4.0. Thus, together with agronomic knowledge, the value of this technology results in an ongoing process of assessment, interpretation, and targeted operations [8, 9]. The maps generated using these sensors characterize the analyzed areas, allowing the creation of management zones.

The pH, MO and electrical conductivity (EC) sensors embedded in agricultural equipment have been used for a longer time in the US, generating useful maps for fields management [10, 11]. The soil pH (active acidity) interferes with the availability of several essentials chemical elements for plant development. This fact is due to membrane cells selectivity present in root which does not allow the absorption of some chemical forms of the elements [12]. Thus, the pH sensors have great potential for decision-making regarding the liming or not of areas in order to raise the soil pH to the fields ideal value, aiming to raise productivity rates. However, in Brazil this technology still needs further studies for validation and implementation, mainly due to the variability of soils and management between countries.

2. Method and Results

2.1 Active Acidity (Soil pH)

The active acidity corresponds to the concentration of H^+ ions in the soil solution and its measurement is expressed by the pH index, given by the negative logarithm of the molar concentration of H^+ in the solution, according to the equation [12]:

$$pH = -\log [H^+]. \tag{1}$$

Solutions at pH = 7,0 are neutral because the concentration of H^+ is equal to the concentration of OH^-, that is 1×10^{-7} M [13]. In the soil, pH affects nutrient availability, microorganism activity and root growth [14]. In addition, at pH values below 5,5, Al^{3+} is available in the soil solution. This is considered toxic to plants, making the application of lime in crops necessary, in order to raise the pH [15]. To quantify the lime needed to raise the pH to the appropriate value for the growth of crops, the potential acidity of the soil must be determined in the laboratory, using the SMP index. Potential acidity (H + Al) estimates the soil buffering capacity exerted through the release of H^+ ions adhered to hexahydrate colloids or Al into the solution, causing the soil pH withstand small variations [4, 16].

2.2 Standard Laboratory Method for Soil pH Determination

The soil's active acidity is usually estimated in the laboratory using the standard method, by determining the pH in the soil solution. This method consists of the electrochemical measurement of the activity of H^+ ions in the soil-water suspension, in order to generate fertility reports [17].

To this end, the soil collection in the field is necessary. The soil sample preprocessing consists of drying it in an oven (45 °C) and grinding it in a hammer mill passing through a 2 mm diameter mesh sieve. After this process, soil and water are added in polypropylene bottles in a 1:1 ratio, whose 10 mL of soil and 10 mL of distilled water. The samples are stirred with a glass rod and laid to rest for 30 minutes. Then, a stirring is carried out again to determine the pH in a bench peagometer. This equipment consists of an ion selective membrane electrode for H^+ ions and a thermocouple. The peagometer calibration is done with commercial buffer solutions of pH 4.0 and 7.0 before taking measurements [4, 18].

2.3 pH Sensor

The pH sensor is composed of two wear-resistant antimony (Sb) electrodes with good relative calibration stability, ideal for mapping the field in motion [3, 19, 20]. The sensor provides pH values obtained through an oxidation-reduction reaction of a film consisting mainly of antimony trioxide (Sb_2O_3) with the H^+ ions present in the sample [21]. Thus, it can be called ion-selective electrode (ISE) of Sb [3] for pH measurements in soil according to the reaction presented by the equation below (BATES, 1961):

$$Sb_2O_3 + 6H^+ + 6e = 2Sb + 3H_2O \quad (2)$$

The conversion of the electrical signal obtained in the oxidation-reduction reaction into a pH value is presented below [21]:

$$Ec = Ec\ (0) - R.T.In10/F\ pH \quad (3)$$

Although robust, the Sb electrodes can be damaged if their surface dries out, when going through an inadequate storage period. This factor impairs its functioning due to the presence of pores in the Sb film and makes hydration in KCl solution important. Thus, the electrodes were immersed in a KCl solution (50% buffer pH 4,0 and 50% KCl 1 mol^{-1}) for hydration during four days. After this process, the sensor's calibration was possible and, later, the soil samples were analyzed.

As in the case of the bench peagometer, the sensor's calibration must be done before performing measurements. In this regard, buffer solutions of pH 4,01 and 7,01, supplied by Veris Technologies, were used. When the sensor-check option (calibration) is selected, the buffer solution of pH 4,01 is requested, and later, that of pH 7,01. The cleaning with distilled water and drying with a paper towel of the electrodes between the solutions is necessary.

2.4 pH Sensor Evaluation in Bench

The pH sensor (Sb electrodes) was evaluated in two phases: (a) bench (laboratory tests), and (b) field (tests in crops areas). The work carried out in the bench phase, in a static structure, required a system composed of two electronic boxes, a power source and a virtual terminal (panel), connecting the sensor by cables. A method was created for bench measurements: soil samples were collected and placed in a tray soon after, in order to measure the pH in conditions as close as possible to the natural ones. The pH sensor was directly inserted into the freshly collected samples.

The pH values displayed on the panel were manually noted. The soil samples collected showed pH values in a range from 4,0 to 6,0, in order to verify the sensor's accuracy in terms of pH variation in the region's soils. The data from the first samples analyzed using this method created with the sensor and compared to the standard method are shown in Table 1.

Table 1. Soil pH values obtained using the Sb ISE under natural humidity conditions and the standard laboratory method

Sample	Sensor			Standard			Agreement (%)
	Average	SD[1]	RSD[2] (%)	Average	SD	RSD (%)	
Soil reference	4.9	0.07	1	5.3	0	0	92
1	4.6	0.35	8	4.8	0.05	1	96
2	4.3	0.07	2	5.6	0.01	0	77
3	4.1	0.14	3	5.1	0.03	1	80
4	4.2	0.28	7	5.1	0.04	1	82
5	4.5	0.28	6	5.3	0.01	0	85
6	4.1	0.14	3	6.4	0.03	0	64
7	6	0.14	2	7.2	0.01	0	83
8	5.5	0.28	5	6.4	0.04	1	86
9	4.7	0.14	3	4.6	0.04	1	98
10	4.2	0.14	3	5	0.06	1	84
11	4	0.14	4	4.7	0.05	1	85
12	5.55	0.071	1	5.8	0	0	96
13	6.3	0.283	4	7	0	0	90

[1] SD = Standard deviation. [2] RSD = Relative standard deviation.

The values obtained with the pH sensor were lower to those obtained with the standard method. The accuracy of the measurements with the sensor was related to the reference sample, showing a concordance of 91%. The evaluation of the other samples showed that the concordance between the methods ranged from 64 to 98%, demonstrating a low correlation. Regarding the precision of the results, expressed by the relative standard deviation (% RSD), the Veris sensor showed a maximum variation of 8%.

In a scatter plot generated with the values obtained using the sensor and the standard method, it is possible to state that the sensor underestimated the pH values, developing a low correlation ($R^2 = 0.55$) between the methods. However, if sample 6 (4,1; 6,4) is removed from the evaluation, the correlation rises to $R^2 = 0.77$, indicating the presence of an atypical data, also known as 'outlier'. Outliers can be removed using mathematical formulas because they are abnormal values, thus increasing the correlation between the methods used.

The moisture assessment was carried out, aiming to verify a possible interference of the amount of water in the soil during the measurement. Thus, the sensor analyzed four soil samples with increasing additions of distilled water (0–20 mL) in 20 mL of soil. Using the data obtained, a graph was created from one of the samples (Fig. 1). It represents the stabilization of pH values from the addition of 8 mL of distilled water to the soil (20 mL) highlighted by the rectangle. Furthermore, in this moisture condition, the values are consistent with the reference value (98–101%).

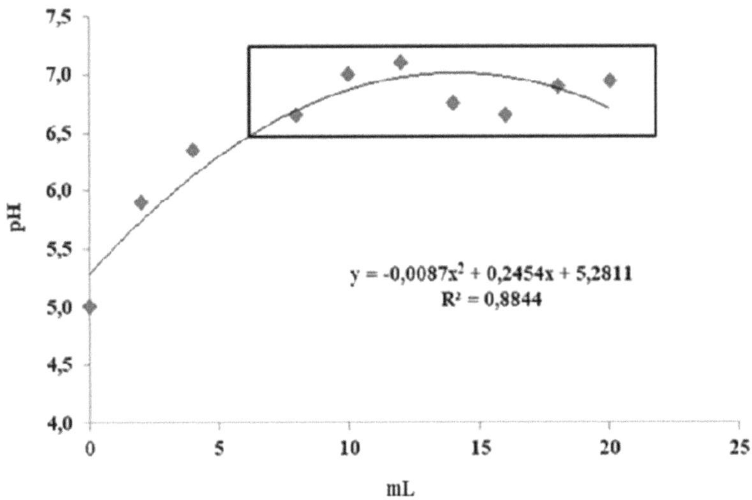

Fig. 1. Evaluation of moisture interference in pH measurements using the sensor.

Due to the results obtained with the determination of pH using the sensor, it was possible to identify the need for moisture in the soil during the measurement, to generate consistent values with the results obtained using the standard method. The presence of 8 mL of water in 20 mL of soil represents a volumetric moisture content of 24%. This classifies the soil in a friable condition, indicating management to be carried out in the field.

More pH measurements were carried out, totaling 50 soil samples, in order to obtain a greater precision of the results. The statistical program "The Unscrambler 3.5®" was used to evaluate the results. Thus, a correlation of 0,91 was observed between the data obtained by Veris on the bench compared to the standard method of analysis. The results obtained using the sensor underestimate the pH values with around 10%, i.e., it indicates a more acidic soil pH comparing to the standard method.

2.5 Evaluation of the pH Sensor in Field

Traditionally, in the southern region of Brazil, soil sampling and analysis have been carried out using methods that ensure the collection of a composite sample by area. Therefore, the property must be initially divided into areas considered 'homogeneous' and approximately 15–20 sub-samples are collected to compose the composite sample that is representative of an area of up to 20 ha. Using the results of laboratory analyses, a characterization of the average levels of soil attributes was done. This will guide the management decisions, which will be identical for the entire area.

The introduction of precision agriculture (PA) and the use of georeferenced samples made possible the identification of the spatial variability of soil chemical attributes (fertility). The georeferenced soil sampling follows a predefined collection pattern, where sampling points are regularly placed over the area. Regular sampling networks, also known as sampling grids, are normally used to guide the systematic arrangement of sampling points. This procedure is used in most areas managed with PA.

Defining the size of the sampling grid is one of the most critical points in the soil sampling process due to the direct correlation between the number of samples collected and the quality of the results obtained [22-24]. Some studies seek to define the best sampling grid in accordance with the spatial dependence of physical and chemical attributes in Brazilian soils [22-25]. However, at least one sample per hectare is needed for the reliable mapping of the main fertility attributes [26]. In general, the definition of these grids dimensions is basically based on economic and practical reasons, affecting the accuracy of the results obtained.

Due to the challenges associated with the spatiotemporal characterization of soil attributes, the sensing technologies make it possible to monitor the agricultural soils at low cost, even with high sampling resolution and satisfactory accuracy [26]. The use of dynamic sensors, which enables the determination of soil attributes on site, allow an increase of the sampling point's number and a better identification of the variability of these attributes, such as pH.

2.5.1 Validation of Dynamic Soil pH Sensor: Comparison with Precision Agriculture Methodologies

The use of sensors in the field and its applicability at scale requires validation regarding the precision and accuracy of soil attributes. This validation was carried out by comparing the values obtained from sensors with the values obtained by standard/reference methods.

The most used soil sampling methodology, in terms of PA system, in southern Brazil is grid collection. It is characterized by sub-samples collected in a radius close to the central point of this grid [27]. This methodology was considered as a standard (for comparison purposes) in a first evaluation of the sensors, as it is the most used. Therefore, the antimony sensor was sought to be evaluated regarding the results obtained by grid collections used in the methodology used in PA.

The selected assessment site was an area of 10,5 hectares used for agricultural production, under the management of a consolidated no-tillage system. The agricultural area is located in the physiographic region of the Planalto Médio (the Middle Plateau), Rio Grande do Sul, Brazil, 28°39′13.99″ S and 53°06′31.23″ W.

In order to carry out the pH measurements using the ISE of Sb and the soil collections from the area for laboratory pH determination, the area was divided into 0,5 ha sampling grids (totaling 22 points) in a diamond pattern for a better identification of the fertility of the area. The center of each grid was highlighted with a stake inserted in a georeferenced point. At each point of the grid, ten sub-samples were collected within a radius of 17 meters from the stake and after homogenization they created a sample analyzed using the standard laboratory method. The pH field measurements were carried out using a Sb ISE sensor, coupled to the commercial equipment (Veris Technologies, Salina, KS, USA). Six measurements were taken at 3 points near the center of the grid. The pH value for each grid was obtained calculating the arithmetical mean of these measurements.

The comparison of the pH obtained using the standard method, with the sensor, using a concordance test. The concordance analysis obtained by dividing the pH value of the Sb ISE sensor with that of the Standard pH and multiplying the result by 100; the closer to 100%, the greater the concordance.

The results obtained showed an average variation of concordance greater than 10% between the methodologies for pH determination. Despite the initial results showing low correlation, it should be noted that the soil collection methodology followed a grid sampling standard (representing the mean value of the grid). Although close, the readings were performed in different places from the standard method, where sub-samples with spatial variability (vertical and horizontal) of the soil were collected. This factor could have contributed to the difference in pH values obtained between the sensor and the standard method, resulting an inadequate concordance.

The homogenization of the soil of the 0–10 cm layer for standard determination led to an average pH value for this depth layer, lower than the pH of the most superficial layer (3–5 cm), which is the location of the (Sb) ISE sensor measurements. According to [28], the spatial variation of soil pH can occur due to different soil management and, according to [29], due to the application of agricultural correctives only on the surface and without incorporation.

However, the performance evaluation of the Sb ISE sensor carried out on bench (Section 2.4) demonstrated the influence of soil moisture during the measurements. Thus, a new field evaluation was carried out in order to verify if the moisture influenced the low concordance obtained between methods. The new evaluation consisted of the standard determination of pH in soil samples collected in the exact measurement point of the field equipment using the (Sb) ISE sensor.

A second evaluation, consisting in two readings, was carried out with the Sb ISE sensor in a location close to the stake (less than 1.5 m radius from the central point of the grid). The first pH reading was carried out under the conditions of natural soil moisture, and then, 75 mL of distilled water was added in the same location and a new reading was carried out a few moments later. For comparison, a soil sample was collected at the exact location where the Sb ISE sensor touched the soil for standard laboratory analysis.

The comparison of the pH obtained with the Sb ISE sensor using the standard method (Table 2) shows mean correlations close to 100, indicating a good concordance between methods.

Table 2. pH values from the exact location of the reading carried out by the Sb ISE set, obtained using standard laboratory methodology, compared to the pH results obtained by the Veris® U3 equipment's antimony ISE set

Point	pH lab. location	Sb ISE set (no water addition)	Agreement	Sb ISE set (with water addition)	Agreement
1	6.81	7	103	7.13	105
2	5.82	6.05	104	6.37	110
3	6.34	6.56	103	6.91	109
4	6.29	6.86	109	6.75	107
5	6.53	6.67	102	6.82	104
6	6.72	6.77	100	6.76	101
7	6.37	5.63	88	5.65	89
8	6.56	6.65	101	6.79	104
9	6.41	6.19	96	6.27	98
10	5.78	6.29	108	6.12	106
11	6.6	6.87	104	6.76	103
12	6.4	6.73	105	6.6	103
13	6.32	6.74	107	6.85	108
14	6.86	6.68	97	6.76	99
15	6.62	6.91	104	6.97	105
16	6.37	6.75	106	6.85	108
17	6.09	6.5	107	6.52	107
18	6.28	6.45	102	6.54	104
Average			103		104

In this evaluation, the pH values obtained by the Sb (ISE) in the field did not depend on the soil moisture as there was no significant difference between measurements in the soil under natural moisture condition and with the addition of 75 mL of water. It should be highlighted that in both cases, the soil moisture during the measurements was in a friable condition, suitable for carrying out agricultural operations. Similar results were obtained by [3] who did not observe significant differences in pH readings through Sb electrodes, within the typical moisture range of field soils. However, Silva [30] observed an increase in pH units as soil moisture increased.

These results show the good efficiency of the Sb ISE sensor in estimating the active acidity (soil pH) and that the variations in results observed in previous evaluations were caused by the natural spatial variability of the soil in relation to its chemical, physical and biological parameters. Soils are heterogeneous and their properties resulted mainly from pedogenetic processes can vary widely. Variations of these parameters can occur both in the vertical direction of the soil profile and horizontally on the area's surface [31].

2.5.2 Data Filtering

A wide variation in the amplitude of concordance was obtained for all methodologies, although the mean values of concordance are closer to 100, compared with the analysis made from the collection of soil from the 10 cm layer. Therefore, graphs were generated between the data from the Sb ISE set and the laboratory data (standard methodology) of the exact location where the sensors touched the ground, seeking to verify the coefficient of determination in relation to the observed values. Figure 2 presents the graphs referring to the comparisons between the methodologies and the standard method.

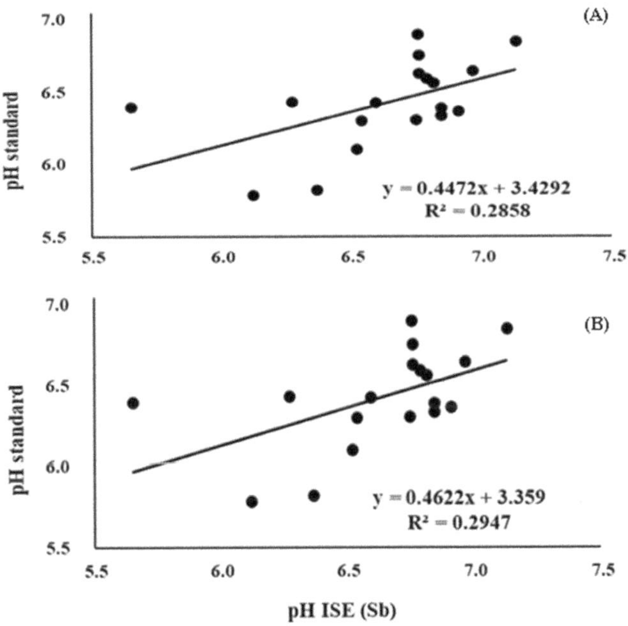

Fig. 2. pH determination by the antimony ion selective sensor set, without addition of water (A), and with addition of 75 mL of water (B), compared to the laboratory result of the sensors' exact location contact with the ground.

The pH results obtained by the equipment's sensors and the soil analyses obtained using the standard methodology present low values of R^2. Eitelwein et al. [33] present a correlation greater than 0.8 R^2, found by Veris enterprise for soil pH determinations between the SpeedRead™ pH sensors system and the laboratory standard. In this case, 14 experimental fields were evaluated in 5 different American states (Kansas, Nebraska, Iowa, Illinois and Wisconsin), always working with a dense data grid.

One of the factors that could have led to unrepresentative R^2 values is the number of points sampled. The main advantage of using sensors is the ability to map with a dense grid of samples, which did not happen for this evaluation. Thus, with a low amount of observed values, the possibility of discrepancy between values is greater, decreasing its significance. Eitelwein et al. [33] observed in soil pH spatial distribution maps determined by different methods, that when maps with greater number of reading are generated the results showed a greater similarity, demonstrating that there is a gain in information quality as the sample density increases and thus justifying the low correlation of the study.

Using equipment similar to the one in the present study, having the same operating principle and antimony electrodes, Eitelwein et al. [33] observed that there were different scales of variability in pH when comparing soil samples in the laboratory and analyzes in the field. The authors attributed three reasons that may have interfered with the readings: the influence of the basic pH (8.25) of the water used to wash the electrodes in the field; the variable soil temperature; and the ion concentration dilution in the residual water from washing the electrodes.

Given these doubts, new tests were carried out, aiming to evaluate the same chemical parameter of the soil with the Sb ISE sensor, in order to verify factors and hypotheses that justify the results found, in addition of seeking new methodologies and ensuring accurate results generated by the sensor.

Evaluations were carried out in two experimental areas, targeting a greater amount of data and testing the efficiency of the Sb ISE in performing readings in different soil layers: the crop area (previously evaluated) and the parcels of an (long term) experiment which evaluates the results of lime and gypsum doses. Trenches 0.20 m deep were dug in both evaluated areas and in each trench the pH value was determined with the Sb ISE and using the standard method. The readings were taken at the average depth of the layers (0–0.05; 0.05–0.10; 0.10–0.15; and 0.15–0.20 m). The values for the layers from 0.0 to 0.10 m and from 0.10 to 0.20 m were also used. For the layers of 0–0.10 m and 0.10–0.20 cm, obtained in the equipment, the arithmetical means of the values were calculated for the layers 0–5 cm; 5–10 cm for the 0–10 cm layer, and 10–15 cm; and 15–20 cm for the 10–20 cm layer.

As the antimony (ISE) is fixed to the equipment, it takes readings to a depth of 0.05 m, it was removed to take the readings in different layers. The equipment was placed in manual mode to take the readings and to wash the electrodes. The equipment was positioned close to the trench and the antimony ISE was inserted in the central position of each layer, waiting about 10 seconds until the values stabilized.

The pH determination through the standard method was carried out in soil samples extracted from the trench wall, where the pH was determined with the antimony ISE. With the aid of a spade, 5 cm thick soil slices were extracted and divided into layers (0–0.05; 0.05–0.10; 0.10–0.15; and 0.15–0.20 m) for further analysis using the standard method.

In pH laboratory determination, the samples are processed in duplicate in order to maintain the accuracy and quality of the results. When these values do not consist with each other, it is redone. The use of two data filtering forms (methodologies) was proposed based on this methodology and understanding that a classification of the determinations carried out by the two sensors that compose the equipment (ISE Sb 1 and ISE Sb 2) should be applied.

The first filtering method evaluated was the exclusion of values for which the sensors showed a difference greater than 0,5 pH units. The other filtering method proposed was the concordance between the electrodes (Sb ISE 1 and Sb ISE 2) that compose the sensor, where the pH values were discarded when the results differed from the range of 95–105% of concordance.

The data acquired in the field were subjected to comparisons with the values obtained using the laboratory method. This comparison was carried out using linear regression. From the unfiltered data, graphs were created for each depth (Fig. 3).

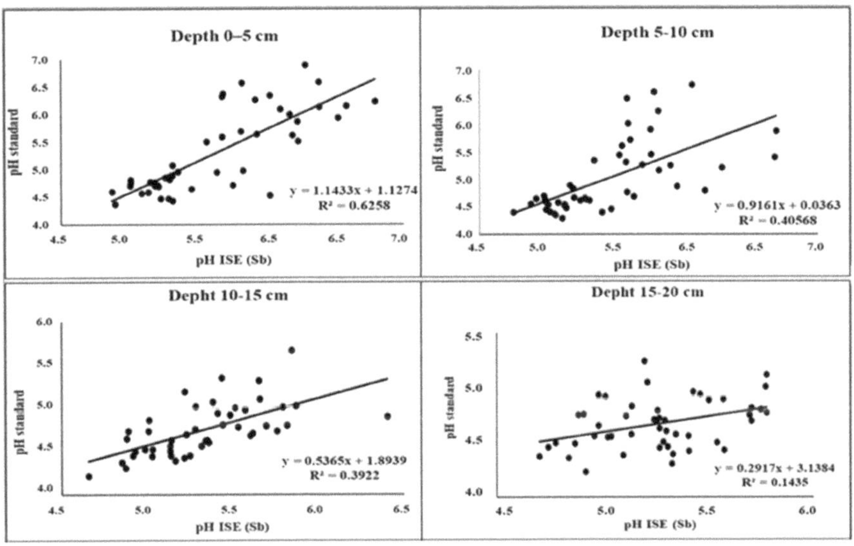

Fig. 3. Comparative graphs without filtering for pH values in the field (Sb ISE) and in the laboratory (pH - standard), prepared for each layer evaluated.

The analyses show an inversely proportional relationship between depth and the correlation coefficient (R^2), i.e., the shallower the sampled layer, the greater its data reliability. The trend was maintained, when data filtering was carried out using the methodologies described above, showing the sensors' best performance in more superficial layers.

The results presented confirm the good performance of the pH reading system in the 0–5 cm layer, which is also the depth range during the operation of the equipment under its working conditions.

Analyzing the entire set of data generated by the evaluations in different soil layers after performing the filtering process (Fig. 4), it can be observed that the R^2 results were superior to those obtained in the grid analysis (Fig. 2).

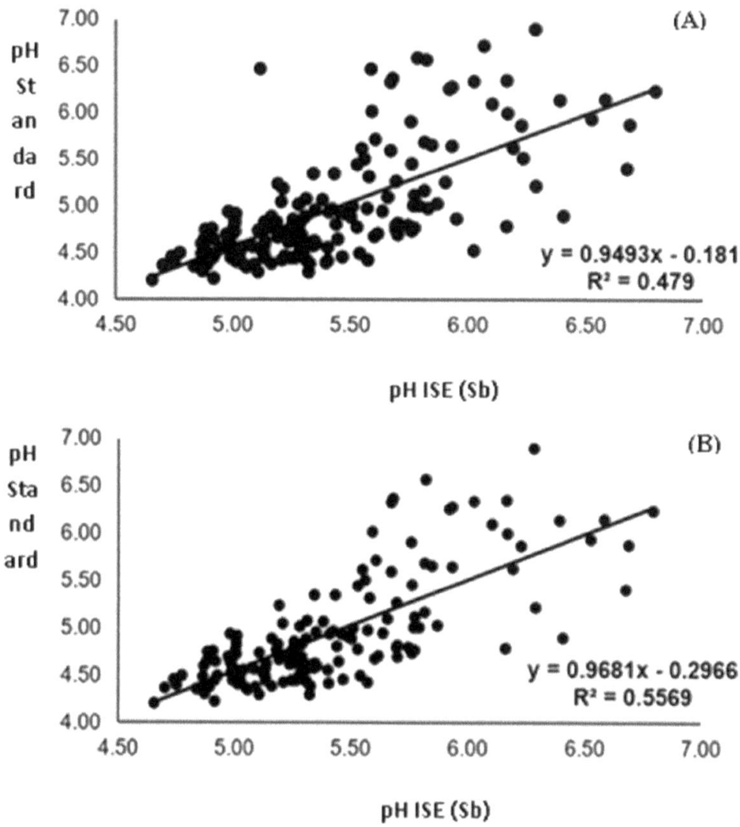

Fig. 4. Comparative graph containing data from all depths, obtained with two data treatment methods: (a) using the difference in values, (b) between their concordances, between pH values in the field (pH ISE-Sb) and in the laboratory (pHLab).

The R^2 results improvement is possibly influenced by a greater number of samples, which in this case had a total of 192 duplicate samples. It can be observed that filtering the data by the concordance method showed better results, although it excludes a larger number of data comparing to other method analyzed.

In field tests, Adamchuk et al. [34] found an R^2 of 0.66 significance between antimony electrodes comparisons and laboratory tests, under normal working conditions with the ISE-Sb set mounted on the equipment, establishing an ideal contact condition ground × electrode. In more recent studies, Adamchuk et al. [32] present an index with greater significance (0.8 R^2).

The results obtained from the different performed evaluations indicate that the Sb ISE set obtains results more similar to the standard method when it analyzes the soil layer of 0–0.05 m. It was also highlighted that the data filtering methodology through concordance provided higher R^2 values.

In order to observe the sensors' performance results under the best conditions tested, the graph below was generated (Fig. 5). Through the result of R^2 of 0.8596, it demonstrates the efficiency in determining the active pH of the soil in comparison with the standard method.

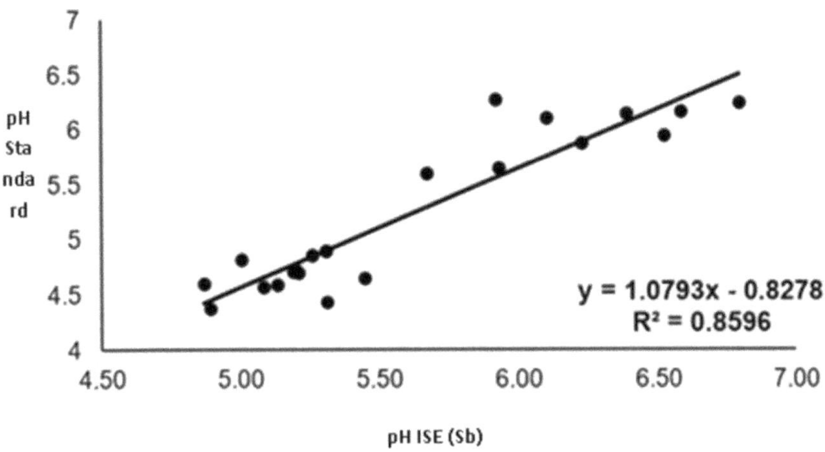

Fig. 5. Comparative graph of the 0–5 cm layer between pH values in the field (pH Sb ISE) and in the laboratory (pHLab) for those who received readings and soil collection at the same point.

2.6 pH Fertility Maps

After measuring the antimony ISE efficiency in determining the active pH in the soil, an evaluation was carried out with the aforementioned sensor in the experimental area of the field described above. This evaluation met the equipment manufacturer's recommendation for use, which involves crossings in the field every 15 m, with pH sampling point collection every 65 m in the crossing line. Therefore, an active soil pH data was collected every 975 m^2 (65 × 15 m).

Making a comparison with precision agriculture techniques that uses grid sampling, for the southern region of Brazil, the sampling grids that have been adopted range from approximately 10000 m^2 (100 × 100 m) to 30000 m^2 (173 × 173 m). This definition of the sampling grid size is generally based on economic and practical criteria, often neglecting geostatistical principles, which could not demonstrate all the existing variability in the area [27]. Analyzing precision agriculture techniques in the United States of America, Kerry et al. [35] describes that sample grids of 100 × 100 m (one sample per ha) are insufficient to detect different scales of variation in soil attributes.

The sensors use can increase the number of points analyzed per area, reducing the sampling grid size, improving the variability observation of soil chemical attributes, such as active pH, as presented in Fig. 6.

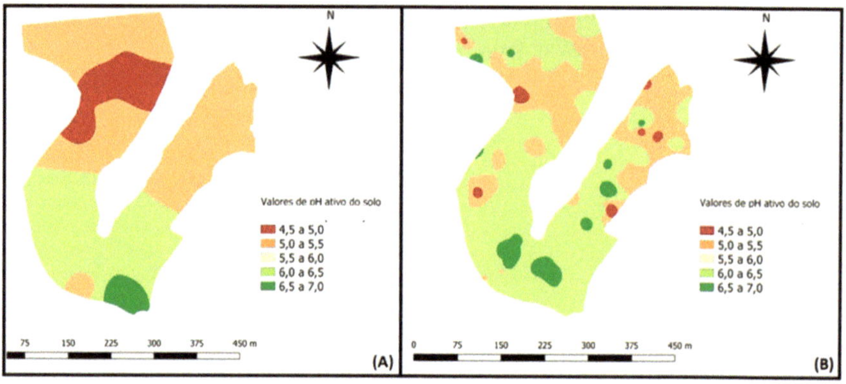

Fig. 6. Active soil pH map generated with a 0.5 ha sampling grid (density of 22 sampling points) and laboratory analysis (A), and soil pH map generated with data collected by the Sb sensors set, crossings at a distance of 15 m and readings taken every 65 m (density of 103 sampling points) (B), for an agricultural area of 10.5 ha.

Figure 6 presents a better detailing of the soil active pH spatial variability, when a higher density of samples per area is used. The results corroborate what was described by [24], who state that the sampling grid size reduction increases the accuracy in characterizing the soil attributes spatial variability through thematic maps and that soil fertility maps become more dissimilar as the sampling grid size increases.

The generation of maps using sensing allows a better detailing of the fertility map in relation to the PA carried out with conventional soil analysis, in addition of being sustainable, less costly and faster [26]. The sensors efficiency in generating more accurate fertility maps can replace laboratory analysis; however, the soil type and the sensor calibration must be observed [20]. With the data collection for this work and as Benhossi & Reynaldo [20] claim, the greater detail of the maps generated by the (Sb) ISE sensor can be observed in concordance with the standard method, allowing agronomic management decisions.

3. Conclusions

For proper functioning, the Sb electrodes must be stored properly or immersed in a solution of 50% KCl 1 mol L^{-1} and 50% buffer of pH 4,0 in order to hydrate. It is essential to calibrate the sensor before taking the measurements, as well as cleaning and drying the electrodes between their immersion in the different buffer solutions. The pH measurements carried out by the sensor composed of Sb electrodes require soil moisture to obtain values consistent with the standard laboratory method for agronomic indications; however, it can still underestimate the values.

The field-level evaluation of the Sb electrodes indicates that the reading carried-out by the sensor obtained a result consistent with the laboratory analysis using the standard method. The pH values obtained by the Sb ISE, under the evaluated moisture conditions, the ISE Sb sensor demonstrated adequate efficiency and precision for the results of active acidity (soil pH) in the 0–5 cm layer.

The data statistical analysis was made from the pH results obtained, aiming to exclude 'outliers', whose variation was greater than 0.5 pH unit, or if the concordance between the methods was outside the range of 95–105%, which increases the accuracy of the results.

The pH data collection obtained with the Sb ISE allows a greater sampling density, when compared to soil collection in grids ≥1 ha. After proper calibration and data filtering, it provides the generation of more detailed maps of the area, in addition to being more sustainable and faster.

References

1. Molin, J.P. and Rabello, L.M. (2011). Estudos sobre a mensuração da condutividade elétrica do solo. *Eng.* Agrícola. *Jaboticabal*, 31(1), 90–101. jan./fev.
2. Gao, F.L., He, Q.S., Zhang, Y.D., Hou, J.H. and Yu, F.H. (2021). Effects of soil nutrient heterogeneity on the growth and invasion success of alien plants: A multi-species study. *Frontiers in Ecology and Evolution,* 8, 525.
3. Schirrmann, M., Gebbers, R., Kramer, E. and Seidel, J. (2011). Soil pH mapping with na On-The-Go Sensor. *Sensors*, 11, 573–598.
4. Sociedade brasileira de ciência do solo (SBCS) (2016). Manual de adubação e calagem para os estados do Rio Grande do Sul e de Santa Catarina. Comissão de Química e Fertilidade do Solo. SBCS-NRS: Brasil (10ª Edn.) Porto Alegre, 376 pp.
5. Villafuerte, A., Valadares, F.G., Campolina, G.F. and Silva, M.G.P. (2018). Agricultura 4.0: Estudo de inovação disruptiva no agronegócio brasileiro. *ISTI/SIMTEC, Aracaju/ SE.*, 9(1), 150–162.
6. Massruhá, S.M.F.S. and Leite, M.A. de A. (2017). AGRO 4.0: Rumo à agricultura digital. JC na Escola Ciência, Tecnologia e Sociedade: Mobilizar o Conhecimento para Alimentar o Brasil. Embrapa Informática Agropecuária.
7. Silva, J.M.P. and Cavichioli, F.A. (2020). O uso da agricultura 4.0 como perspectiva do aumento da produtividade no campo. *Interface Tecnológica*, 17, (2).
8. Kitchen, N.R. (2008). Emerging technologies for real-time and integrated agriculture decisions. *Computers and Electronics in Agriculture*, 61, Elsevier B.V.
9. Ferraz, M.N. (2009). Investigação preliminar sobre o uso do Veris P4000 Soil Force/ Eca/Spectral Profiler para predição de densidade de carbono em solos australianos. Universidade de São Paulo Escola Superior de Agricultura "Luiz de Queiroz" Departamento de Engenharia Rural – Setor de Máquinas Agrícolas.
10. O'Halloran, J. (2020). *Soil Mapping Technologies. Precision Agriculture in Vegetable Systems*. Department of Agriculture and Fisheries. Jan.
11. Manitoba Agriculture, Food and Rural Initiatives. (2008). *Soil Management Guide*.
12. Malavolta, E., Vitti, G.C. and Oliveira, S.A. (1997). Avaliação do estado nutricional das plantas; princípios e aplicações. *Piracicaba, POTAFOS*, 319 pp.

13. Bohnen, H., Meurer, E.J. and Bissani, C. (2006). Solos ácidos e solos afetados por sais. *In:* Egon José Meurer (Ed.), *Fundamentos de Química do Solo* (3° Edn.) Porto Alegre: Evangraf.
14. Taiz, L., Zeiger, E., Møller, I.M. and Murphy, A. (2017). *Fisiologia e Desenvolvimento Vegetal.* (6ª Edn.). Porto Alegre: Artmed.
15. Gatelli, V. (2014). Calagem: Importância e aplicação. Unifertil Universal de Fertilizantes S.A. Maio, 006(Ano 03).
16. Salet, R.L. (1998). Toxidez de alumínio no sistema plantio direto. Tese de doutorado, UFRGS. Porto Alegre.
17. Silva, L.S. da, Gatiboni, L.C., Escosteguy, P.A.V., Scivittaro, W.B. and Spagnollo, E. (2016). Métodos de análise de solo e de plantas. Manual de calagem e adubação para os Estados do Rio Grande do Sul e de Santa Catarina/Sociedade Brasileira de Ciência do Solo-Núcleo Regional Sul. – [s. l.]: Comissão de Química e Fertilidade do Solo - RS/SC. (11ª Edn.), 51–62.
18. Tedesco, M.J., Gianello, C., Bissani, C.A., Bohnen, H. and Volkweiss, S.J. (1995). *Análises de solo, plantas e outros materiais* (2ª Edn.). Porto Alegre: UFRGS/ Departamento de Solos, 174. (Boletim Técnico, 5).
19. Adamchuk, V.I. and Lund, E.D. (2008). On-The-Go Mapping of Soil pH Using Antimony Electrodes. Paper No. 083995; ASABE: St. Joseph, MI, USA.
20. Benhossi, G. and Reynaldo, E.F. and Machado, T.M. (2021). Differences between laboratory and sensor analyses for soil attributes. *Pesq. Agropec. Trop., Goiânia*, 51.
21. Bates, R.G. (1961). Electrodes for pH measurement. *J. Electroanal. Chem.*, 2, 93–109.
22. Cherubin, M.R., Santi, A.L., Eitelwein, M.T., Ros, C.O. and Bisognin, M.B. (2014). Sampling grids used to characterise the spatial variability of pH, Ca, Mg, and V% in Oxisols. *Rev. Ciênc. Agron.*, 45, 659–672.
23. Cherubin, M.R., Santi, A.L., Eitelwein, M.T., Menegol, D.R., Da Ros, C.O., Pias, O.H.C. and Berghetti, J. (2014). Eficiência de malhas amostrais utilizadas na caracterização da variabilidade espacial de fósforo e potássio. *Ciência Rural*, 44, 425–432.
24. Cherubin, M.R., Santi, A.L., Eitelwein, M.T., Amado, T.J.C., Simon, D.H. and Damian, J.M. (2015). Dimensão da malha amostral e a acurácia na caracterização da variabilidade espacial de fósforo e potássio. *Pesq. Agropec. Bras.*, 50, 168–177.
25. Nanni, M.R., Povh, F.P., Demattê, J.A.M., Oliveira, R.B.D., Chicati, M.L. and Cezar, E. (2011). Optimum size in grid soil sampling for variable-rate application in site-specific management. *Scien. Agric.*, 68, 386–392.
26. Molin, J.P. and Tavares, T.R. (2019). Sensor systems for mapping soil fertility attributes: Challenges, advances, and perspectives in Brazilian tropical soils. *Engenharia Agrícola, Jaboticabal*, 39(Special Issue), 126–147.
27. Cherubin, M.R., Santi, A.L., Pias, O.H.C., Eitelwein, M.T., Basso, C.J., Della Flora, L.P. and Melo, J.D. (2016). Amostragem de solo na agricultura de precisão. *In*: Santi, A.L., Sebem, E., Giotto, E., Amado, T.J.C. (Eds.), *Agricultura de precisão no Rio Grande Do Sul* (1ª Edn.). Santa Maria. Cespol. 309 pp.
28. Teixeira, I.R., Souza, C.M. de, Borém, A. and Silva, G.F. da. (2003). Variação dos valores de pH e dos teores de carbono orgânico, cobre, manganês, zinco e ferro em diferentes profundidades em argissolo vermelho-amarelo, sob diferentes sistemas de preparo do solo. *Bragantia, Campinas*, 62(1), 119–126.

29. Gatiboni, L.C. (2003). Alterações nos atributos químicos de solo arenoso pela calagem superficial no sistema plantio direto consolidado. *Cienc Rural*, 33, 283-290.
30. Silva, F.C. de S. (2013). Mensuração direta do pH do solo em campo por meio de eletrodos íon-seletivos. Dissertação de mestrado apresentada no curso de Engenharia de Sistemas Agrícolas na Escola Superior de Agricultura Luiz de Queiroz, Universidade de São Paulo, Piracicaba.
31. Artur, A.G., Oliveira, D.P., Costa, M.C.G., Romero, R.E., Silva, M.V.C. and Ferreira, T.O. (2005). Variabilidade espacial dos atributos químicos do solo, associada ao microrrelevo. *R. Bras. Eng. Agríc. Ambiental*, 18, 141–149.
32. Adamchuk, V.I., Lund, E.D., Sethuramasamyraja, B., Morgan, M.T., Dobermann, A. and Marx, D.B. (2005). Direct measurement of soil chemical properties on-the-go using ion-selective electrodes. *Comput. Electron. Agric.*, 48, 272–294.
33. Eitelwein, M.T., Trevisan, R.G., Colaço, A.F., Vargas, M.R.I. and Molin, J.P. (2016). On-the-go measurements of pH in tropical soil. *Proceedings of the 13th International Conference on Precision Agriculture Monticello*. International Society of Precision Agriculture.
34. Adamchuk, V.I., Lund, E.D., Dobermann, A. and Morgan, M.T. (2002). *On-the-go Mapping of Soil Properties Using Ion-selective Electrodes*. ResearchGate, Dez.
35. Kerry, R., Oliver, M.A. and Frogbrook, Z.L. (2010). Sampling in precision agriculture. *In*: Oliver, M.A. (Org.). *Geostatistical Applications for Precision Agriculture*. Heidelberg: Springer-Verlag, 35–63.

CHAPTER

11

Image Processing-based via Agridrones for Agriculture Fields

Awais Khan Jumani[1*][0000-0001-9468-0446], Asif Ali Laghari[2][0000-0001-5831-5943], Rashid A. Laghari[3][0000-0002-9710-7538], Abdullah A. Khan[2][0000-0003-2838-7641], Vania V. Estrela[4][0000-0002-4465-7691], Lin Teng[5][0000-0001-6572-4032], Andrey Terziev[6][0000-0002-7069-367X], G.G. de Oliveira[7][0000-0001-7164-2068] and Yuzo Iano[7][0000-0002-9843-9761]

[1] Department of Computer Science, ILMA University Karachi, Sindh, Pakistan
[2] Department of Computer Science, Sindh Madressa-tul-Islam University, Karachi, Sindh, Pakistan
[3] College of Mechanical and Electrical Engineering, Nanjing University of Aeronautics and Astronautics, Nanjing 210016, China
[4] Department of Telecommunications, Federal Fluminense University (UFF), RJ, Brazil
[5] Software College, Shenyang Normal University, Shenyang 110034, China
[6] TerziA, Bul "Vazkresenie" 103, 1504, Sofia, Bulgaria
[7] Faculty of Electrical and Computer Engineering, UNICAMP, Campinas, Brazil

Some of the best sustenance resources come from agriculture, and most nations' economies hinge on it. Technology enacts a vital task in every expertise field. This book part debates the four agribusiness revolutions, each with an associated technology level, focusing on the latest applied science and open issues. Farming can benefit from "information and communication technologies" (ICT) like image processing, cloud/fog support, Blockchain, artificial intelligence (AI), aviation, and the "Internet of Things" (IoT) for a myriad of gains, e.g., meeting farmers' predictions and expectations, accomplishing fungi management, and precluding pest attacks that affect one-third of the planet's annual harvest. What will AI mean for the fourth revolution, Agriculture 4.0 (Agri4)? Agriculturalists are pressured daily to boost their yield by nations demanding it and to furnish proper food worldwide. Farmers already exploit such ICT types in their harvest. A top-grade agribusiness high-tech solution is the agridrone (alias agri-UAV), to safeguard living organisms and soil state, identifying matters from

* Corresponding author: awaisjumani@yahoo.com

photogrammetry and remote sensing (RS) imageries and making husbandry more impressive and productive without losing farmers' hope in the time crops. This chapter pins down the up-to-the-minute Agri4 applications, emphasizing agridrone technology. Agriculturalists can cut labor with agridrones to protect their crops and diminish fertilizer utilization while easily analyzing the land's imageries. Agridrones can easily follow plants' diseases and stress, delivering possible solutions without damage. Furthermore, they can render digital scenes to visualize crops' health while assessing the ground and creating a 3D field map to inform farmers when planning seed-planting patterns.

1. Introduction

Agriculture arranges for some of the best alimentary articles, paramount to most nations' markets. In all spheres, technology is indispensable. This chapter covers four agribusiness revolutions and their respective technology levels, emphasizing present-day studies and unanswered demands. "Information and communication technologies" (ICT), notably "image processing" (ImP), "artificial intelligence" (AI), Blockchain, avionics, and the "Internet of Things" (IoT), can sustain many farming practices that satisfy agriculturalists' predictions and expectations.

ImP treats an image digitally, performing certain operations to enhance imagery or acquire valuable facts from imaging inputs, e.g., a still-picture frame, video, or voxels. The output may contain image-related characteristics. The ImP system normally uses set techniques to handle images by identical symbols or representations. ImP is a fast-expanding technology in today's world for engineering, applied math, computer science, and graphic projects, establishing the research core. ImP mimics the "human visual system" (HVS) system within digital imaging equipment, most precisely. The HVS does not perceive the real world correctly, with different frequency bands as cameras with more sound effects and frequency bandwidths. There will be substantial distinctions between human and digital reconnaissance detectors, particularly in the translation/transforming stages. Image editing should be scientifically handled to replicate and verify human discoveries [1] via documenting, reporting, and treating for appropriate control images.

ImP can get or extract valuable information from agronomic imageries, where the input is a static image or video frame, and the outcomes may partake in image-related features. Usually, an ImP system involves predefined signal processing techniques to handle 2-D, 3-D, 4-D, etc., images. Preprocessing functions embrace those procedures typically necessary before data breakdowns and information extraction, usually classified as radiometric or geometric adjustments. Correcting sensed data anomalies and undesired (or ambient) noise and correctly representing the perceived radiation reflected/emitted and recorded includes radiometric correction. Figure 1 portrays scientific compensation for Earth-sensing geometry mismatches and translating Earth's surface data to real-world coordinates (i.e., latitude and longitude). Combining ImP, automation, drones, and AI leads to "precision agriculture" (PA) or "smart farming" (SF).

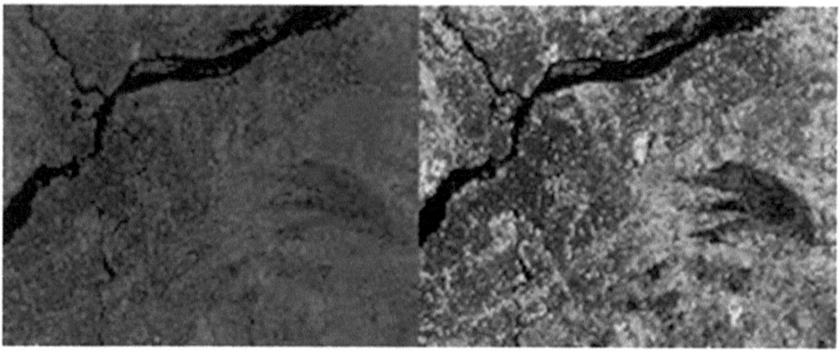

Fig. 1. Geometrically sensing images [2].

The second set of ImP functions enhances or normalizes imagery's look to support visual discernment and analysis. Image transformations are associated with image-enhancing techniques, albeit typically combining information processing from several spectral bands. Image-enhancing operations usually operate simultaneously on a single data channel. Merging and translating original bands to 'new' images are done in arithmetic operations (i.e., subtracting/adding (combining) elements, multiplying (scaling) images, and splitting a scene into sub-images).

Figure 2 shows that digitally identifying and classifying pixels in the data relies on image classification and analysis procedures. Classification typically takes place in multi-channel sets of information (A). Each of them, established on the statistical features of the pixels' brightness, is allotted to a distinctive image class or theme (B). Digital categorization is based on several methods. ImP extracts evidence or elaborates helpful traits from imagery. Signal processing procedures aid in obtaining liable imageries' analyses, viz., a harvest image or

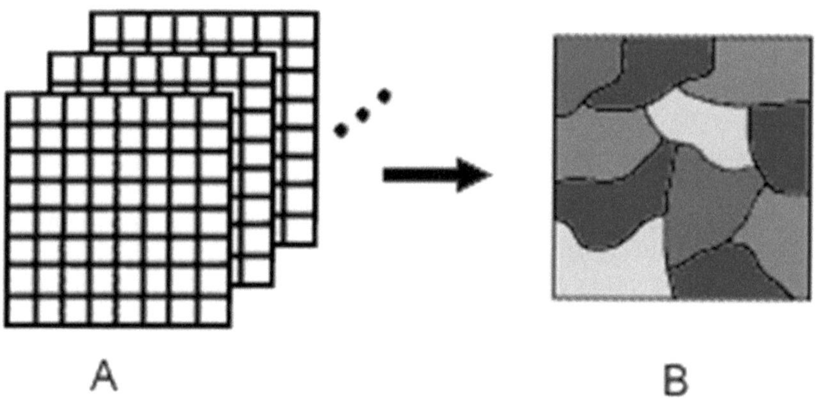

Fig. 2. Digital identification/organization of images [2].

a terrain's associated topographies. ImP is now amid fast-growing technology, constituting engineering and IT control's primary study field. There are three steps concerning ImP:

- Detection tool development to import or gather an image,
- The image can be explored and manipulated, and
- A picture can be scrutinized to build a better-quality version.

The gathered or generated data are generally crude, implying that they are not functional for many potential application reasons. One must appraise, process, and then utilize data [2].

2. Image Processing Steps

ImP has a few steps where problems can be identified through image acquisition and processed in parallel, whose actions appear in Fig. 3. These are well elucidated below.

Image Acquisition: First, ImP detects particular elements, such as an element from an interior layout of an authentic scenario, intended for processing, collecting, accumulating, printing, and visualizing data. Image acquisition may vary from a modest smartphone to cutting-edge equipment.

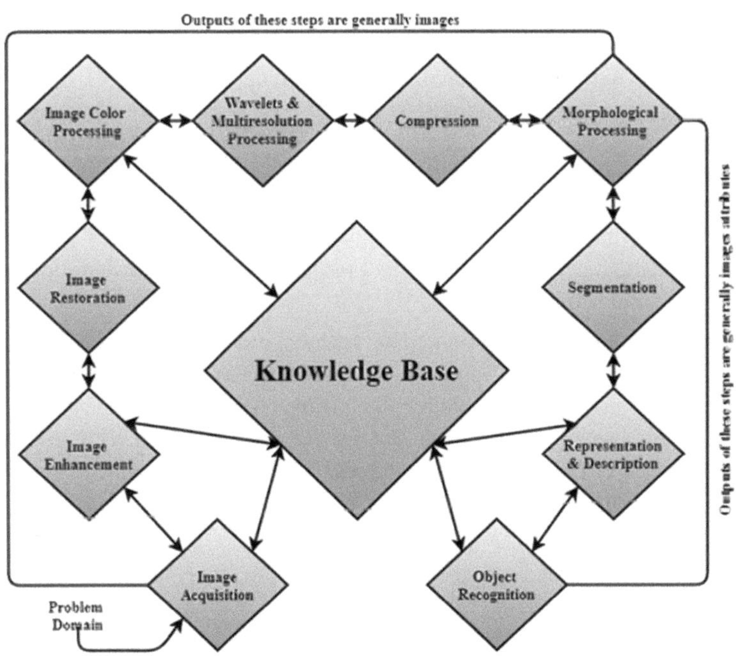

Fig. 3. Framework illustrating Image processing steps [3].

Image Enhancement: It addresses more displaying/rendering modalities or numerous analytical outcomes, viz., muting videos, sharpening, or switching other visualization types to facilitate essential characteristics identification.

Image Restoration: This service transforms and simplifies a noisy or defective image allowing usage, improvements, and measurements from a spoiled image. Image restoration tactics aim to minimize noise and allow for decision-making in, e.g., action blurring, sounding, and camera focus.

Coloring Image Processing: Color, the processing of images, needs knowledge of light physics and vision physiology, e.g., how the HVS perceives hue and handles quotidian properties, construction materials, repasts, locations, and time. Color also helps the image separation process.

Wavelets and other Multiple Processing Solutions: When decorating the image environment, one may utilize different tool sets established on the topography. Wavelets aid in representing ad examining signals efficiently. The wavelet transform facilitates many image solutions' investigations.

Image Compression: Compression shrinks information storage space and broadcasting expenses. Processes may also gain knowledge, get visual advantages, and comprehend other data impacts.

Morphological Processing: Images may be extracted for shape representation and description. Erosion and dilation amount to the staples of morphological processes whose processing outputs (imagery features) also appear in the block diagram above.

Character Recognition: "Optical character recognition" (OCR) amounts to obtaining knowledge or computer-readable text from scanned images and turning this material into accessible records, whether from a paper, bill, bank statement, income receipt, business card, printed document, or email. Digitalized documents can be sorted, searched, and saved by electronic systems involved in machine conversions and online displays. Text-to-voice conversion, key data collection, OCR, and text mining often produce immense knowledge, patterns, and other computer outcomes.

Representation and Description: The segmentation stage output is typically raw pixel data, representing either a zone's border or all this region's points. Representation and portrayal practically always follow. Previously to turning raw facts into a usable presentation for further processing, many more factors must be considered. Attributes that result in quantitative information or are crucial for individualizing one class of items from another may be extracted via description.

Segmentation: It breaks up imagery into smaller, more controllable pieces, albeit notoriously challenging. A robust segmentation entails identifying distinct objects more quickly.

Knowledge Database (KNB): Knowledge may be as easy as identifying areas in a picture where the chosen information is, thus dropping the time and effort indispensable to find it. For example, a materials inspection problem's knowledge base might include an interconnected list of all relevant prospective faults, or a program to tag changed regions could encompass a high-resolution satellite depictions' databank of an area [3, 59–61].

3. Image Processing in Agriculture

ImP has imaging tools to amass and scrutinize images that farmers can utilize efficiently. In particular, a logical management strategy spares time and capital to automate analyses. Agricultural ImP is an utmost creative and important topic. Figure 4 exhibits a comprehensive perspective of plant recognition. Research becomes paramount because several subjects are connected with it [4].

Fig. 4. Plant recognition from drone analysis [4].

A. **Population is growing:** In the upcoming decades, the global number of inhabitants will possibly augment by 33%. Populace estimates may reach about 10 billion by 2050 and 11.2 billion by 2100. Meanwhile, the planet's diet undergoes modifications because of ever-changing demographics that call for valuable animal/vegetal protein. Urbanization, rising wages, and natural population expansion make up for this predisposition [5].
B. **Urbanization is increasing:** Macro-urbanization may conduct a 2.4 billion people net increase in towns and cities up to 2050. Urbanization encourages infrastructural upgrades such as cold chains that allow commerce in perishable commodities. As a share of comprehensive nutritional shifts, it also lifts wages and augments demands processed vegetal/animal/minerals foodstuff and other resources. The planet's pastoral land is becoming unfit for production: 25% of all farmland is already considered severely degraded by some dealings, while a further 44% is moderately or mildly declining.

Over 40% of the Earth's rural population subsists in water-scarce regions with severe apprehension. The land has long been assumed as an infrequent resource. Formerly, old farms would be replaced with new uncultivated land. Such spaces are scarce now, and what is recurrently left cannot be cultivated sustainably. Land shortages prompted intensive tiny farm usage, reduced output per person, and more landlessness, all contributing to rural impoverishment. Agronomy is a main source and indirect victim of agronomic degradation, with innumerable elements prompting this process from multiple standpoints [6]. Besides other factors, over-cultivation, animal over-grazing, harsh rain, terrain exposure by strong airstreams, poorly managed water flow, poorly done agricultural rotations, and anthropogenic activities all erode the soil. Unbalanced application of fertilizers to restore yield leads to nutritional imbalance. Agricultural concerns drive around 80% of the planet's deforestation. Deforestation ultimately causes hydric supplies destruction. Though irrigation expertise has enriched terrestrial and hydric usage efficiency, increasing populations are genuinely alarmed about hydric resources' safety and availability. The required irrigation management in emergent nations alone will be $1 trillion by 2050 [7].

Moreover, rural populations are maturing quickly, bringing significant penalties for the labor force, production arrangements, land tenure, rural/urban social structure, and cost-effective growth in general. Due to the planet foodstuff supply organization, the Western dietary demands may boost the food price for exports to developing nations, besides crop displacement difficulties dictated by ostracized, unattended populational groups, viz., indigenous people. Moreover, famine is not only an issue "somewhere else" in the UK. In the USA, over 1 million people use a food bank yearly, whereas 40 million USA inhabitants undergo food vulnerability. Eco-friendly foodstuff waste is astonishing. Cultivating, while handling unattained food, requires massive land and the ability to handle deforested areas, extinct species, and indigenous people who relocated and eroded soil, among other doomed practices that degraded food production or triggered huge farming losses. But there is a drawback to wealthier diets, particularly excessive meat intake. Infant obesity and an astonishing number of persons suffering from chronic illnesses (e.g., diabetes) stem from shortages of fresh meals, reliance on fast foods (many based on meat), and processed nourishments. Chronic disease epitomizes nearly half the global load, which crafts a double burden in combination with infectious illnesses, which are still the leading disease cause in underprivileged countries. The increasing meat production consequences on the environment matter too: Livestock raising exploits about one-fourth of the worldwide agricultural water, accounting for approximately 18% of human "greenhouse gas" (GHG) emissions. The environmental effect is ultimately unsustainable. In short, more people equals more demand, which implies higher production. The UN "Food and Agriculture Organization" (FAO) affirms that by 2050, agrarians will produce 70% more foodstuff adapted to the escalating urban population demands, covering the whole value agronomical chain. While agricultural investments and innovations improve productivity, income growth has decelerated to rates too low for comfort. The Agribusiness labor force amounts to a pressing issue. Even as alimentary requirements and demands rise, the rustic population declines.

Major changes are underway, but first one must scrutinize the past. The first agricultural technological breakthrough was impressive: Cereal yields in East Asia increased by 2.8% per year or nearly 300% from 1961 to 2004, crafting new and more productive crops with modern techniques, including irrigation, fertilizers, and pesticides (World Bank) [7–9]]. But there are falling efficiency improvements: yield gains have slowed. And difficulties are graver: By 2050, the world will engender 70% more sustenance spending little energy, fertilizer, and pesticide to help reduce GHG levels and cope with weather alterations. Old and innovative technology must blend in. The Agri4 must be green, with science and technology at its core [6]. Agri4 will try to solve the alimentary shortage equation (which comprises demand and supply chain components) with technological innovations, enhancing and meeting customers' genuine requirements and reengineering value chains. Modern and conventional farmsteads will operate differently mostly because of advances, comprising sensors, gadgets, machinery, and ICT. In future agriculture, robotics, sensors (e.g., pH, temperature, plant nutrient contents, and humidity), actuators (e.g., pumps, warnings, etc.), aerial images, and GPS will target several purposes ubiquitously. These developments will make agribusinesses more lucrative, efficient, safer, and ecologically approachable. Agri4 no longer has to rely on water, fertilizers, and pesticides in whole fields. Farmers will instead employ negligible quantities or dismiss them from the supply chain, leading to dry zones' cultivation and drawing on plenty of clean assets for food production, like the sun and marine waters. Fortunately, these technical and digital developments may take over the segment, enriching the whole alimentary value chain. According to AgFunder, companies in agricultural technology have augmented by more than 80% yearly since 2012. Startups from Agritech flourish with entrepreneurs and investors demonstrating a ravenous appetite for the niche [10]. Agri4 may remediate food shortage. One can discern three broad technological disruptions with particular solutions:

1. Introduce new schemes,
2. Utilize novelty to approach food production, humanizing the whole food chain's efficiency, and
3. Combine technological advances to fit many industries' needs [11].
4. Precision Agriculture Rise.

4. Precision Agriculture Rise

There exist ambitious plans to feed the globe with decreasing GHG emissions. The UK's National Farmers' Union (NFU) stated it plans to neutralize British agriculture by 2040 by introducing a series of initiatives to enhance land management, raise farming efficiency, and promote a far-reaching bioeconomy. Although unsettled, the NFU has argued for working 'smarter' to reduce direct pollution from farming by providing the same value with lower emissions. PA techniques appear in Fig. 5. "Unmanned aerial vehicles" (UAVs), are not yet globally mainstream in agriculture. Still, they play a progressively significant "precision agriculture" (PA) role, helping agriculturalists toward sustainable agronomy and preserving and boosting profitability. "Global positioning systems" (GPSs) and "global imaging system" (GIS) tools constitute a substantial SF component for in-field monitoring, yield mapping, and acquiring crop parameters data. These present more intensive and effective cultivation forms that can aid farmers in modifying fertilizer prescriptions or detecting/anticipating crop illnesses. More data empower farmers to deliberate economic and environmental variables, and optimize fertilizer treatment, applying the appropriate quantity timely [12].

Fig. 5. Precision in agriculture with IoT [12].

5. Embracing Drones

Agridrone usage gradually augments as an efficient, sustainable agriculture management approach. This tactic enables agronomists, agri-technicians, and farmers to improve their operations through strong analytic data to obtain useful crop insights. Crop monitoring via drone data facilitates planning correctly and constantly improves, employing ditches and changing fertilizer applications. Produces may be tracked accurately and stepwise from farmstead to fork and vice-versa with GPS coordinates rather than via more orthodox, labor-intensive data acquisition. UAVs help especially thoroughly monitoring vast agricultural expanses to discover the most applicable seed prescriptions envisaging variables such as slope and elevation. The technique also comprehensively examines plant growth and population since more precise data may contribute to replantation choices, thinning and cutting, and the crop models' development. Critical factors embrace high-resolution data to evaluate crop fertility, enabling farmers to utilize fertilizer more precisely, minimize waste, and plan and solve irrigation glitches. Figure 6 suggests a device to handle natural catastrophes' aftermaths akin to floods to assist farmers in evaluating land damage not easily reached on foot [13].

Fig. 6. Drone adoption setting in husbandry [13].

Taking Drones Further: Advancing Agridrones even more: UAVs' potential for sustainable SF is enormous. The agribusiness drone market is already projecting huge amounts of money, a sign the sector realizes the advantage over more conventional techniques, e.g., ground mapping. Given the large surveillance field, drones may improve efficiency so that users may collect images of high quality faster than other techniques. Appraising yearly profits may assist in guiding decision-making and managing expectations, particularly in unpredictable market state of affairs. Furthermore, UAVs are increasingly regarded as a safer alternative for mapping regions like uneven or vast terrain that may harm operators mainly compared to on-the-foot terrestrial methods. When historically employed for satellites and crewed aircraft farm surveillance, Agri-UAVs can be indeed a more precise and cost-effective alternative, offering a better precision and resolution even on cloudy days. Although conventional land-dwelling may delay projects in challenging meteorological circumstances, precise crop well-being inspections may take place throughout the year exploiting UAVs [14, 15].

5.1 Applications of Image Processing in Agronomy

In this radiation source, gamma-ray imaging, X-rays, and other means are needed to improve an individual sensor's incoherence and imprecision. Indeed, agronomic applications incite soil scrutiny data that provoke a crop intensity and production regional growth through digital and analog techniques. Image analysts utilize many theoretic and hands-on principles, mainly when utilizing certain image methods. AI may either liberate us from repetitive effort and trigger enormous increases in productivity or engender widespread rumors about tyrannical, dystopic unemployment and doomed automation. Concerning agronomy, some academics, entrepreneurs, and politicians believe AI's and other sophisticated technologies' impacts are as significant as to encourage Agri4. Given the potentially revolutionary implications of future technology on both good and bad agriculture, one must rest and contemplate multiple scenarios before Agri4 implementations. It must function for everyone, whether planters, landowners, rural/urban personnel, rural/urban communities, i.e., the general public, notwithstanding the agribusiness size. However, political figures, media, and legislators have presumed Agri4 a largely good outcome without emphasizing negative outcomes in recent studies [55]. The first agri-revolution befell when people began husbandry circa 12,000 years ago. The subsequent restructuration ascended in the 17th century when European feudalism collapsed. And finally, introducing chemical fertilizers, pesticides, new high-yield crops, and heavyweight machinery in the 1950s and 1960s, alias the Green Revolution [16].

Agri4 hinges on new technological developments, especially AI, to make better planning choices and autonomous power robots, much like Agri4. Such smart robots (agribots) may be utilized for weeding, cultivation, picking crops, milking, and agrochemicals dispensation (e.g., by drone). New kinds of gene editing are other agrarian technologies for developing better yields, disease-resistant plants, vertical farms, and synthetic meat produced in labs. These technologies draw enormous capital and effort to intensify sustenance production and minimize forthcoming environmental damage [17] with media attention. Nonetheless, many other agrarian novelties, such as the first "genetically modified" (GMO) crops and chemicals, viz., the now banned pesticide dichlorodiphenyltrichloroethane (DDT), also encountered similar hype before leading to more debate. Uncontrolled or blind techno-optimism is undesirable due to the broader controversy around emerging novelties, viz., nanotechnology and autonomous vehicles [18]. One cannot presuppose that all these new agri-technologies are implemented without overcoming some obstacles. Precedent evidence tells that advantages are unlikely to be evenly spread in society and that some will get the worst of it. Robotic milking may be effective but it brings in new stresses.

Robotic milking of cows shown in Fig. 7 has improved the work vs. life balance by alleviating dexterous chores. Yet, owing to the information flood and the impression that the farmer must monitor data 24/7, they have also generated

Fig. 7. Robotic milking of cows [19].

a "different stress." The "National Farmers' Union" (NFU) believes innovations may entice younger, more technically trained agrarian workers. Such advances may stimulate more people to join in agribusinesses by removing backward preconceptions by using equipment more effectively. However, the potential for change will sadden current agrarian employees who may be exchanged by machinery or whose expertise is unsuitable for a new farming method. They may not expend time in open-air and can become more computer-dependent.

5.2 Power Inequality

Possible authority imbalances may arise in Agri4, albeit some farmers are enthusiastic about the upcoming agri-technology. Nonetheless, others question whether those with less capital, low internet availability, limited IT skills, and less contact with guidelines/technical support may profit from novelties. History signposts that technological firms and bigger farms are frequently the winners of such changes and that advantages do not unavoidably go to smaller family farms. Agriculturalists may not access/own innovations in the Agri4 framework, or will have to count on supplementary businesses to preserve progressively more sophisticated equipment.

The dispute over GMO crops (Fig. 8) arises from other species' DNA introduction, an open reminder of eventual lack of public trust and guarantee about new technologies adoption. Likewise, if the public perceives gene editing (which includes tiny, controlled changes in a living body's DNA) as equivalent to GM. Wearable livestock technology advocates vow well-being enhancement, albeit public opinion may deem these devices as treating creatures like machines. One must determine the Agri4 advantages and drawbacks and for whom, rather than naïve optimism. An all-encompassing individuals' diversity must exist to aid in fostering socially acceptable ideas for the agrarian future.

Fig. 8. Advanced machinery can bridge agriculturalists and tech firms [19].

5.3 Usage of Corporate Technologies and Applications

High-quality productivity will expand as PA grows and farms become closer together in the next several years. More than 75 million agricultural IoT equipment was used by 2020: the typical farm in 2050 will produce 4.1 million "data facts" daily. Granting the cumulative number of linked HW/SW structures delivers great visions for foodstuff producers, it also upsurges complexity. The answer exploits cognitive technologies for understanding, learning, reasoning, interaction, and efficiency. Some are more forward-looking than others, but the evolvements are quite promising [20].

Internet of Things (IoT): Digital change is disrupting agriculture. IoT technology offers insights into food production via structured and unstructured data interrelationships. IoT platforms, e.g., IBM's Watson (Fig. 9) can adapt agridrone-sensed data into actual AI-reliant outputs.

Fig. 9. IoT and husbandry possible scenario [21].

Automation of skills and workforce: The UN projects a reduction of 2/3 of the planet's rural population. New technologies will alleviate farmers' workload, such as remotely performed operations, automated procedures, risk prevention, and environmental/agrarian problems mitigation. The future farmer's abilities will comprise ICT more and more with biology than conventional agronomy employing few machines and lots of anthropological labor [22].

Data-Driven Farming: Farmers will make better-informed choices by evaluating and connecting weather, seed type, soil quality, disease likelihood, historical data, market trends, and pricing.

Chatbots: AI-fueled chatbots back up retail, leisure industry, media-related endeavors, and insurance activities. However, in agriculture, they may assist farmers in responding and suggesting particular issues [23].

Drone Technology: Per estimates, drone-powered solutions prices in all relevant sectors may exceed $127 billion. And husbandry is exceptionally favorable where drones can solve significant caveats. Drone technology is a high-tech makeover for agriculture and it may be inserted sixfold:

1. Soil and Field Analysis: Drones may be a part of planning seed planting and collecting data for irrigation and nitrogen management by engendering accurate 3D maps for initial soil analysis [24].
2. Planting: Startups have developed drone planting solutions that lessen planting expenditures by 85%. These systems shoot soil pods containing seeds and fertilizers that provide all nutrients needed for crops' growth.
3. Crop Spraying: Agri-UAVs may scan the Earth and spray it for real-time uniform coverage. The deduction is that aerial spraying is five times faster than conventional equipment.
4. Crop Monitoring: Inefficient monitoring of crops is an enormous barrier. With drones, time-series animations may highlight crop growth and inefficiencies in production, consequently allowing amended management.
5. Irrigation: Sensor UAVs can tell apart a field's dry from improved areas [25].
6. Health Assessment: Drone-carried sensors can monitor plant changes, indicate their health via crop scanning with visible and near-infrared (NIR) light, and notify agriculturalists of illness. One day, UAVs may work as independent swarms for data collection, and other missions. The greatest barrier to making this reality is that sensors can gather high-quality data and numerical crushing software (SW) that can make this high-tech fantasy come true [26].

Blockchain and Securing Agriculture: Blockchain, distributed Bitcoin, and other cryptocurrency ledger technology enables secure digital transactions and record-keeping. Albeit, Blockchain aimed primarily at virtual currencies, it may also help agronomic transactions to shorten inefficiencies and fraud, enhance food security, and pay for farmers and time for transactions. By ameliorating supply chain traceability, authorities can rapidly identify the source of tainted foodstuffs and define their range during instances of contamination. In addition, by identifying bottlenecks, the system may cut waste in the supply chain contributing to discarded food. Blockchain's openness may also assist in combating food fraud. With demand from consumers for organic, GMO-free, and antibiotic-free foods rising, the news is replete with false labels. Minor transactions on the farm, in warehouses, and in plants are efficiently IoT monitored and transmitted with sensors and "radio frequency identification" (RFID) tags across the supply chain. Maersk, a shipping and logistics business, has supply networks intra-continental involving dozens of people and hundreds of contacts. Blockchain may save billions by lessening fraud and human errors. All honest market actors gain from openness. Blockchain may inhibit price extortions and delayed payments while simultaneously removing intermediaries and reducing transaction costs. These traits may result in fairer prices and empower smallholder farmers to get a greater share of the value of their crops [27].

Nanotechnology and Precision Agriculture: Tactless pesticides and chemical fertilizers usage led to the loss of soil biodiversity, and an increase in resilience to diseases and pests was the driving force of the 20th-century Green Revolution. The next revolution is nanotechnology-driven PA [11], with nanoparticles and sophisticated PA biosensors supplied to plants. The gradual and continuous release of nutrients and agrochemicals from traditional fertilizers, herbicides and insecticides resulting in plants getting the correct dose. Nanotechnology PA advantages include:

- Approximately 60% of fertilizers used are lost in the environment creating pollution,
- Nano fertilizers enable the gradual and sustained agrochemicals' release, which leads to accurate doses,
- Enhanced plant protection and disease treatment, and
- Biosensors can detect pesticides in plants, leading to better decision-making [28].

Food Sharing and Crowd Farming: Finally, the sharing economy and crowdsourcing may assist in preventing food waste. Technology permits communities to exchange products and services. First popular for transportation sharing and home sharing and now adapted to many industries, including food. The use of drones is rapidly increasing in virtually every area of the economy, but the use of drones in agriculture is exploding. In a few years, drone usage will escalate widely on big and small farms, from reconnaissance to security. The information collected by drones on farms is frequently used to influence agricultural choices better and is part of a PA system. UAVs have already become an important component of large-scale SF in several fields. The information gathered from the drone fields helps farmers plan their plantations and treatments to get the greatest potential returns. Some studies show that SF may improve yields by up to 5%, a substantial gain in the sector with usually small profit margins [29].

Monitoring Plant Health: A drone imaging application can successfully monitor plant health. Detailed color information is used by drones equipped with sophisticated imaging technology known as "normalized differential vegetation index" (NDVI) to assess plant health [57]. This enables farmers to watch crops as they develop so that any issues may be solved quickly enough to preserve harvests. This image shows just how NDVI works. Drones with 'normal' cameras can also perform agricultural health monitoring. Many farmers currently use satellites to monitor crop growth, density, and staining, but access to satellite data is expensive and sometimes less useful than closer drone footage. Because drones fly near fields, the cloud cover and low light conditions are less important than satellite imagery. Satellite imagery can provide precision to the meter, but drone imagery can provide an exact image position up to the millimeter. Figure 10 shows the booth gaps can be identified and replaced as necessary after plantation, and disease or insect issues may be detected and addressed immediately.

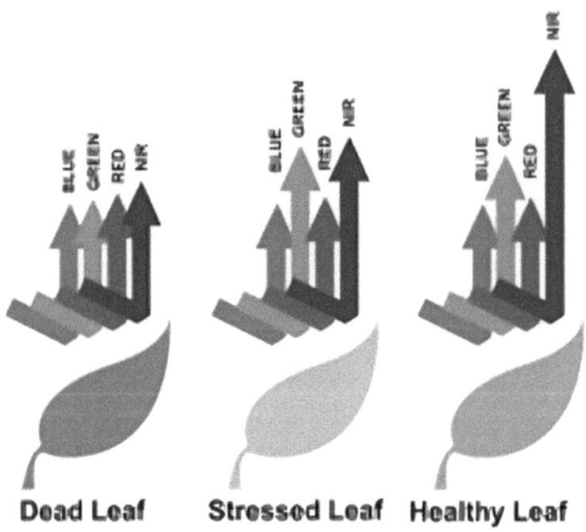

Fig. 10. Different forms of plant conditions [30].

Monitoring Field Conditions: Drone reconnaissance is also utilized for soil and the health of related resources. Drones can offer meticulous field mapping, including elevation information, supporting farmers in detecting land anomalies. Field elevation data help identify drainage patterns and wet/dry areas, enabling more effective watering methods. Some agricultural drones and service providers additionally check the nitrogen level using better soil sensors, making fertilizer usage possible precisely, excluding weak growth areas and taming soil health [31, 32].

Planting and Seeding: A novel and less common UAV application is seed sowing. Automated drone seeds are mostly now for the forestry industry, but the prospect of planting with drones implies that places may be extremely difficult to access without harming employees. They can also plant with two operators and 10 drones, generating 400,000 trees a day far more effectively [33].

Spray Application: South-East Asia widely utilizes spraying agridrones to administer treatments. In South Korea, spraying drones handle roughly 30% of its crops. These agri-UAVs can traverse extremely difficult locations, viz., high-altitude steep tea plantations, preventing employees from navigating over fields harmful to their well-being. Sprayers perform exceptionally precise spray applications with only the necessary quantity of chemicals to optimize productivity in certain regions, albeit laws vary significantly across nations. Currently, Canada vetoes them since more study will better determine the spray drift effect. Some proposed regulations suggest that only qualified professionals be given spraying drones, not selling but renting them from licensed operators (Fig. 11).

Fig. 11. Drone spray on plants [34].

Security: Drone safety is a rapidly expanding business besides agriculture, but it is also very helpful for farming. Drone cameras can follow agronomic activities all day long to guarantee smooth operations and the location of equipment [35, 36]. Instead of hiring additional security people, safety drones may be used to monitor fences and perimeters of more expensive crops such as cannabis. Figure 12 shows that drone cameras are being utilized innovatively to safeguard farm animals by finding lost or wounded herd animals in faraway pastures. Remote regions that required hours to traverse may now be supervised in a few minutes.

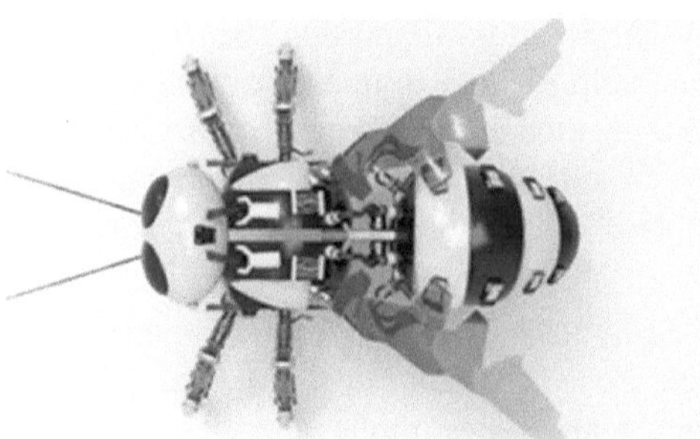

Fig. 12. Secured sensing robot for plant [37].

Drone Pollination: Some novel agronomic UAV applications continue to be tested and developed. Pollination is one of the most popular and frequently dramatized application realms. Dutch and Japanese researchers are crafting tiny UAVs to pollinate florae without destroying them. Developing independent pollination drones covers the next stage, functioning and checking crop health without continuous operator instructions [38, 39].

Drone AI: AI is essential to help smallholder farmers in underdeveloped countries. Today's drones are more suited for monitoring popular crops such as maize developed in broad monoculture field patterns. Drone monitoring systems flop to identify fields with greater crop variety, lesser recognized products, and grains that appear similar throughout their development phases. To train AI systems requires extra effort to have less frequent crops with more varied plantation patterns [40–42].

Drone Irrigation: Intriguing drone usage possibilities arise as climate changes, affecting drought conditions further and demanding more effective irrigation methods [56]. With microwave sensing, drones may get extremely precise information on the soil's health, including moisture level, without interfering with the plants [43] to successfully disperse water within an area to preserve resources.

Growth of Agricultural Use: The agronomy sector's drone usage will only expand. Recent Markets Report projects the "unmanned aerial systems" (UASs) usage in agriculture industry at approximately $5.7 billion internationally by 2025. UASs are cyber-physical systems (CPSs) containing drones among their virtual and real parts [55]. A fast-growing global populace and greater technical improvement of UAVs drive this upsurge. Their enormous flexibility, low cost, and saving possibilities are the key industries for adopting and benefiting from UAS capabilities. So, UK farmers are increasingly using UAS technology, ever altering how crops are produced and raising animals, with agrarian cases continuing to grow. Field mapping utilizing many SF drones can monitor crop health in various soil types, scrutinize these data and project future outputs. As a result, UAV usage improves productivity and crop output and may also contribute to delivering fertilizer, fungicide, and pesticides via accurate GPS routes. Regarding pesticides, some drones can better spray crops than a conventional tractor, reducing overall expenses and exposing employees to chemicals less frequently. Nevertheless, British drones cannot deliver pesticides [44].

Global Agridrone Trendsetters: The growing agridrone usage in North America, places the USA at the forefront of this sector. Agri-UAVs' enhancement have become part of the national plan, particularly given the decrease in Asian "Total Factor Productivity" (TCP and the UAVs' acceptance by the "Federal Aviation Administration" (FAA). Europeans expect to continue the UAV usage, viz., as a major (wine-producing) industry [45].

Agri-UAVs and COVID-19: Throughout a global health emergency, the UK experienced problems about picking fruits and vegetables by workers who could be sick. With the British government asking for assistance on the 2020 harvests, apprehensions about agri-UAVs optimizing agribusiness efficiency emerged. While UAS fruit-picking technology is still under development, some technology companies are already experimenting with prototypes that can collect hard fruit, as it is the case with peaches, apples, apricots, and pears. Current UAS applications have already evidenced noteworthy time and cost gains [46, 47].

UAS Agronomic Efficiency: Drones have historically functioned in many agribusiness sectors (e.g., dairy farming, planting, spraying, horticulture, and tractors), augmenting production intensively. They offer extremely precise real-time data for improved earth and crop advance investigations with fertilization and irrigation evidence. Agridrones can rely on computer vision (CV) to discriminate between plants and weeds [59–61]. They can even remove weeds by transferring that information on the ground to an unmanaged car. Since drones take on more physically arduous duties and perform them faster, they enable better efficiency in many agricultural sectors with higher returns for agrarians. Market predictions point toward agridrones revolutionizing SF. Drones and technological innovation [48] will be continuously utilized to feed the increasing population with difficulties stemming from climate change and the industry's severe weather.

Usage Reasons: A camera solution to augment overall efficiency emerges in Fig. 13. With drones becoming an essential component of SF cultivation, farmers encounter numerous paybacks and caveats. Most recompenses stem from eradicating guesses and shrinking uncertainty. Agricultural success usually relies on various variables, like temperature, weather, Earth, rainfall, etc. Their adaptability translates into effectiveness and fundamentally rests on trustworthy, nearly real-time information that can alter the game [49]. With massive agrarian databases, agriculturalists may augment crop yields, expedite tasks, cut budgets, and answer back with unparalleled precision. Adaptation is indispensable, and agriculturalists will need new-generation technology to handle the expanding population and worldwide changing weather conditions.

Fig. 13. Agridrone-sensing cameras [50].

6. Smart Farming with Drone Technology

A deeper look at drone technology helps better understand its usage in agriculture. An agri-UAV construction normally encompasses propulsion and navigation frameworks, GPS, sensors (e.g., cameras), programmable controllers, and automated flying equipment. The UAV technology for agriculture is designed to gather more precise information than aircraft and satellites (Fig. 14). Agritech SW can analyze the acquired records and render them easily and human-friendly.

Fig. 14. Aerial view with agridrone sensing.

Few agridrones logical steps:

- Specifying flight parameters: Drone navigation can send the evaluation data through surveillance.
- Autonomous flights: With preset flying parameters, an agri-UAV flies as per a devised pattern to collect data.
- Data upload: The agridrone data collected go to a processing and analysis center on ground.
- Information output.

Agrarians get the treated data visually represented to judge facts. Profound, precise data analyses allow for better administrative choices. Agridrone exploitation important aspects are pondered more acutely below.

Best Agricultural Practices: Drones are gaining PA traction since they can provide the most up-to-date information quickly and effectively. Agridrone SW development with affordable cost also augments applications' echelons. How agri-UAVs may aid agriculture follows.

Estimating Soil Condition: Data-driven SF lets farmers espouse measures based on precise soil conditions information. This data extraction required prior physical field trips and the manual collection of measurements. Equipped with PA sensors, drones can gather and provide farm data more rapidly and accurately.

Planting Future Crops: Instead of obsolete planting methods, the prepared cultivation land can receive seeds from an agri-UAV. The Drone Seed startup can deliver robots to grow crops and more. In addition, uncrewed aviation can spray meadows with the right water, fertilizer, and pesticide quantities over fields, saving expenses, human labor, and time.

Fighting Pests: Not only can agri-UAVs gather thermal, multispectral, and hyperspectral bands' facts about soil conditions. Similarly, they can identify regions troubled by weeds, maladies, and pests. Founded on this information, farmers may decide the necessary, precise quantities to exterminate problems, dropping costs and improving produce fields [51].

Agriculture Spraying: SF also utilizes drones to spray on crops, which assists in reducing human interaction with fertilizers, insecticides, and other potentially hazardous constituents. Agri-UAVs can execute this occupation more quickly and efficiently than cars and tripulated aircraft, being a fantastic option for farms operating manually. Agri-UAVs are determinant for spot-on treatment. One can identify and act on diseased regions using sensors, especially cameras, while the healthy field portion remains untouched, saving time, boosting security, and dropping expenditures.

Crop Surveillance: Agricultural fields are spread over vast zones, and the total produce status can be challenging to assess. With drones for mapping agriculture, farmers may keep informed on the plants' conditions in a certain region and identify which meadows need care. Drones examine the field through in-front cameras to gage the crop status, as well as estimating light absorption rates. Agribusiness vendors may go on to enhance cultivated area conditions anywhere, centered on clear-cut, real-time information.

Livestock Monitoring: UAVs can watch faunas grazing on pastures and lessen people's need for horseback and foot dislocations in livestock production. Agridrones can catch missing livestock, detect wounded or ill animals and compute their precise figures utilizing thermal sensing. Admittedly, agridrones can better monitor animals than dogs [52].

Developing Agridrone Solutions: Albeit accessible and realistically cost-effective, before adopting solutions, list essential factors to a target agribusiness. The procedures to convert a farmstead into a drone-enhanced rural property include these five facets:

1. *Determining Goals:* Agridrones have many usages: monitoring fields and crops, sowing seeds, monitoring livestock, etc. What does one want to accomplish with them? Determine the key agricultural operations that may reap the most from agri-UAVs.
2. *Legislation:* Contracting a lawyer for commercial drones' legal issues is a better alternative. Laws diverge consistent with a nation or residency place. Register a drone and research what kinds of drones one may operate lawfully in a region. Also, read the safety guidelines carefully while flying an agri-UAV to evade mishaps.

3. *Deciding on Hardware:* The hardware (HW) to equip a drone hinges on its tasks, e.g., a drone will require an infrared (IR) camera to verify crop well-being. For soil assessments, one needs suitable sensors, etc. Briefly, a drone's capacity (payload) depends on the desired activities to be performed.
4. *Deciding on Software:* Agridrones' data must be treated by agronomic SW. Depending on the agridrone, one will probably require navigation and flight controlling packages, terrain mapping, statistical and analytical data suites, and ImP SW. In certain instances, ready-made, off-the-box SW will not function, and custom-built solutions will be crucial [53].
5. *Open Agri4 Challenges:* Originally, the evidence-gathering, exploration, and human visualization procedures must be acknowledged. One must study plant physiology profoundly to train algorithms to detect specific patterns. These blueprints help diagnose plant fitness (resembling soil conditions or livestock behavior patterns). Agridrones can modernize husbandry in various forms. Still, farmers must be systematically prepared to determine their business intents and decide on the correct HW package and agridrone SW characteristics to benefit completely. Developing agri-UAVs is intricate, requiring partnerships with reliable third-party HW and SW dealers.

Farming: A-Farming: Aerial services empower farmers to manage their property adaptively. Agridrone imaging delivers efficient, rapid land/crop surveying, triggering clever choices to save costs and amplify the output. Producers may:

- Upsurge yields through monitoring various problems influencing plant fitness and growth by agri-UAV services.
- Find and react speedily to diseases, invasions, and scant irrigation imperceptible by earth control stations. Testing schemes like NDVI aid in zeroing in, relating, and revising conditions to optimize yields.
- Use thermal and spectral mapping with topographic knowledge to monitor and evaluate crop health more quickly and accurately over vast regions.
- Fulfill soil and field analyses to evaluate the levels of humidity and fertility.
- Identify irrigation indexes and the needed nitrogen levels to cut down waste.
- Plant more effectually by amassing observable facts via landscape maps, topography, and soil information.
- Administer irrigation occupying an integral, precise land layout. Inflight images let identify slopes and drainage, besides infrared (IR) cameras, to detect dry and wet soil, enabling irrigation to be adjusted as required.

Smarter Equipment for Smarter Farmsteads: PA or SF focuses on integrating sophisticated crop and animal managing tools to expand production without compromising quality. The necessity to balance cost and quality made SF technology developers especially attentive to inexpensive agridrones not relying on much pilot training. In addition, UAVs may carry "remote sensing" (RS) equipment, which has necessitated the utilization of full-size, human-crewed aircraft or satellite connections.

Keeping Watch: Environmental vicissitudes occur rapidly, impairing vast fields' surveillance for farmers. The same occurs to cattle. Particularly, big sheep may be hard to watch without a fauna panorama. As it turns out, the drones make great babysitters soaring. Agri-UAVs can fly high quickly, using surveillance to create animations that allow farmers to manage and remediate crop failures and lost or injured animals in real-time. Fortunately, intelligent agridrones assist in diminishing environmental impacts. These specialist UAVs are fitted with sprayers and other technology types, such as ultrasonic echoing systems and lasers, for accurate distance assessments. Afterward, the total spraying and the groundwater chemical levels are expressively lessened. And drones can do a spraying job 5 times quicker than old-school techniques.

Optimized planting for efficiency. An amazing newest intelligent application is MIT's planting systems, which may reduce planting expenditures by roughly 85% and deploy customized seed- and nutrient-packed pods in the soil below. Likewise, many agriculturalists now exploit 3-D mapping through UAVs to extract Earth and other environmental analysis data to simplify planning/planting patterns. These means assist in irrigation management and the amount of soil nitrogen throughout the crop cycle, enriching irrigation the most. Like said before, experts believe that water constraints would certainly agitate the agricultural future.

Nonetheless, SF may comprise high-tech drones with RS (viz., multispectral, hyperspectral, and thermal bands discoveries) to detect the driest field portions, allowing them to distribute water supplies considerably more cheaply to dry and less wet regions. UAVs permit vigorous yields since they act like tiny, airborne triage guardians, exploiting visible and 'near-infrared' (NIR) light sensors and monitoring equipment to check crop and animal health. During field flights, drones may gather sufficient precise facts to compute the crop's vegetation index. One notices minute variations in a NIR and green light reflected in the crop, generating multispectral diagnostic images.

- Take inventories of tree height and canopy volume;
- Screen tree vigor and quality;
- Management of water, nutrients, pests, and recurrent diseases;
- Estimates of fruit/nozzle production and yield;
- Marketing tools conception for orchards and tree nurseries (e.g., videos promoting orchards or tree/fruit sales) [54].

7. Conclusion

Most nations' economies are heavily agriculture-reliant since it provides some of the finest alimentary goods. In today's humankind, technology is essential in every industry. This chapter scrutinizes the most recent innovations and unanswered questions linked with the four major agribusiness technological waves. One-third of the biosphere's twelve-monthly harvest is squandered due to fungi management and pest attacks, which ICT can remediate, e.g., via image processing, AI, avionics, the IoT, and Blockchain. How will AI technology impact the Agricultural Revolution 4.0 (Agri4)? Farmers keep facing their governments' demands to meet escalating global food needs. Harvesting already benefits from this kind of ICT use by agriculturalists. By guarding florae, soil, and other ecological conditions, spotting prospective caveats and risks from evidence taken from photos/video, and enhancing agriculture effectiveness and productivity, the agridrone is the finest solution for yield hitches, allowing agriculturalists to preserve their harvests optimism despite obstructions. This manuscript looked at contemporary agridrone technology uses. Protecting crops with agridrones, reducing fertilizer usage, simply analyzing collected photographs of the land, and collecting soil data may all be undertaken with fewer human struggles. Agridrones simplify plant illness detection and offer non-destructive remedies. They may also examine the soil and craft a 3-D meadow map to support farmers with seed-planting patterns and render scenes/scenarios for crops' healthiness visualization.

The cumulative agrarian provisions demand steered folks to develop more production to fulfill market needs. Issues impeding plant development have been acknowledged. Most agrarians utilize ICT in harvesting. The drone is one of the best solutions for harvest problems. It makes our harvest more effective and productive without losing farmers' hope in the time crops. Agrarians can lessen labor to safeguard their crops with agri-UAVs that can easily analyze the land through their captured images and blend, e.g., soil data for supplementary processing. The drone applications can easily identify the disease's weight on plants and provide possible damage-free solutions. Furthermore, the agridrones can help graphically exhibit the crops' health. This chapter has discussed the latest Agriculture 4.0 technological innovations and trials.

Agriculture 5.0 will require further adjustments to conciliate AI and human well-being [58. 61, 62].

References

1. Karim, S., Zhang, Y., Laghari, A.A. and Asif, M.R. (2017). Image processing based proposed drone for detecting and controlling street crimes. *2017 IEEE 17th International Conf. Communication Technology* (*ICCT*), 1725–1730. IEEE.
2. Digital Image Processing. Retrieved from: https://www.nrcan.gc.ca/maps-tools-and-publications/satellite-imagery-and-air-photos/tutorial-fundamentals-remote-sensing/image-interpretation-analysis/digital-image-processing/9279.
3. Retrieved from: https://www.bharathuniv.ac.in/colleges1/downloads/courseware_ece/notes/BEC007%20%20-Digital%20image%20processing.pdf.
4. Image Processing in Agriculture. Retrieved from: https://www.mathworks.com/company/events/webinars/upcoming/image-processing-in-agriculture-3355302.html.
5. Demirhan, M. and Premachandra, C. (2020). Development of an automated camera-based drone landing system. *IEEE Access*, 8, 202111–202121.

6. Bennett, M.K., Younes, N. and Joyce, K. (2020). Automating drone image processing to map coral reef substrates using Google Earth engine. *Drones*, 4(3), 50.
7. Liu, H., Wei, Z., Chen, Y., Pan, J., Lin, L. and Ren, Y. (2017). Drone detection based on an audio-assisted camera array. *2017 IEEE Third International Conf. Multimedia Big Data (BigMM)*, 402–406. IEEE.
8. Manohar, A., Sneha, D., Sakhuja, K., Dwivedii, T.R. and Gururaj, C. (2017, May). Drone based image processing through feature extraction. *2017 2nd IEEE International Conf. Recent Trends in Electronics, Information, & Communication Technology (RTEICT)*, 155–160. IEEE.
9. Kim, B., Min, H., Heo, J. and Jung, J. (2018). Dynamic computation off loading scheme for drone-based surveillance systems. *Sensors*, 18(9), 2982.
10. Zhao, H. (2018). *A New Image Processing Algorithm for Geological Structure Identification of Rock Slopes Based on Drone-Captured Images*. Ph.D. Dissertation. Colorado School of Mines.
11. Kulbacki, M., Segen, J., Knieć, W., Klempous, R., Kluwak, K. et al. (2018, June). Survey of drones for agriculture automation from planting to harvest. *2018 IEEE 22nd International Conf. Intelligent Engineering Systems (INES)*, 000353–000358. IEEE.
12. Gashi, E.M.F., Ngjeliu, M. and Ferati, B. (2016). *Developing of Land Information System for Supporting and Managing of Agricultural Sector in Viti Municipality.* Geo-SEE Institute, 6, 94–107. ISSN: 1857-9000, EISSN: 1857-9019.
13. Retrieved from: https://www.harveynorman.com.au/dji-phantom-4-multispectral.html.
14. Saha, A.K., Saha, J., Ray, R., Sircar, S., Dutta, S. et al. (2018, January). IOT-based drone for improvement of crop quality in agricultural field. *IEEE 8th Computing and Communication Workshop and Conference (CCWC)*, 612–615. IEEE.
15. Bassine, F.Z., Errami, A. and Khaldoun, M. (2018). Real-time video processing using RGB remote sensing by drone. *Proc. 2018 International Conf. Electronics, Control, Optimization and Computer Science (ICECOCS)*, 1–5. IEEE.
16. van der Merwe, D., Burchfield, D.R., Witt, TD., Price, K.P. and Sharda, A. (2020). Drones in agriculture. *Advances in Agronomy*, 162, 1– 30.
17. Tsouros, D.C., Bibi, S. and Sarigiannidis, P.G. (2019). A review on UAV-based applications for precision agriculture. *Information*, 1(11), 349.
18. Daponte, P., De Vito, L., Glielmo, L., Iannelli, L., Liuzza, D. et al. (2019). A review on the use of drones for precision agriculture. *IOP Conference Series: Earth and Environmental Science*, 275(1), 012022. IOP Publishing.
19. Hillsdon, M. (2019). Dutch farmers plot a greener revolution by sowing biodiversity. Retrieved from. https://theconversation.com/the-fourth-agricultural-revolution-is-coming-but-who-will-really-benefit-145810.
20. Rachmawati, S., Putra, A.S., Priyatama, A., Parulian, D., Katarina, D. et al. (2021). Application of drone technology for mapping and monitoring of corn agricultural land. *Proc. 2021 International Conf. ICT for Smart Society (ICISS)*, 1–5. IEEE.
21. Naqvi, K. (2021). Smart Farming with IoT and Cloud in Malaysia. Retrieved from: https://www.technologytimes.pk/2021/08/30/smart-farming-with-iot-and-cloud-in-malaysia/
22. Colloredo-Mansfeld, M., Laso, F.J. and Arce-Nazario, J. (2020). Drone-based participatory mapping: Examining local agricultural knowledge in the Galapagos. *Drones*, 4(4), 62.

23. Dharmaraj, V. and Vijayanand, C. (2018). Artificial intelligence (AI) in agriculture. *International J. Current Microbiology and Applied Sciences*, 7(12), 2122–2128.
24. Paliwal, N., Vanjani, P., Liu, J.W., Saini, S. and Sharma, A. (2019). Image processing-based intelligent robotic system for assistance of agricultural crops. *International J. Social and Humanistic Computing*, 3(2), 191–204.
25. Dutta, P.K. and Mitra, S. (2021). Application of agricultural drones and IoT to understand food supply chain during post COVID-19. *Agricultural Informatics: Automation Using the IoT and Machine Learning*, 67–87.
26. Magomadov, V.S. (2019). Deep learning and its role in smart agriculture. *J. Physics: Conference Series*, 1399(4), 044109. IOP Publishing.
27. Bermeo-Almeida, O., Cardenas-Rodriguez, M., Samaniego-Cobo, T., Ferruzola-Gómez, E., Cabezas-Cabezas, R. et al. (2018). Blockchain in agriculture: A systematic literature review. *International Conf. Technologies and Innovation*, 44–56. Cham.: Springer. Cham.
28. Panday, U.S., Pratihast, A.K., Aryal, J. and Kayastha, R.B. (2020). A review on drone-based data solutions for cereal crops. *Drones*, 4(3), 41.
29. Raoufat, M.H., Dehghani, M., Abdolabbas, J., Kazemeini, S.A. and Nazemossadat, M.J. (2020). Feasibility of satellite and drone images for monitoring soil residue cover. *J. Saudi Soc. Agricultural Sciences*, 19(1), 56–64.
30. Kamle, M., Mahato, D.K., Devi, S., Soni, R., Tripathi, V. et al. (2020). Nanotechnological interventions for plant health improvement and sustainable agriculture. *3 Biotech.*, 10(4), 1–11.
31. Al-Thani, N., Albuainain, A., Alnaimi, F. and Zorba, N. (2020). Drones for sheep livestock monitoring. *2020 IEEE 20th Mediterranean Electrotechnical Conference (MELECON)*, 672–676. IEEE.
32. Liang, M. and Delahaye, D. (2019). Drone fleet deployment strategy for large scale agriculture and forestry surveying. *2019 IEEE Intelligent Transportation Systems Conference (ITSC)*, 4495–4500. IEEE.
33. Meivel, S. and Maheswari, S. (2021). Remote sensing analysis of agricultural drone. *J. Indian Soc. Remote Sensing*, 49(3), 689–701.
34. Dou, H., Zhang, C., Li, L., Hao, G., Ding, B. et al. (2018). Application of variable spray technology in agriculture. *IOP Conference Series: Earth and Environmental Science*, 186(5), 012007. IOP Publishing.
35. Devi, Y.S.S., Prasad, T.K.D., Saladi, K. and Nandan, D. (2020). Analysis of precision agriculture technique by using machine learning and IOT. *In*: Soft Computing: Theories and Applications, 859–867. Singapore: Springer.
36. Srivastava, K., Bhutoria, A.J., Sharma, J.K., Sinha, A. and Pandey, P.C. (2019). UAVs technology for the development of GUI-based application for precision agriculture and environmental research. *Remote Sensing Applications: Society and Environment*, 16, 100258.
37. Retrieved from: https://www.croptracker.com/blog/drone-technology-in-agriculture.html.
38. Basri, R., Islam, F., Shorif, S.B. and Uddin, M.S. (2021). Robots and drones in agriculture—A survey. *In*: Computer Vision and Machine Learning in Agriculture, 9–29. Singapore: Springer.
39. Kamilaris, A., Kartakoullis, A. and Prenafeta-Boldú, F.X. (2017). A review on the practice of big data analysis in agriculture. *Computers and Electronics in Agriculture*, 143, 23–37.

40. Sharma, R., Kamble, S.S. and Gunasekaran, A. (2018). Big GIS analytics framework for agriculture supply chains: A literature review identifying the current trends and future perspectives. *Computers and Electronics in Agriculture*, 155, 103–120.
41. Goel, R.K., Yadav, C.S., Vishnoi, S. and Rastogi, R. (2021). Smart agriculture: Urgent need of the day in developing countries. *Sustainable Computing: Informatics and Systems*, 30, 100512.
42. Gupta, S.K., Kumar, R., Limbalkar, O.M., Palaparthi, D. and Divte, P.R. (2019). Drones for future agriculture. *Agriculture & Food: e-newsletter*, 16.
43. Grossi, M., Berardinelli, A., Sazonov, E.S., Beccaro, W. and Omaña, M. (2019). Sensors and embedded systems in agriculture and food analysis. *J. Sensors*, 1-6808674: 2.
44. Kadethankar, A., Sinha, N., Hegde, V. and Burman, A. (2021). Signature feature marking enhanced IRM framework for drone image analysis in precision agriculture. *ICASSP 2021 IEEE International Conf. Acoustics, Speech, and Signal Processing (ICASSP)*, 2385–2389. IEEE.
45. Awais, M., Li, W., Cheema, M.M., Hussain, S., Shaheen, A. et al. (2021). Assessment of optimal flying height and timing using high-resolution unmanned aerial vehicle images in precision agriculture. *International J. Environmental Science and Technology*, 1–18.
46. Kumar, A., Sharma, K., Singh, H., Naugriya, S.G., Gill, S.S. and Buyya, R. (2021). A drone-based networked system and methods for combating coronavirus disease (COVID-19) pandemic. *Future Generation Computer Systems*, 115, 1–19.
47. Kamilaris, A. and Prenafeta-Boldú, F.X. (2018). Deep learning in agriculture: A survey. *Computers and Electronics in Agriculture*, 147, 70–90.
48. Islam, N., Rashid, M.M., Wibowo, S., Xu, C.Y., Morshed, A. et al. (2021). Early weed detection using image processing and machine learning techniques in an Australian chilli farm agriculture. *Agriculture*, 11(5), 387.
49. Fanigliulo, R., Antonucci, F., Figorilli, S., Pochi, D., Pallottino, F. et al. (2020). Light drone-based application to assess soil tillage quality parameters. *Sensors*, 20(3), 728.
50. Retrieved from: https://www.russellwebster.com/british-prison-shields-itself-against-drug-drones/
51. Namani, S. and Gonen, B. (2020, March). Smart agriculture based on IoT and cloud computing. *2020 3rd International Conf. Information and Computer Technologies (ICICT)*, 553–556. IEEE.
52. da Silva, M.V., Pandorfi, H., Lopes, P.M.O., da Silva, J.L.B., de Almeida, G.L.P. et al. (2020). Pilot monitoring of Caatinga spatial-temporal dynamics through the action of agriculture and livestock in the Brazilian semiarid. *Remote Sensing Applications: Society and Environment*, 19, 100353.
53. Mahroof, K., Omar, A., Rana, N.P., Sivarajah, U. and Weerakkody, V. (2021). Drone as a Service (DaaS) in promoting cleaner agricultural production and circular economy for ethical sustainable supply chain development. *J. Cleaner Production*, 287, 125522.
54. Newton, J.E., Nettle, R. and Pryce, J.E. (2020). Farming smarter with big data: Insights from the case of Australia's national dairy herd milk recording scheme. *Agricultural Systems*, 181, 102811.
55. Barrett, H. and Rose, D.C. (2022). Perceptions of the fourth agricultural revolution: What's in, what's out, and what consequences are anticipated? *Sociologia Ruralis*, 62, 162–189. doi: 10.1111/soru.12324.

56. Tingey-Holyoak, J.L., Fenemor, A. and Syme, G.J. (2022). Enhancing the value of water: The need to start from somewhere else. *Australasian J. Water Resources*, 26, 1–6.
57. Aroma, R.J., Raimond, K., Estrela, V.V. and de Jesus, M.A. (2023). A coastal band spectral combination for water body extraction using Landsat 8 images. *Int. J. Environ. Sci. Technol.* https://doi.org/10.1007/s13762-023-05027-z.
58. Monteiro, A.C.B., Franca, R.P., Estrela, V.V., Fernandes, S.R., Khelassi, A. et al. (2020). UAV-CPSs as a test bed for new technologies and a primer to Industry 5.0. *In*: Estrela, V.V., Hemanth, J., Saotome, O., Nikolakopoulos, G., Sabatini, R. (Eds.), *Imaging and Sensing for Unmanned Aircraft Systems*, 2(1), 1–22. London, UK: IET,. doi: 10.1049/PBCE120G_ch1.
59. Deshpande, A., Patavardhan, P., Estrela, V.V. and Razmjooy, N. (2020). Deep learning as an alternative to super-resolution imaging in UAV systems. *In*: Estrela, V.V., Hemanth, J., Saotome, O., Nikolakopoulos, G., Sabatini, R. (Eds.), *Imaging and Sensing for Unmanned Aircraft Systems*, 2(9), 177–212. London, UK: IET. doi: 10.1049/PBCE120G_ch9
60. Aroma, R.J., Raimond, K., Razmjooy, N., Estrela, V.V. and Hemanth, J. (2020). Multispectral vs. hyperspectral imaging for unmanned aerial vehicles: Current and prospective state of affairs. *In*: Estrela, V.V., Hemanth, J., Saotome, O., Nikolakopoulos, G., Sabatini, R. (Eds), *Imaging and Sensing for Unmanned Aircraft Systems*, 2(7), 133–156, London, UK: IET. doi: 10.1049/PBCE120G_ch7.
61. Estrela, V.V., Saotome, O., Loschi, H.J., Hemanth, D.J., Farfan, W.S., Aroma, R.J. et al. (2018). Emergency response cyber-physical framework for landslide avoidance with sustainable electronics. *Technologies*, 6, 42. doi:10.3390/technologies6020042.
62. Khan, A.A., Laghari, A., Gadekallu, T.R., Shaikh, Z.A., Javed, A.R., Rashid, M., Estrela, V.V. and Mikhaylov, A. (2022). A drone-based data management and optimization using metaheuristic algorithms and blockchain smart contracts in a secure fog environment. *Computers and Electrical Engineering*, 102, 108234.

CHAPTER

12

Composting Concepts, Legislation and Panorama: From Solid Waste to Organic Fertilizer in Agriculture 4.0

Ingrid de Souza Siqueira[1*][0009-0009-4636-4604], **Alef da Silva Sousa**[2][0009-0007-7492-1506], **Gabriela Petroceli Mota**[1][0009-0001-9744-1173], **Glacielen Ribeiro de Souza**[1][0009-0003-0773-1092], **Analyze Villanueva Gaete**[1][0000-0003-1471-7443], **Gabriel Quintanilha-Peixoto**[3][0000-0002-7476-5845], **Carlos Eduardo de Souza Teodoro**[2][0000-0001-8591-0502] **and Aline Chaves Intorne**[1,4*][0000-0001-8015-6926]

[1] Laboratory of Physiology and Biochemistry of Microorganisms, Universidade Estadual do Norte Fluminense Darcy Ribeiro (UENF), Campos dos Goytacazes, RJ, 28013-602, Brazil

[2] Biotechnology Laboratory, Universidade Federal Fluminense (UFF), Volta Redonda, RJ, 2629-5000

[3] Laboratory of Molecular and Computational Biology of Fungi, Federal University of Minas Gerais (UFMG), Belo Horizonte, MG, 31270-901, Brazil

[4] Laboratory of Chemistry and Biology, Federal Institute of Education, Science and Technology of Rio de Janeiro (IFRJ), Volta Redonda, RJ, 27213-100, Brazil

Qualified specialists and other human stakeholders must fulfill "solid waste management" (SWM), given the growing worldwide population's worries. Mexico, South, and Central America account for 12% of worldwide garbage generation due to ongoing increased consumption/consumerism, urbanization, and lifestyle changes. Brazil generates 50% of its "organic solid waste" (OSW). These detrita pollute landfills, limiting their viability. This efficacious, low-priced biotechnological process controls biological waste and yields fertilizers with added value while supporting the circular economy. Soil microbes degrade "organic materials" (OMs) into bioavailable nutrients, improving nutrient-deficient soils, reducing erosion, and acting as composting bio-inoculants. Composting by-products can replace chemical fertilizers, shrinking soil, and

* Corresponding authors: ingrids.uenf@gmail.com; aline.intorn@ifrj.edu.br

hydric pollution. Composting mechanisms drop food debris along the supply chain while recovering and fostering better soil usage, sustainable communities, and remediating weather transitions, all in agreement with the UN Sustainable Development Goals (SDGs). This chapter examines composting fundamentals and the possibility of producing and applying composting-based organic fertilizers in Agriculture 4.0, intending to advance edible production in a sustainable and modern form.

1. Introduction

Poor "solid waste" (SW) handling or no "SW management" (SWM) harms humankind. Economic development and demographics also escalate as urbanization develops. The corresponding "urban solid waste" (USW) [1, 2] grows as well, especially food and green waste (FGW), aka "organic waste" (OW). In line with [3], the globe's population engenders a USW average of 2.01 billion tons per annum. This value estimate will reach 2.59 billion tons (t) in 2030, with 3.4 billion t in 2050. As per the World Bank (in absolute terms), the region encompassing 37 countries (the Pacific and East Asia) engenders an immense SW quantity (23%). Comparatively, all Africa (69 nations) responded for 15% [3] (Fig. 1).

Of the total USW, 31% is estimated and deemed not managed carefully, thus engendering severe health populational hazards [3]. The natural environment also endures ground, hydric, and air pollution. Locally, the accumulation of garbage generates liquid discharges that can infiltrate and contaminate land and hydric resources. Globally, open dumps trigger most GHGs [1, 4] (Fig. 2).

As per the "Outlook Report on Waste Management in Latin America (LA) and the Caribbean" [4], the USW composition may vary consistently with the inhabitants' income. Dissimilar consumption patterns can interfere with waste composition [2]. In high-GDP countries, 32% of garbage is FGW, while 51% is dry, viz., glass, paper, metallic pieces, and plastic. FGW in low- and middle-income nations is within the interval 53–57%, and dry materials constitute only 20% of the residues [3] (Fig. 3). OW makes up for 75% of the USW composition when seeing only low-income nations [4].

Income impacts the "waste management" (WM) (Fig. 4), too. Controlled landfills exist almost exclusively in middle- and high-income nation-states. Customarily, low-GDP nations discard 93% of their total waste in the open. In middle- and high-GDP nations, this number plummeted to 2% [3].

For 2016, 1.6 billion t of carbon dioxide (CO_2) from USW had been expected (akin to 5% of the Earth's emanations), first and foremost driven by landfills with modern WM or lacking gas collection systems [3, 5]. Without waste management improvement, this data could escalate to 2.38 billion t by 2050. One should realize that around 2.5% of USW emanates from "organic matter" (OM), also called "organic waste" (OW), decomposition [3, 6].

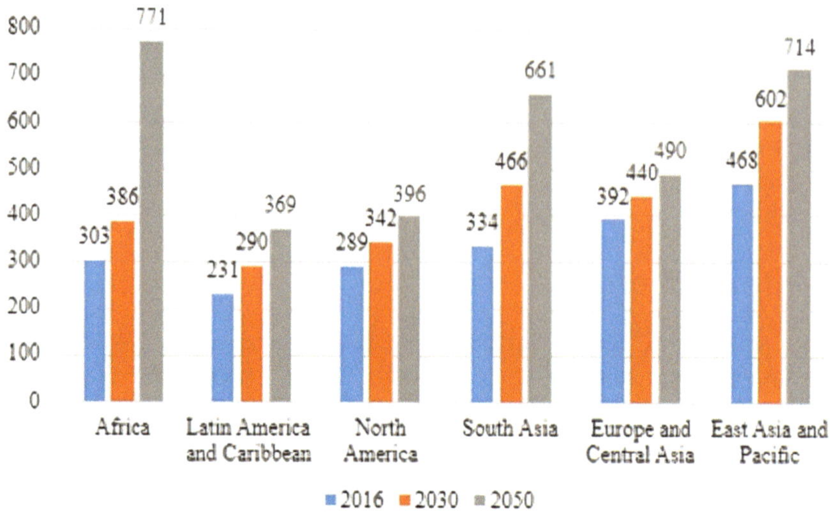

Fig. 1. World waste for 2016 and forecasts for 2030 and 2050 in millions of tons [3].

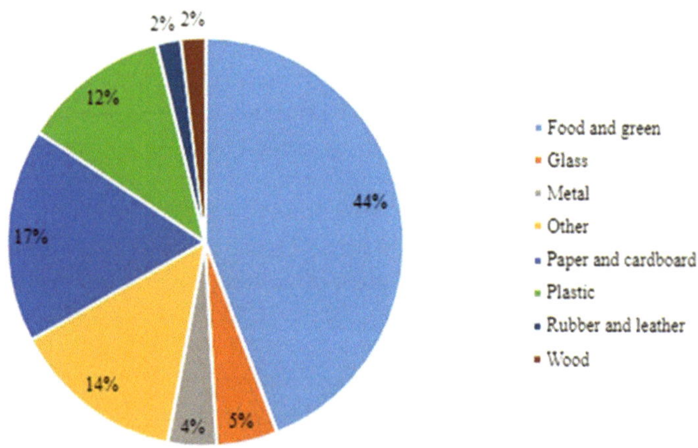

Fig. 2. Global waste configuration [3].

In 2016, Mexico and South/Central Americas got 231 million t of garbage, with an anticipated daily average of 0.99 kg per capita, where 52% of the residues were organic, e.g., food leftovers (Fig. 5). Brazil ranks 4th amongst those with the most waste globally, which can be further highlighted and exemplified (in absolute numbers) after reaching the highest rank in 2020, making 82.5 million t of garbage [7] amidst Mexico and South/Central America nations [3]. As per the population total, most USW outputs came from archipelagic nation-states with active tourism markets. The American Virgin Islands illustrate this point with 4.48 kg of garbage per capita/day, i.e., practically six times the Earth's average of 0.74

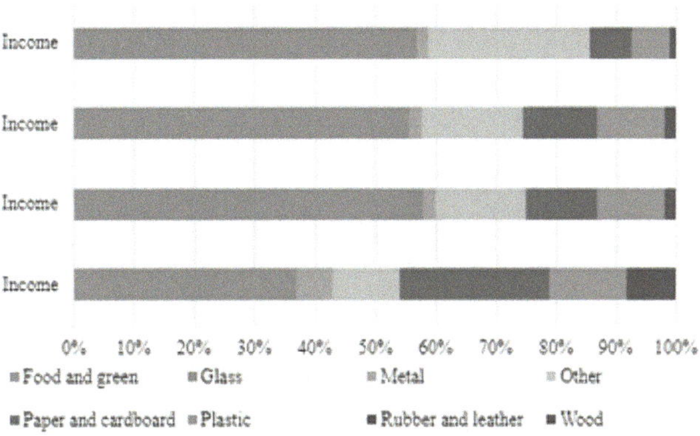

Fig. 3. Waste composition by income [3].

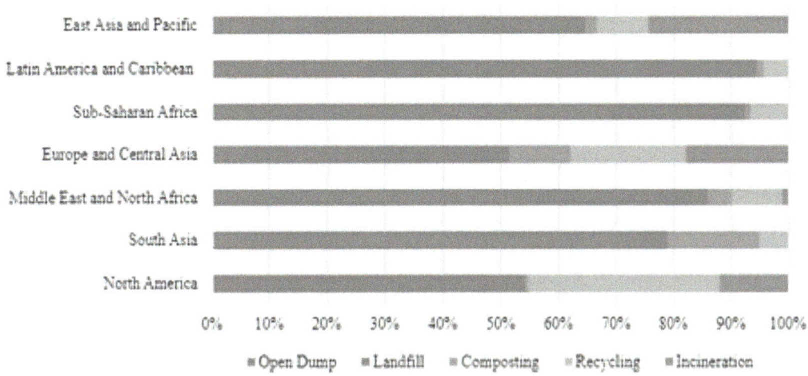

Fig. 4. Worldwide waste management [3].

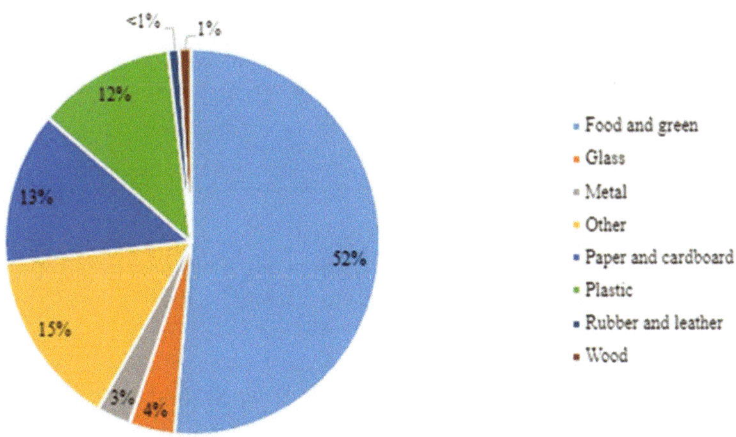

Fig. 5. USW generated in Mexico, South/Central Americas [3].

garbage/per capita/day [3]. The vacation industry impacts and exacerbates USW, predominantly in transnational destinations [2], e.g., Mexico, South and Central American archipelagic regions.

Even though USW collection handling is high in LA's metropolitan zones (84%), averaging 30% in rural areas, of this total, only 4.5% is recycled [3]. About 69% go to landfills, and half are destined for landfills with environmental control [3, 4] (Fig. 4). In Uruguay and Colombia, SW collection reaches up to 95%, while in Haiti, it is around 12%, which discloses some inequalities also present within the countries [3]. Only 10 Brazilian states count on a collection coverage rate above 90%: Sao Paulo, Rio de Janeiro, Distrito Federal (DF, Brasilia), Parana, Santa Catarina, Rio Grande do Sul, Goias, Espírito Santo, Amapá, and Mato Grosso do Sul. The lowest collection coverage rates emerged in the North and Northeast states: Ceará (80.1%), Rondônia (78.9%), Pará (76.7%), Piauí (69.2%), and Maranhão (63. 9%) in the bottom five places [7]. Furthermore, LA's informal sector has robustness, with many garbage collectors (4,000 active laborers per nation collecting recyclable materials on average).

For 2019, the "Food Waste Index Report 2021" projected 931 million t of food depletion, where 61% would come from family units, 26% from food service suppliers, and 13% from retail. This high number embodies 17% of the world's annual food production [8]. LA single-handedly responds to nearly 20% of globally lost food, in view of the thorough production chain up to the consumer (Table 1) [9]. Table 1 also highlights colossal energy, hydric, and further natural resources' depletion not counting money [4, 9].

Table 1. Estimated domestic foodstuff waste in some LA and Caribbean nations compared to the world [8]

Country	kg/per capita/year	t/ano
Belize	53	20,564
Brazil	60	12,578,308
Colombia	70	3,545,499
Mexico	94	11, 979,364
Global	74	563 (million tons)

Only 19% of the planetary USW will likely experience recycling or composting [3]. Thus, public policies encouraging these measures are necessary to hand over farm duties to different actors toward a steady circular economy shift [2, 4]. Despite national variations, LA is subjected to incipient composting/recycling initiatives [10–14, 128, 131], and many cities/villages are committed to recuperating waste. Cities like Montevideo (in Uruguay) and Bogotá and Medellín (in Colombia) reprocess more than 15% of waste, and municipalities like Mexico City and Rosario in Argentina exploit compost, exceeding 10% of waste [3].

With the composting strategy, natural resources would be transformed into raw materials for fabricating goods, generating minimal waste to be pulled together, treated, and recycled. Then, forming new inputs and feeding them back into the process will permit better energy use and nature protection [4, 131]. Moreover, the composting relates straight to the UN's 2030 Agenda (i.e., "Sustainable Development Goals" (SDGs)) as below:

- **Zero starvation and sustainable husbandry:** By augmenting agricultural productivity, warranting a sustainable sustenance production system, and improving food security;
- **Good well-being:** Through tropical diseases withdrawal, decline of deaths and illnesses due to chemicals, hydric, and soil contamination;
- **Drinkable water and sanitation:** By cutting chemicals released into the water;
- **Sustainable towns and communities:** Employing inclusive, sustainable urbanization whereas lessening the negative environmental impact per capita supports sustainable environmental relationships in urban areas;
- **Global climate transformation combat:** Via total GHG emissions' integration measures;
- **Terrestrial life:** By renovating degraded properties and their diverse soil types.

Therefore, using the composting technique as a correct destination for OW brings enormous local, national, and international benefits. Since LA yields 14% of worldwide food and around half of this region's jobs are agri-food related, the economic and agronomic achievements are even more significant [4, 15].

In Mexico, South and Central Americas, three global phenomena stem from agrarian development: climate change, augmented food demand, and "information and communication technologies" (ICT) disruptive application [16]. These drivers will influence agricultural development in the approaching years and have different relevancy and traits per country; however, they will mark the sectorial development agenda in the coming years [17].

Agriculture 4.0 refers to ICT innovation processes, viz., Artificial Intelligence (AI), sensors to probe biological and inanimate entities, actuators to remediate or stabilize problems, e.g., tractors and drones (alias uncrewed aerial vehicles or UAVs). The Internet of Things (IoT) supports local/personal gadgets, virtualization, and so on, combining data from field equipment with real-time reception, simplifying collaboration and decision-making among different farming industries and value chains [18, 19, 128, 129]. Designing ICT strategies for LA's agrarian sector remains incipient. However, some initiatives, e.g., originating laboratories and agribusiness that rely strongly on ICT, boosted Brazil, Chile, Argentina, and Uruguay, which may represent engagement openings and escalate the sector's ICT transformation [17].

2. Composting

It ensues from controlled aerobic OM decomposition into a stable outcome akin to humus, known as compost [15], and essentially means the same as natural decomposition. The difference is its enhancement and acceleration by mixing OW with other elements to optimize microbial development [20]. The subsequent expression enlightens the process [21]:

$$OW + Microorganisms + O_2 \rightarrow H_2O + CO_2 + Compost + Heat.$$

Composting is a sustainable technology for treating organic SW (OSW), encompassing physical-chemical and biotic factors [22]. This process' phases are threefold: (i) mesophilic, (ii) thermophilic, and (iii) maturation [23] (allude to Fig. 6), where microorganisms play a paramount role in decomposing and stabilizing OM [22, 24].

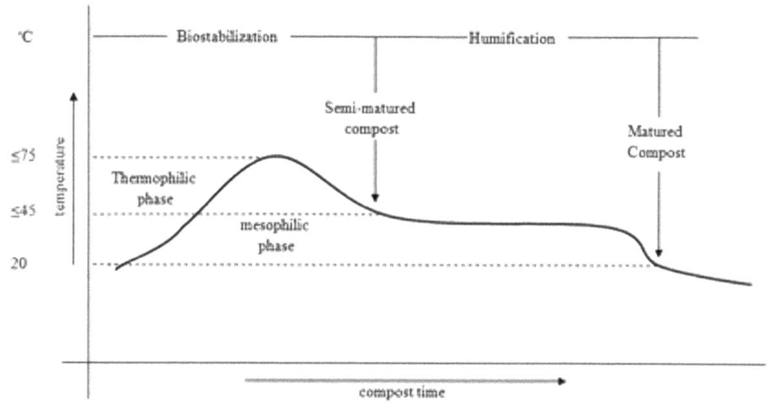

Fig. 6. The composting evolution [23, 25].

In the brief mesophilic stage, microorganisms subsist in milder temperatures (<45 °C) [26], metabolizing the simplest molecules, viz., free amino acids, carbohydrates, etc. [23, 27]. A temperature augmentation (up to 75 °C) with humidity reduction induces the thermo-tolerant microorganisms' appearance, typifying the thermophilic phase [23]. At this stage, thermophilic microorganisms cut down proteins, polysaccharides, and fats. Heat eradicates pathogenic microorganisms from OM [27]. The CONAMA Resolution 481/2017 from Brazil institutes that in open composting structures, the temperature must go beyond 65 °C for 3 days (or exceeding 55 °C for a 2-week minimum), whereas in closed systems, above 60 °C for a 3-day minimum period [28].

Microorganisms rely on organic carbon mineralization throughout the thermophilic phase, liberating carbon dioxide (CO_2) molecules and nitrogen and lessening the carbon-to-nitrogen ratio (C/N) [22, 27]. Compound maturation [29, 30] partakes in decreased microbial activity, lower temperature, low pH levels, and the absence of phytotoxic chemicals. Research [22, 27, 29, 30] proposes optimal C/N ratios within the interval 15-20, and NH_4^+/NO_3^- ratio below 1.

C/N ratio, humidity, temperature, oxygen rate, pH, OW particle size, and compaction degree [6, 27] impact the composting process efficiency and quality. The ideal range of these parameters may differ with the initial residues' characteristics, the operational processes' conduct, and the final compound distillation [31]. Studies recommend these compounds to immobilize potentially toxic elements and to enrich the land [24, 27, 32, 33].

Nevertheless, these components must abide by the restrictions prescribed by a nation's existing agricultural protocols. Many LA and Caribbean countries have established regulations with particular criteria for determining the quality of compounds and their permissible applications. In the context of Brazil, the "Ministry of Agriculture, Livestock and Supply" (MAPA) has enforced guidelines through normative instructions and CONAMA – Resolution n° 481 [34] (Table 2) for illustration. Chile has compound classes A and B, further characterized by the incidence/absence of usage limits, particularly for metals.

Composts can lead to pathogenic bacteria like *Salmonella spp.*, *Escherichia coli*, and Listeria monocytogenes, along with other microorganisms, including fungi, viruses, and parasite eggs transmittable to consumers via eating contaminated food [41, 42]. The compost for cultivating short-stemmed and green vegetables, besides fruits, must be free from pathogens and signs of fecal contamination. Elevated temperatures can control these factors, effectively adjusting time and temperature parameters throughout the production process, which can be crucial [43, 44].

2.1 Soil Microbiology

Regarding microorganisms, bacteria and fungi prevail in soil, exhibiting significantly higher biomass than protists, archaea, and viruses, with a ratio of 104:1 [45]. The microbiota can be affected by innumerable biotic and abiotic parameters, including pH levels, organic carbon concentration, oxygen and redox circumstances, moisture levels, N and K availability, soil texture/structure, vegetal species, temperature, and predation [45, 46]. Soil OM has direct connections to its microbial community. Organic substrate feedback can impact the C/N relationship and transform microorganisms' structure, diversity, and ground functions [47, 48].

Microorganisms decompose OM and control key biogeochemical processes since they are preponderant in energy and nutrient cycles [49]. Multiple species, like *Arthrobacter*, *Streptomyces*, and *Lysinibacillus*, can to enhance carbon mineralization and facilitate OM degradation [47]. These microbial communities' significance necessitates exploring their management as a prospective solution to several global difficulties. These challenges comprise, albeit not restricting, the expanding population's sustenance, alleviating terrain degradation and low fertility, preventing biodiversity deficits, and managing climate change [45, 49].

Table 2. Organic fertilizers regulation in different LA countries

Parameters	Brazil[a]	Chile[b]	Mexico[c]	Colombia[d]
Dry matter (%)	≥50	-	-	-
Organic matter (%)	-	≥25	≥15	-
Particle size (mm)	≤40	≤15	-	-
N total (%)	≥0,5	≥0,8	-	≥2
Organic carbon (%)	≥15	10-24	-	15
C/N Ratio	≤20	10-25	<12	as declared
pH	as declared	5.0-7.5	6.5-8	as declared
Water storage	≥60	-	≥70	≥35
Arsenic (mg/kg)	≤20	10	5	41
Cadmium (mg/kg)	≤3	1	1	39
Lead (mg/kg)	≤150	50	5	300
Hexavalent chromium (mg/kg)	≤2	60	5	1200
Mercury (mg/kg)	≤1	1	-	17
Nickel (mg/kg)	≤70	10	5	420
Selenium (mg/kg)	≤80	6	-	-
Thermo-tolerant coliforms (NMP/g of dry matter)	<1000	<1000	<1000	<1000
Viable helminth eggs (no. per 4 g of ST.)	1	Absence	<10	Absence
Salmonella sp.	Absence in 10 g of mat. dry	Absence	< 3	Absence in 25 g of product
Glass, plastic, metal in dry mass	(>2 mm) ≤ 05%	(>5 mm) ≤ 0.05%	(>2 mm) 0.5%	(>2 mm) 0.2%
Stones in dry dough	(>5 mm) ≤5%	(>5 mm) ≤ 5%	(>2 mm) 0,5%	(>2 mm) 2%

[a] National Environmental Council for Composting [35–37]
[b] Chilean Compost Standard NCh 2882/04 [38]
[c] Standard NTA-006-SMA-RS-2006 for OW soil improvers [39]
[d] Standard NTC 5167/04 – Organic products utilized as fertilizers and land corrections [40]

Organic compounds can stimulate the natural microbiota and insert other microorganisms to intensify benefits with less energy and time [47, 50]. Likewise, they stimulate the microbial populace that solubilizes phosphorus (P) and the terrain enzymatic activity, transforming P with augmented productivity of rice and corn [48, 51]. Adding carbon and nitrogen changed the soil's microbiome's composition, increasing phyla *Actinobacteria* and *Proteobacteria* and reducing the *Chloroflexi* and *Verrucomicrobia* [52]. The constitution and variety of Guizhou's yellow soil fungi were altered with the OM input and reduced pathogens [53].

2.2 Composting Methods

Crop residues, manure from animal farms, agro-industrial processing residues, USW, and unsold agrarian products can be outstanding for composting [54–58]. In general, the wastes cited above possess characteristics that facilitate their degradation and offer nutritional and fertilizer properties. So, it is vital to steadily insert less compressible materials with advantageous structuring capabilities, such as wood shavings, pruning residues, and straw (aka bulk materials), to foster proper composting. These components can give porosity to the dough and facilitate timely oxygen diffusion, which is decisive for assisting the aerobic metabolic processes of microorganisms [15]. The primary concerns when opting for a composting technology include [44]:

- Processing time;
- Space requirements;
- Necessary hygienic safety;
- Preliminary material (i.e., without or with the animal-origin material presence);
- Local weather conditions (strong winds, torrential rain, freezing temperatures, or other acute meteorological conditions).

Methodologies can be: (i) closed systems (made in containers), and (ii) open systems (outdoors). Composting methods differ in the decomposition duration and the compost's stability and maturity [44, 59]. Table 3 portrays different compost schemes.

Continuous compost research optimizes the process to encounter the lowest time to obtain compost, seeking to shrink contamination threats and GHG emanations [66, 67]. Process optimization can rely on changeable aeration that nourishes microorganisms, besides putting in biochar and bioinoculants alongside specific microorganisms [68–72]. Therefore, optimizing the composting process and complying with current legislation can contribute to agronomic OW valorization and a circular LA/Caribbean economy consolidation.

2.3 Composting Uptakes and Benefits

These results are related to the increased concentration of macro and micronutrients, augmented carbon concentrations, greater microbial activity, and correction of the soil's physical-chemical properties (temperature, pH, humidity, oxygen) [31, 75, 82]. Therefore, due to low-price production and investment costs, organic compounds can increase fertilizer-use efficiency, maximize crop yields, promote sustainable husbandry, and raise income. OW discarded in landfills produces undesirable GHG amounts [6, 73] and occupies land that may have agronomic usages. Thus, composting has aroused interest as an eco-friendly alternative to

- manage OSW [31],
- alleviate weather change impacts by employing carbon sequestration,
- lower fossil fuel consumption [74, 75],

Table 3. Composting methods [60–65]

Method	Description
Windrow composting	The process involves arranging the raw materials in elongated and thin formations, commonly named piles or windrows, and periodically agitating the material. Combining several materials eases aeration within the assumed arrangement. The maintenance can be costly; nonetheless, this system's expeditiousness and ability to effectively preserve thermal energy.
Container bin or in-vessel composting	The term "enclosed composting" pertains to composting within a confined space, for instance, a container or vessel. Implementing different techniques is contingent upon the vessel varieties, aeration devices, and mechanical revolving tactics employed within a given setting.
Bokashi composting	The process exploits OM, molasses, effective microorganisms, and H2O anaerobically. Numerous studies have acknowledged the positive fallouts of enhancing fertility, augmenting crop profits, and fostering plant growth. Furthermore, as an organic substance, bokashi can boost the earth's retaining water capability.
Vermicomposting	Worms decompose biodegradable OM by breaking down many organic compounds during feeding. The waste, called 'castings', presents a raised nitrate concentration and promptly obtainable elements (like P, K, Ca, and Mg) to heighten land fertility.
Static composting	To promote aerobic decomposition, aeration can be: (i) passive, and (ii) sporadic/static. A simple, time-consuming procedure that stores unprocessed OM with a small staff and equipment preconditions.
Leaf composting	Decomposing leaf debris and other OW fosters nutrient-rich composts. Garden surplus, wild plants, foliage, and foodstuff types can be mulched thinly upon the soil surface. Instead of piling up stuff, the terrain is worked, and allowed to cheaply deteriorate until the plantation starts again with unsophisticated, readily manageable strategies.
Berkley rapid composting	Aka hot composting, employs aerobic treatment, utilizing abundant microorganisms (e.g., bacteria, fungi) to break OW into healthy, humus-rich materials.

- bioremediate metals, pesticides, and soon-to-appear contaminants [6],
- balancing nutrients in soils [31], and
- safeguard nourishment for an escalating world population [75].

A Kean University study (USA) had compost GHG emissions 31% lower than in landfills. During growth, the plants treated with the organic compound dropped GHG emissions by 9%, pollution by 20%, eutrophication by 100%, and fossil fuel usage by 30%. These outcomes highlighted composting benefits when contrasted to chemical fertilizers [73]. The economic aspects of the compost system were also evaluated, leading to a profit of $13,200 per year by trading compost-grown vegetables, and this income could increase to $23,550 when adding the educational and environmental aspects.

Moreover, applying compost in agribusiness can upgrade the quality of land and crops [76] and perform phytopathogens' biocontrol [77, 78], diminishing factory-made fertilizers and pesticides. Composts can upsurge soil OM, essential for maintaining high agricultural productivity, as it contains C, K, N, P, and other essential crops' nutrients, as well as ameliorating the hydric retention and lowering soil density [77, 78]. OM holds high humic substances' (HSs) amounts that can enrich and balance ground germination, rooting, overall harvest development, native flora/fauna, productivity, and nutrient availability [77].

Associating inorganic with organic fertilizers augmented the corn yield by 10–15%, with no difference when completely replacing the synthetic compound [79]. An old Chinese assessment had shown that this blend heightened the soil's OM content and yield with a 14% average in rice, wheat, and corn production compared with solely inorganic fertilizer [80]. An olive tree (Cornelius) study handled with organic fertilizer obtained productivity nearly 55% higher than without fertilizers and significantly increased the leaf area index [81]. According to the authors, intermixing organic and inorganic fertilizers leads to greater canopy density and superior olive tree yield than conventional handling. Compost usage furthered the yield of vegetables (endive, eggplant, cauliflower, onion, tomato, melon, and fennel) similar to or greater than artificial chemical fertilizers [33]. These fallouts stem from the

- increased concentration of macro and micronutrients,
- augmented carbon concentrations,
- greater microbial activity, and
- corrected topsoil's properties (viz., temperature, pH, humidity, oxygen) [31, 75, 82].

Therefore, organic compounds can increase fertilizer use efficiency, make the most of crops, recommend sustainable husbandry, and increase income due to reduced production and investment costs.

3. Agriculture 4.0 Composting

Industry 4.0 has spawned significant advances in computer science, joining production chain processes and elements at all stages, times, and sectors [83–87]. These production management enablement and optimization can be applied to agronomy (Agriculture 4.0), where these gains arrive gradually. Agriculture 4.0 comprises innovations that optimize and monetize productivity by reducing errors [84]. Food production in terms of calories will have to grow by 56% by 2050, compared to 2010, but it must be more sustainable, producing less GHGs, and employing less water, energy, and additives. Achieving this depends even more on scientific and technological development [88]. Integrating Agriculture 4.0 and relying solely on the necessary quantities for specific regions will be feasible, avoiding waste [89]. This scenario is a rich, sustainable approach, furthering the entire farming production chain, from planting to managing OW. The organic liquid or solid fertilizer from compost bins has various nutrients, e.g., slurry is a nutrient-rich liquefied biofertilizer [90]. The nutrients found in the compost bin products will affect the OM and the composting approach, customarily rich in C, N, K, and P [91, 92]. In addition to the physical-chemical analysis, microbiological breakdowns are prerequisites to endorse large-scale fertilization [93]. Compost products' phytotoxicity inspection before employing them for planting is paramount for healthy farming [94]. Agriculture 4.0 can help solve the high global food demand from the composting viewpoint threefold, as described below [89].

3.1 New Production Methods

Agriculture has been a conservative activity, albeit a robust economic pillar. However, this sector faces challenges like food security, climate change, hydric contamination, and soaring energy demands, which call for fresh applied science results to tackle these drawbacks [95]. The latest Agriculture 4.0 production means would be divided into (i) mature technologies embracing hydroponics, algae-related materials, plus bioplastics and (ii) near-future agribusiness-related knowledge [89] akin to desertic and marine farming.

Plants produced without soil but water while supplying minerals to plants via nutrient solutions are called hydroponics. They surpass traditional agronomy when (i) it comes to control of soil pathogens, (ii) production is independent of the soil quality, (iii) controlled growth is according to the available nutrients, (iv) reusing nutrient solution, and others. The system suits vertical cultivation, optimizing metropolitan spaces [96], and biofertilizers can sustain hydroponic plants [97, 98], favoring these foods' production.

Algae and microalgae have fed fish, other marine entities and even humans for years [99–101]. These organisms can be raised in tanks and fed with liquid biofertilizers from the compost bin [102]. Bioplastics can be placed in compost bins according to their produced materials [103–105].

3.2 New Technologies Invigorating Food Chains' Efficiency

Vertical and urban farmsteads stand as mature translational technologies [89]. Near-term technologies embrace 3D foodstuff sector applications, genetic modifications, livestock, and crops [89]. The liquid fertilizer from composting can irrigate vertical farming [106, 107]. Since Agriculture 4.0 also cuts down chemicals [84, 89], using composting-derived solid fertilizer is also viable [108–110]. Furthermore, this fertilizer insertion recovers nutritional and natural fauna/flora factors, besides other pivotal issues for maximizing efficiency

(Table 4), with remarkable urban uptakes in metropolitan areas lacking some aspects due to the large urban occupation [111, 112]. One can add microbiological bioinoculants to fertilizers for process optimization, magnifying crop nutrient assimilation [111].

Table 4. Benefits for the soil of applying fertilizer from a compost bin [111, 113–115]

Aspects	Benefits
Physical	The compost particles create porosity, facilitating hydric retention and gas exchange; Increased vegetation cover; Improves soil structure; Increases to a litter of only fertile; Augments nutrients' movements; and Soil safeguards from shifting meteorological conditions.
Chemical	OM improves soil fertility; OM stabilization and mineralization; Humic substances' precursors' production; and Soil pesticide concentration reduction by creating bonds with organic compost molecules.
Microbiological	Microbiological microorganisms supply that help soil formation and nutrient movements; Augmented growth of the rhizosphere's microorganisms and roots; Microorganisms-based bioremediation to degrade toxic agents; Favorable microorganisms' source to control vegetal pathogens; and The terrain microbial populace and its enzymatic activity can be soil quality change markers.

3.3 Incorporating Technologies

Industry 4.0 has mature technologies applicable to agronomy (e.g., nanotechnology, AI, food division/sharing, crowd/fog computing agri-business are some near-future technical expertise) [89, 116].

Data analytics and intelligent tools afford essential knowledge for agriculturalists that can help expand yields, reduce production inputs, and enhance environmental well-being, delivering more foodstuff on existing terrain, thus reducing land conversion to crops [117].

Plentiful ICT usage is already implemented in worldwide crops. By 2050, 4.1 million agriculture-related data are expected daily [89]. An Indian city wants to employ IoT for automated waste separation, joining forces en route for intelligent, sustainable cities/villages [118] and optimizing the OM separation as well. IoT can be applied to composting to expedite it up to 2.35 times faster with the technology [119].

Soil fertility degradation is also a major agrarian problem. "Geographic Information Systems" (GISs) fertility maps provide land fecundity statuses and aid as a decision-making tool for cultivation. These maps contain viable artificial/biological fertilizer recommendations based on soil assessments for sustainable, intensive specific crop production in peculiar locations [120, 121].

Moreover, drones can facilitate the (a) geographic assessment of planted areas, (b) strategic points' assessment for soil breakdowns and irrigation tactics, (c) dissemination of seeds with portions of nutrients, (d) spraying in needy soils, (e) crop monitoring, and (f) plant health analysis [89]. The University of Sao Paulo has conceived an autonomous, AI-driven, drone-reliant system for pesticide spraying [122]. Literature depicts liquid composting-achieved biofertilizers for foliar spraying [123–125] and natural pesticides to repel and prevent pests [123, 126]. Thus, it could replace toxic agrochemicals sprayed by drones.

Only a few LA countries have developed Agriculture 4.0 strategies. In Brazil, policies are undergoing formulations and reviews regarding significant developmental and feasibility topics via an "Agro 4.0 Chamber". This effort counted with ministries, the "Agricultural National Confederation" (CNA) and the "Organization of Brazilian Cooperatives" (OCB) [17]. Colombia has also modernized agronomic ICT utilization by directives from the "Ministry of Information and Communication Technologies" (MINTIC) and the "Bogotá Chamber of Commerce" [40].

The El-Salvadorian "Ministry of Agriculture and Livestock" has announced an "Agricultural Intelligence Unit" to monitor productive areas through multispectral and hyperspectral imageries [127, 128]. The Chilean "Ministry of Agriculture" initiated, in 2020, an Agriculture 4.0 line of attack at the ministerial level. It instituted a scientific counseling board, an advisory discussion table with the corporate sector, and working committees [17].

4. Discussion

Treatment with composting takes know-how, time, expertise, equipment, and work. The method and product drawbacks must be weighed against the benefits of composting. Composting is ideal for unstable, non-uniform, hard-to-store organic by-products. Manures, biosolids, and food-handling residuals must be warehoused periodically. After composting, manure becomes drier, more homogeneous, and biologically stable, making it useful beyond land application. Farmers and feedlot owners value composted manures more. Composting degrades random materials like yard cuttings into homogeneous organic mulch. Biosolids dry out as composts and are easier to land-apply. Pathogen-containing by-products are safer following high-temperature composting. Compost products are slow-release fertilizers with a more elevated carbon-to-nitrogen ratio than original end-products.

Composting by-products have limitations, viz., site preparation and equipment expenditures, a long treatment period, targeting the final compost product, and ecological difficulties like odors and dust. Composting may require weeks to stabilize, contingent on the procedure (technology). A good compost market is paramount to substantiate the extra work demands to generate and manage it, viz. biological procedures entail odors if not appropriately controlled. Slow-release compost nutrients require larger application rates to match plant development response than the original composting outcome. Grander application rates demand more material and field trips than the existing by-product supplies, opening an avenue to an Agriculture 4.0 Composting-as-a-Service (CaaS) paradigm.

Composting paybacks and caveats help producers estimate the cost vs. return proportion necessary for initiating processes at any scale. The final treatment outcome and associated benefits from by-product usage may call for environmental adjustments. The suitable by-product to co-compost or utilize with composts can produce value-added yields that can be traded far from their composting source site. The aerobic breakdown of organic waste mixtures in co-composting produces compost for fertilizer or soil enhancement.

5. Conclusion

Composting specialists and other parties undergo multiple "solid waste management" (SWM) obstacles thanks to the snowballing global population. Hence, Latin American/Caribbean alterations in consumption patterns, urbanization, and lifestyle prompt 12% of the world's waste, of which 50% is "organic solid waste" (OSW) in Brazil alone. Other BRICS countries mirror this scenario and escalate globally once a country's GDP sinks. The contamination of landfills by these residues may diminish their viability. Composting using biotechnology is a cost-effective method of managing organic waste that simultaneously provides valuable fertilizers and strengthens the circular economy. Soil microbiota recasts "organic matter" (OM) into bioavailable nutrients, fortifying soils, lessening erosion, and performing as bio-inoculants for decomposition. It mitigates both ground and hydric pollution through the substitution of chemical fertilizers as per the United Nations SDGs. Composting reduces food waste, restores a portion of the ecological equilibrium of the soil, fosters sustainable communities, and mitigates climate change. This essay examined Agriculture 4.0 sustainable and practicable compost elements: (i) biotechnological fundamentals, and (ii) viability of forging and employing this offshoot using Industry 4.0 tools, viz., remote sensing, sensors/actuators, epigenetics, the Internet of Things, novel energy practices, among other possibilities.

References

1. Magazzino, C., Mele, M. and Schneider, N. (2020, July). The relationship between municipal solid waste and greenhouse gas emissions: Evidence from Switzerland. *Waste Manag.*, 15(1)13, 508–520.

2. Zambrano-Monserrate, M.A., Ruano, M.A. and Ormeño-Candelario, V. (2021). Determinants of municipal solid waste: A global analysis by countries' income level. *Environ. Sci. Pollut. Res.*, 28, 62421–62430.
3. Kaza, S., Yao, L., Bhada-Tata, P. and Van Woerden, F. (2018). What a waste 2.0: A global snapshot of solid waste management to 2050. *Urban Development*, Washington, DC: World Bank. CC BY 3.0 IGO.
4. Savino, A., Solórzano, G., Quispe, C. and Correal, M.C. (2018). Waste Management Outlook for Latin America and the Caribbean Waste Management Outlook for Latin America and the Caribbean. UN Environment Programme, Latin America and the Caribbean Office; Panama City, Panama: 2018.
5. Devadoss, P.S.M., Pariatamby, A., Bhatti, M.S., Chenayah, S. and Shahul Hamid, F. (2021). Strategies for reducing greenhouse gas emissions from municipal solid waste management in Pakistan. *Waste Management & Research*, 39, 914–927.
6. Cerda, A., Artola, A., Font, X., Barrena, R., Gea, T. and Sánchez, A. (2018). Composting of food wastes: Status and challenges. *Bioresource Technology*, 248(Pt A), 57–67.
7. ALBREPE. Panorama dos Resíduos Sólidos no Brasil, 2021. https://abrelpe.org.br/panorama. Accessed Oct. 2023.
8. FAO, United Nations Environment Programme. (2021). *Food Waste Index Report 2021*. Nairobi.
9. FAO, United Nations Environment Programme. (2019). *The State of Food and Agriculture: Moving Forward on Food Loss and Waste Reduction*. Rome. CC BY-NC-SA 3.0 IGO.
10. https://smia.munlima.gob.pe/uploads/documento/927d34ddcb1597e8.pdf pg 50 e 51.
11. http://localhost:8080/xmlui/handle/123456789/363.
12. https://www.prefeitura.sp.gov.br/cidade/secretarias/subprefeituras/amlurb/noticias/?p=169047.
13. https://zerowasteworld.org/wp-content/uploads/Suchitoto_espan%CC%83ol.-1.pdf.
14. Avila, K.S. (2020). Planta de Compostaje de Puente Nacional "Un Compromiso de Todos." *Repositorio Institucional UNAD*. https://repository.unad.edu.co/handle/10596/39008.
15. Pérgola, M., Persiani, A., Palese, A.M., Meo, V.D., Pastore, V. et al. (2017). Composting: The way for a sustainable agriculture. *Applied Soil Ecology*, 123, 744–750.
16. Trivelli, C. and Berdegué, J.A. (2019). Rural transformation. Looking towards the future of Latin America and the Caribbean. 2030: Food, agriculture and rural development in Latin America and the Caribbean, No. 1. Santiago. FAO. 76 pp.
17. Sotomayor, O., Ramírez, E. and Martínez, H. (2021). Digitalización y cambio tecnológico en las mipymes agrícolas y agroindustriales en América Latina. Documentos de Proyectos, 46965, UN, CEPAL.
18. Pogorelskaia, I. and Várallyai, L. (2020). Agriculture 4.0 and the role of education. *J. Agricultural Informatics*, 11(1), 45–51. ISSN 2061-862X.
19. Razmjooy, N. and Estrela, V.V. (Eds.) (2019). *Applications of Image Processing and Soft Computing Systems in Agriculture*, pp. 300. Hershey, PA, USA: IGI Global. doi: 10.4018/978-1-5225-8027-0.
20. USDA-NRCS. (2000). *National Engineering Handbook*, Part 637, Environmental Engineering, Chapter 2, Composting. US Department of Agriculture, Natural Resources Conservation Service, Washington, DC. Available at: http://www.nrcs.

usda.gov/Internet/FSE_DOCUMENTS/nrcs143_022229.pdf. Accessed on 26 July 2023.

21. Neugebauer, M. and Sołowiej, P. (2017). The use of green waste to overcome the difficulty in small-scale composting of organic household waste. *Journal of Cleaner Production*, 156, 865–875.
22. Biyada, S., Merzouki, M., Dėmčėnko, T., Vasiliauskienė, D., Urbonavičius, J. et al. (2020). Evolution of microbial composition and enzymatic activities during the composting of textile waste. *Applied Sciences*, 10(11), 3758.
23. Biyada, S., Merzouki, M., Dėmčėnko, T., Vasiliauskienė, D., Ivanec-Goranina, R. et al. (2021). Microbial community dynamics in the mesophilic and thermophilic phases of textile waste composting identified through next-generation sequencing. *Sci. Rep.*, 11, 23624.
24. Lopez, R., Antelo, J., Silva, A., Bento, F. and Fiol, S. (2021). Factors that affect physicochemical and acid-base properties of compost and vermicompost and its potential use as a soil amendment. *J. Environmental Management*, 300, 113702.
25. D'Almeida, M.L.O. and Vilhena, A. (2000). *Lixo municipal: Manual de gerenciamento integrado* (2nd Edn.). São Paulo, IPT, CEMPRE, Brazil.
26. Storey, S., Chualáin, D.N., Doyle, O.P., Clipson, N.J. and Doyle, E. (2015). Comparison of bacterial succession in green waste composts amended with inorganic fertiliser and wastewater treatment plant sludge. *Bioresource Technology*, 179, 71–77.
27. Onwosi, C.O., Igbokwe, V.C., Odimba, J.N., Eke, I.E., Nwankwoala, M.O. et al. (2017). Composting technology in waste stabilization: On the methods, challenges and future prospects. *J. Environmental Management*, 190, 140–157.
28. https://www.ibama.gov.br/component/legislacao/?view=legislacao&legislacao=137380.
29. El Fels, L., El Ouaqoudi, F.Z., Barje, F., Hafidi, M. and Ouhdouch, Y. (2014). Two culture approaches used to determine the co-composting stages by assess of the total microflora changes during sewage sludge and date palm waste co-composting. *J. Environmental Health Science and Engineering*, 12.
30. Biyada, S., Merzouki, M., Imtara, H. and Benlemlih, M. (2019). Assessment of the maturity of textile waste compost and their capacity of fertilization. *European J. Scientific Research*, 154(3), 399–412.
31. Chiarelotto, M., Restrepo, J.C., Lorin, H.E. and Damaceno, F.M. (2021, Dec.). Composting organic waste from the broiler production chain: A perspective for the circular economy. *J. Cleaner Production*, 329, 129717.
32. Palansooriya, K.N., Shaheen, S.M., Chen, S.S., Tsang, D.C., Hashimoto, et al. (2019). Soil amendments for immobilization of potentially toxic elements in contaminated soils: A critical review. *Environment International*, 134, 105046.
33. Morra, L., Bilotto, M., Baldantoni, D., Alfani, A. and Baiano, S. (2021). A seven-year experiment in a vegetable crops sequence: Effects of replacing mineral fertilizers with Biowaste compost on crop productivity, soil organic carbon, and nitrates concentrations. *Scientia Horticulturae*, 290, 110534.
34. CONAMA, Resolução n° 481/17, 2017. Available from: http://conama.mma.gov.br/?option=com_sisconama&task=arquivo.download&id=702 Acesso em 27 de dezembro de 2021.
35. http://conama.mma.gov.br/?option=com_sisconama&task=arquivo.download&id=702.

36. https://www.gov.br/agricultura/pt-br/assuntos/insumos-agropecuarios/insumos-agricolas/fertilizantes/legislacao/in-sda-27-de-05-06-2006-alterada-pela-in-sda-07-de-12-4-16-republicada-em-2-5-16.pdf.
37. https://www.in.gov.br/en/web/dou/-/instrucao-normativa-n-61-de-8-de-julho-de-2020-266802148.
38. http://www.ingeachile.cl/descargas/normativa/agricola/NCH2880.pdf.
39. https://sma.edomex.gob.mx/sites/sma.edomex.gob.mx/files/files/sma_pdf_ntea_006_sma_rs_2006.pdf.
40. UN. https://es.scribd.com/doc/138627365/NTC-5167-Apartes1-1-Norma-Para-Abonos-Organicos-y-Fertilizantes.
41. Wichuk, K. and McCartney, D. (2007). A review of the effectiveness of current time-temperature regulations on pathogen inactivation during composting. *J. Environmental Engineering and Science*, 6, 573–586.
42. Lepesteur, M. (2021). Human and livestock pathogens and their control during composting. *Critical Reviews in Environmental Science and Technology*, 52, 1639–1683.
43. Wichuk, K., Tewari, J.P. and McCartney, D. (2011). Plant pathogen eradication during composting: A literature review. *Compost Science & Utilization*, 19, 244–266.
44. FAO (2013). https://www.fao.org/3/i3388s/I3388S.pdf.
45. Fierer, N. (2017). Embracing the unknown: Disentangling the complexities of the soil microbiome. *Nat. Rev. Microbiol.*, 15, 579–590. https://doi.org/10.1038/nrmicro.2017.87.
46. Luo, S., Chen, L., Lu, Y., Daia, S., Sun, D. et al. (2021). The microbial community structure of rhizosphere soil was influenced by different sugarcane varieties with different ratooning abilities. *Sugar Tech.*, 23, 1306–1316. https://doi.org/10.1007/s12355-021-01019-w.
47. Rao, D.L.N., Aparna, K. and Mohanty, S.R. (2019). Microbiology and biochemistry of soil organic matter, carbon sequestration, and soil health. *Indian J. Fertil.*, 15, 124–138.
48. Huang, Y., Dai, Z., Lin, J., Li, D., Ye, H. et al. (2021, Dec.). Labile carbon facilitated phosphorus solubilization as regulated by bacterial and fungal communities in *Zea mays*. *Soil Biology and Biochemistry*, 163, 108465.
49. Ma, B., Stirling, E., Liu, Y., Zhao, K., Zhou, J. et al. (2021). Soil biogeochemical cycle couplings inferred from a function-taxon network. *Research*, 2021. Article ID: 7102769.
50. Bonfim, F.P.G., Honorio, I.C.G., Reis, I.L., Pereira, A. de J. and Souza, D.B. de. (2011). Caderno dos Microrganismos Eficientes (EM). *Instruções práticas sobre uso ecológico e social do EM.* (2nd Edn.).
51. Chatterjee, D., Nayak, A.K., Mishra, A., Swain, C.K., Kumar, U. et al. (2021). Effect of long-term organic fertilization in flooded rice soil on phosphorus transformation and phosphate solubilizing microorganisms. *J. Soil Sci. Plant Nutr.*, 21, 1368–1381.
52. Liu, X.J., Hayer, M., Mau, R.L., Schwartz, E., Dijkstra, P. and Hungate, B.A. (2021, Dec.). Substrate stoichiometric regulation of microbial respiration and community dynamics across four different ecosystems. *Soil Biology and Biochemistry,* 163, 108458.
53. Zhang, M., Liu, Y., Wei, Q. and Gou, J. (2021). Effects of short-term application of Moutai lees biochar on nutrients and fungal community structure in yellow soil of Guizhou. *Environ. Sci. Pollut. Res.*, 28, 67404–67413.
54. Sáez, J., Clemente, R., Bustamante, M.A., Yañez, D.R. and Bernal, M.P. (2017).

Evaluation of the slurry management strategy and the integration of the composting technology in a pig farm: Agronomical and environmental implications. *J. Environmental Management*, 192, 57–67.
55. Forte, A., Fagnano, M. and Fierro, A. (2017). Potential role of compost and green manure amendment to mitigate soil GHGs emissions in Mediterranean drip irrigated maize production systems. *J. Environmental Management*, 192, 68–78.
56. Meng, X., Yan, J., Zuo, B., Wang, Y., Yuan, X. and Cui, Z. (2020). Full-scale of composting process of biogas residues from corn stover anaerobic digestion: Physical-chemical, biology parameters, and maturity indexes during whole process. *Bioresource Technology*, 302, 122742.
57. Abreu, L.M., Hora, H.R.M., Rangel, J.J.A., Erthal, Jr. M., Razmjooy, N, Estrela, V.V. et al. (2020). A multi-criteria modelling for ranking CO_2 emitting G20 countries from the Kaya Identity and their impacts on elderly health. *In*: Iano Y. et al. (Eds.), *Proc. of the 5th Brazilian Technology Symposium* (*BTSym'19*). BTSym. 2019. Smart Innovation, Systems and Technologies, Cham: Springer. doi: 10.1007/978-3-030-16053-1_62.
58. Hameed, M., Bhat, R.A., Pandit, B.A., Ramzan, S., Dijoo, Z.K. and Wani, M.A. (2021). Qualitative assessment of compost engendered from municipal solid waste and green waste by indexing method. *J. The Air & Waste Management Association*, 72, 210–219.
59. Iqbal, M.K., Khan, R.A., Nadeem, A. and Hussnain, A. (2012). Comparative study of different techniques of composting and their stability evaluation in municipal solid waste. *J. Chem. Soc. Pak.*, 34(2), 273–282.
60. Ayilara, M.S., Olanrewaju, O.S., Babalola, O.O. and Odeyemi, O. (2020). Waste management through composting: Challenges and potentials. *Sustainability*, 12(11), 4456.
61. Daud, N.M., Khalid, S.A., Nawawi, W.N. and Ramli, N. (2016). Producing fertilizer from food waste recycling using Berkeley and Bokashi method. *Ponte*, 72(4). doi: 10.21506/j.ponte.2016.4.11
62. Lew, P.S., Nik Ibrahim, N.N.L., Kamarudin, S., Thamrin, N.M. and Misnan, M.F. (2021). Optimization of Bokashi-composting process using effective microorganisms-1 in smart composting bin. *Sensors*, 21(8), 2847.
63. Manyapu, V., Mandpe, A. and Kumar, S. (2018). Synergistic effect of fly ash in in-vessel composting of biomass and kitchen waste. *Bioresource Technology*, 251, 114–120.
64. Xu, Z., Zhao, B., Wang, Y., Xiao, J. and Wang, X. (2019). Composting process and odor emission varied in windrow and trough composting system under different air humidity conditions. *Bioresource Technology*, 297, 122482.
65. Liu, Z., Wang, X., Wang, F., Bai, Z., Chadwick, D. et al. (2020). The progress of composting technologies from static heap to intelligent reactor: Benefits and limitations. *J. Cleaner Production*, 270, 122328. Elsevier.
66. More, A. and More, S. (2019). Enriched rapid composting techniques for agro industrial organic solid waste. International *J. Innovative Technology and Exploring Engineering*, 8(10), 4032–4038. Blue Eyes Intelligence Engineering and Sciences Engineering and Sciences Publication – BEIESP.
67. Shah, G.M., Tufail, N., Bakhat, H.F., Ahmad, I., Shahid, M. et al. (2019). Composting of municipal solid waste by different methods improved the growth of vegetables and reduced the health risks of cadmium and lead. *Environmental Science and Pollution Research Springer Science and Business Media LLC*, 26(6), 5463–5474.

68. Manu, M.K., Kumar, R. and Garg, A. (2017). Performance assessment of improved composting system for food waste with varying aeration and use of microbial inoculum. *Bioresource Technology*, 234, 167–177. Elsevier.
69. Wei, Y., Wu, D., Wei, D., Zhao, Y., Wu, J. et al. (2019). Improved lignocellulose-degrading performance during straw composting from diverse sources with actinomycetes inoculation by regulating the key enzyme activities. *Bioresource Technology*, 271, 66–74. Elsevier.
70. Selvamani, K., Annadurai, V. and Soundarapandian, S. (2019). Improved co-composting of poultry manure with complementary consortium of indigenous Bacillus spp. *Biotech.*, 9(6). Springer.
71. Manyapu, V., Mandpe, A. and Kumar, S. (2018). Synergistic effect of fly ash in in-vessel composting of biomass and kitchen waste. *Bioresource Technology*, 251, 114–120. Elsevier.
72. Pandit, N.R., Schmidt, H.P., Mulder, J., Hale, S.E., Husson, O. and Cornelissen, G. (2019). Nutrient effect of various composting methods with and without biochar on soil fertility and maize growth. *Arch. Agronomy and Soil Science*, 662(2), 250–265. Informa UK Limited.
73. Mu, D., Horowitz, N., Casey, M. and Jones, K. (2017). Environmental and economic analysis of an in-vessel food waste composting system at Kean University in the US Waste Management, 59, 476–486. Elsevier.
74. Bong, C.P.C., Lim, L.Y., Ho, W.S., Lim, J.S., Klemeš, J.J., Towprayoon, S. et al. (2017). A review on the global warming potential of cleaner composting and mitigation strategies. *J. Cleaner Production*, 146, 149–157. Elsevier.
75. Biala, J., Wilkinson, K., Henry, B., Singh, S., Bennett-Jones, J. and De Rosa, D. (2021). The potential for enhancing soil carbon levels through the use of organic soil amendments in Queensland, Australia. *Regional Environmental Change, Springer Science and Business Media LLC*, 21(4).
76. Muscolo, A., Papalia, T., Settineri, G., Mallamaci, C. and Jeske-Kaczanowska, A. (2018). Are raw materials or composting conditions and time that most influence the maturity and/or quality of composts? Comparison of obtained composts on soil properties. *J. Cleaner Production*, 195, 93–101. Elsevier.
77. Guo, X., Liu, H. and Wu, S. (2019). Humic substances developed during organic waste composting: Formation mechanisms, structural properties, and agronomic functions. *Science of the Total Environment*, 662, 501–510. Elsevier.
78. De Corato, U. (2021). Effect of value-added organic co-products from four industrial chains on functioning of plant disease suppressive soil and their potentiality to enhance soil quality: A review from the perspective of a circular economy. *Applied Soil Ecology*, 168, 104221. Elsevier.
79. Van Tung, T., Tran, Q.B., Phuong Thao, N.T., Vi, L.Q., Hieu, T.T. et al. (2021). Recycling of aquaculture wastewater and sediment for sustainable corn and water spinach production. *Chemosphere*, 268, 129329. Elsevier.
80. Wei, W., Yan, Y., Cao, J., Christie, P., Zhang, F. and Fan, M. (2016). Effects of combined application of organic amendments and fertilizers on crop yield and soil organic matter: An integrated analysis of long-term experiments. *Agriculture, Ecosystems & Environment*, 225, 86–92. Elsevier.
81. Roussos, P.A., Gasparatos, D., Kechrologou, K., Katsenos, P. and Bouchagier, P. (2017). Impact of organic fertilization on soil properties, plant physiology and yield in two newly planted olive (*Oleaeuropaea* L.) cultivars under Mediterranean conditions. *Scientia Horticulturae*, 220, 11–19. Elsevier.

82. Paradelo, R., Lerch, T.Z., Houot, S. and Dignac, M.-F. (2019). Composting modifies the patterns of incorporation of OC and N from plant residues into soil aggregates. *Geoderma*, 353, 415–422.
83. Monostori, L. (2014). Cyber-physical production systems: Roots, expectations, and R&D challenges. *Procedia CIRP*, 17, 9–13.
84. Zambon, I., Cecchini, M., Egidi, G., Saporito, M.G. and Colantoni, A. (2019). Revolution 4.0: Industry vs. Agriculture in a Future Development for SMEs. *Processes, MDPI*, 7(1), 6.
85. Aroma, R.J., Raimond, K., Estrela, V.V. and de Jesus, M.A. (2023). A coastal band spectral combination for water body extraction using Landsat 8 images. *Int. J. Environ. Sci. Technol.* doi: 10.1007/s13762-023-05027-z.
86. Khan, A.A., Laghari, A., Gadekallu, T.R., Shaikh, Z.A., Javed, A.R., Rashid, M. et al. (2022). A drone-based data management and optimization using metaheuristic algorithms and blockchain smart contracts in a secure fog environment. *Computers and Electrical Engineering*, 102, 108234.
87. Deshpande, A., Estrela, V.V. and Razmjooy, N. (2021). *Computational Intelligence Methods for Super-Resolution in Image Processing Applications*. Zurich, Switzerland: Springer Nature. doi: 10.1007/978-3-030-67921-7.
88. Searchinger, T., Waite, R., Hanson, C., Ranganathan, J., Dumas, P. and Matthews, E. (2018). Creating a Sustainable Food Future: A menu of solutions to feed nearly 10 billion people by 2050. World Resources Report, ISBN 978-1-56973-953-6 https://research.wri.org/sites/default/files/2019-07/creating-sustainable-food-future2_5.pdf.
89. Clercq, M., De Vats, A. and Biel, A. (2018). *Agriculture 4.0: The Future of Farming Technology*. The World Government Summit, Dubai. https://www.worldgovernmentsummit.org.
90. De Medeiros, M.B. and da Silva Lopes, J. (2006). Biofertilizantes líquidos e sustentabilidade agrícola. *Bahia Agrícola. Salvador, Bahia, Brazil*, 7. http://www.agencia.cnptia.embrapa.br/Repositorio/Biofertilizantes+1_000g76q0gvw02wx5ok0wtedt3kadue0d.pdf.
91. Ciavatta, C., Govi, M., Pasotti, L. and Sequi, P. (1993). Changes in organic matter during stabilization of compost from municipal solid wastes. *Bioresource Technology*, 43(2), 141–145. Elsevier.
92. Sarpong, D., Oduro-Kwarteng, S., Gyasi, S.F., Buamah, R., Donkor, E. et al. (2019). Biodegradation by composting of municipal organic solid waste into organic fertilizer using the black soldier fly (*Hermetiaillucens*) (Diptera: Stratiomyidae) larvae. *International J. Recycling of Organic Waste in Agriculture*, 8(S1), 45–54. Springer Science and Business Media LLC.
93. Riffaldi, R., Levi-Minzi, R., Pera, A. and de Bertoidi, M. (1986). Evaluation of compost maturity by means of chemical and microbial analyses. *Waste Management & Research: The J. Sustainable Circular Economy*, 4(1), 387–396. SAGE Publications.
94. Tiquia, S.M., Tam, N.F.Y. and Hodgkiss, I.J. (1996). Effects of composting on phytotoxicity of spent pig-manure sawdust litter. *Environmental Pollution*, 93(3), 249–256. Elsevier.
95. Yahya, N. (2018). Agricultural 4.0: Its implementation toward future sustainability. *Green Urea*, 125–145. Singapore: Springer.
96. Maucieri, C., Nicoletto, C., Os, E., van Anseeuw, D., Havermaet, R.V. and Junge, R. (2019). Hydroponic technologies. *In*: Goddek, S. et al. (Eds.), *Aquaponics Food Production Systems*, 77–110, Cham: Springer International Publishing.

97. Capulín-Grande, J., Núñez-Escobar, R., Sánchez-García, P., Martínez-Garza, A. and Soto-Hernández, M. (2005). Producción de jitomate con estiércol líquido de bovino, acidulado con ácidos orgánicos e inorgánicos. *Terra Latinoamericana*, 23, núm. 2, abril-junio, 241–247. Chapingo, México: Sociedad Mexicana de la Ciencia del Suelo, A.C.
98. Faruq, H., Novelia, E., Setyaningsih, M. and Nisa, R.A. (2021). The utilization of vegetable waste as a nutrient addition in hydroponic media for the growth of green mustard (*Brassica juncea* L.). *IOP Conference Series: Earth and Environmental Science*, 755(1), 012078. IOP Publishing.
99. An, B.N.T. and Anh, N.T.N. (2020). Co-culture of Nile tilapia (*Oreochromis niloticus*) and red seaweed (*Gracilaria tenuistipitata*) under different feeding rates: Effects on water quality, fish growth, and feed efficiency. *J. Applied Phycology*, 32(3), 2031–2040. Springer.
100. Emblemsvåg, J., Kvadsheim, N.P., Halfdanarson, J., Koesling, M., Nystrand, B.T. et al. (2020). Strategic considerations for establishing a large-scale seaweed industry based on fish feed application: A Norwegian case study. *J. Applied Phycology*, 32(6), 4159–4169. Springer.
101. De Alencar, J.R., Junior, P.A.H. and Celino, J.J. (2010). Cultivo de Camarão Branco Litopenaeus Vannamei (Boone, 1931) com a Macro-alga Ulva Lacuata Linneaus (Chlorophyta) no Tratamento de Efluentes em Sistema Fechado de Recirculação. *Revista de Biologia e Ciências da Terra*, 10(1), 117–137.
102. Postaue, N., Moraes, L.C. and Asmus, R.M.F. (2020). Chorume como fonte de nutriente na produção da biomassa microalgal. *E-xacta*, 12(2), 11. Revista Exacta.
103. Kale, G., Kijchavengkul, T., Auras, R., Rubino, M., Selke, S.E. et al. (2007). Compostability of bioplastic packaging materials: An overview. *Macromolecular Bioscience*, 7(3), 255–277. Wiley.
104. Vaverková, M. and Adamcová, D. (2015). Biodegrability of bioplastic materials in a controlled composting environment. *J. Ecological Engineering*, 16, 155–160. Wydawnictwo Naukowe Gabriel Borowski (WNGB).
105. Ruggero, F., Gori, R. and Lubello, C. (2019). Methodologies to assess biodegradation of bioplastics during aerobic composting and anaerobic digestion: A review. *Waste Management & Research*, 37(10), 959–975. SAGE Publications.
106. Cruz, R.F., Gerude Neto, O.J. de A., Freitas, S.J.N. de, Rodrigues, J.B. and Silva, D.L.L. da. (2019). A aplicabilidade do chorume oriundo do processo de compostagem biofertlizante orgânico para agricultura sustentável. *Nature and Conservation*, 12(3), 37–48. Companhia Brasileira de Producao Cientifica (CBPC).
107. https://www.repository.utl.pt/handle/10400.5/5674.
108. Lucena, L.P. de and Silva, C.E.S.F. da. (2018). Modelos de agricultura urbana para a segurança alimentar: um estudo comparativo entre Singapura e Brasil. *Revista Ibero-Americana de Ciências Ambientais*, 9(3), 379–397. Brazil: CBPC,
109. http://sertaosustentavel.com.br/index.php/home/article/view/14.
110. Pereira, H.R., Alves, R.N., Chaves, J. da S., Matos, S.M. de, Nascimento, J.P.S. do et al. (2020). Compostagem de serragem e adubação da cultura da abóbora no Sul do estado de Roraima. *Research, Society and Development*, 9(11), e929119510.
111. Sánchez, Ó.J., Ospina, D.A. and Montoya, S. (2017). Compost supplementation with nutrients and microorganisms in composting process. *Waste Management*, 69, 136–153. Elsevier.
112. https://www.redalyc.org/jatsRepo/5528/552861694029/552861694029.pdf.

113. Srivastava, V., de Araujo, A.S.F., Vaish, B., Bartelt-Hunt, S., Singh, P. and Singh, R.P. (2016). Biological response of using municipal solid waste compost in agriculture as fertilizer supplement. *Reviews in Environmental Science and Bio/Technology*, 15(4), 677–696. Springer.
114. Kowaljow, E. and Mazzarino, M.J. (2007). Soil restoration in semiarid Patagonia: Chemical and biological response to different compost quality. *Soil Biology and Biochemistry*, 39(7), 1580–1588. Elsevier.
115. Kranz, C.N., McLaughlin, R.A., Johnson, A., Miller, G. and Heitman, J.L. (2020). The effects of compost incorporation on soil physical properties in urban soils: A concise review. *J. Environmental Management*, 261, 110209. Elsevier.
116. Deshpande, A., Estrela, V.V., Patavardhan, P. and Kallimani, G. (2023). Super-resolution in a world of scarce resources for medical imaging applications, *In*: Vyas, S. et al. (Eds.), *Edge-AI in Healthcare: Trends and Future Perspectives*, Chapter 13, London, UK: CRC Press.
117. Loschi, H.J., Estrela, V.V., Hemanth, D.J., Fernandes, S.R., Iano, Y., Laghari, A.A. et al. (2020). Communications requirements, video streaming, communications links and networked UAVs. *In*: Estrela, V.V., Hemanth, J., Saotome, O., Nikolakopoulos, G., Sabatini, R. (Eds.), *Imaging and Sensing for Unmanned Aircraft Systems*, 2(6), 113–132. London, UK: IET, doi: 10.1049/PBCE120G_ch6.
118. Gopikumar, S., Raja, S., Robinson, Y.H., Shanmuganathan, V., Chang, H. and Rho, S. (2021). A method of landfill leachate management using Internet of Things for sustainable smart city development. *Sustainable Cities and Society*, 66, 102521. Elsevier.
119. Pratama, Y.F., Ariyanto, E. and Karimah, S.A. (2019). Prototype of automation of organic fertilizer manufacturing systems based on internet of things. *Proc. 7th International Conference on Information and Communication Technology* (*ICoICT*). IEEE. https://doi.org/10.1109/icoict.2019.8835299.
120. Rawal, N., Acharya, K.K., Bam, C.R. and Acharya, K. (2018). Soil fertility mapping of different VDCs of Sunsari District, Nepal using GIS. *International J. Applied Sciences and Biotechnology*, 6(2), 142–151. Nepal Journals Online (JOL).
121. Razmjooy, N., Estrela, V.V. and Sabatini, R. (2020). Vision in micro-aerial vehicles. *In*: Estrela, V.V., Hemanth J., Saotome O., Nikolakopoulos G., Sabatini R. (Eds.), *Imaging and Sensing for Unmanned Aircraft, Systems*, 1(8), 173–216. London, UK: IET. doi: 10.1049/PBCE120F_ch8.
122. Andrade, R.O. (2016). Drones sobre o campo – Avanços tecnológicos ampliam as possibilidades do uso de aeronaves não tripuladas na agricultura. *Revista Pesquisa FAPESP*, January, 239, Sao Paulo, Brazil. Creative Commons CC-BY-NC-ND. https://revistapesquisa.fapesp.br/drones-sobre-o-campo/?cat=tecnologia/
123. Akanbi, W.B., Adebayo, T.A., Togun, O.A., Adeyeye, A.S. and Olaniran, O.A. (2007). The use of compost extract as foliar spray nutrient source and botanical insecticide in *Telfairia occidentalis*. *World J. Agricultural Sciences*, 3(5), 642–652. ISSN 1817-3047. IDOSI Publications.
124. Romero, C., Ramos, P., Costa, C. and Márquez, M.C. (2013). Raw and digested municipal waste compost leachate as potential fertilizer: Comparison with a commercial fertilizer. *J. Cleaner Production*, 59, 73–78. Elsevier.
125. Vital, M., Howe, A.C. and Tiedje, J.M. (2014 Apr. 22). Revealing the bacterial butyrate synthesis pathways by analyzing (meta)genomic data. *mBio.*, 5(2), e00889. doi: 10.1128/mBio.00889-14.

126. Özer, N. and Köycü, N.D. (2006). The ability of plant compost leachates to control black mold (*Aspergillus niger*) and to induce the accumulation of antifungal compounds in onion following seed treatment. *Biocontrol.*, 51(2), 229–243. Springer. https://doi.org/10.1007/s10526-005-1035-1.
127. Moreno, S.C. and Aedo, M.V.A. (2021). *Análisis de las políticas públicas e iniciativas privadas que apoyan el uso de las tecnologías digitales en las mipymes agrícolas y agroindustriales en El Salvador*. Documentos de Proyectos (LC/TS.2021/64), Santiago, Comisión Económica para América Latina y el Caribe (CEPAL), UN.
128. Aroma, R.J., Raimond, K., Razmjooy, N., Estrela, V.V. and Hemanth, J. (2020). Multispectral vs. hyperspectral imaging for unmanned aerial vehicles: Current and prospective state of affairs. *In*: Estrela, V.V., Hemanth, J., Saotome, O., Nikolakopoulos, G., Sabatini, R. (Eds.), *Imaging and Sensing for Unmanned Aircraft Systems*, 2(7), 133–156, London, UK: IET.
129. Laghari, A.A., Zhang, X., Shaikh, Z.A., Khan, A., Estrela, V.V. and Izadi, S. (2023). A review on quality of experience (QoE) in cloud computing. *J. Reliable Intell. Environ.* doi: 10.1007/s40860-023-00210-y.
130. Estrela, V.V., de Jesus M.A., Aroma, J., Raimond, K., Fernandes, S.R., Andreopoulos, N. et al. (2021). Motion estimation role in the context of 3D video. *International J. Multimedia Data Engineering and Management* (*IJMDEM*), Hershey, PA, USA: IGI Global.
131. Ruiz, Leite, E.C.R., Ruiz, A.M.C. and Aguiar, T. de F. (2005). Educação ambiental e os temas transversais. *Revista de Ciências Humanas da UNIPAR, Akrópolis, Umuarama*, 13(1). jan./mar. Acesso em: 15 out. 2022.

CHAPTER

13

Blockchain and Internet of Things (IoT) Enabled Secure Communication Protocols for Smart Agriculture

Khuda Bukhsh[1,2*][0009-0008-8708-250X], M. Malook Rind[1][0000-0002-3794-1790], Sarmad Shaikh[1][0000-0003-1885-8809], Abdullah A. Khan[1,2][0000-0003-2838-7641], Asif Ali Laghari[1,2][0000-0001-5831- 5943], Vania V. Estrela[3][0000-0002-4465-7691] and Nikolaos Andreopoulos[4][0000-0001-9975-7043]

[1] Faculty of Computer Science, Sindh Madressatul Islam University, Karachi (74000), Sindh, Pakistan

[2] Research Lab of Artificial Intelligence and Information Security, Faculty of Computing Science and Information Technology, Benazir Bhutto Shaheed University Lyari, Karachi (75660), Sindh, Pakistan

[3] Telecommunications Department, Universidade Federal Fluminense (UFF), Rio de Janeiro, Brazil

[4] Computer Science Department, Technological Institute of Iceland Sæmundargata 2, 102, Reykjavik, Ssn. 600169-2039, Iceland

Nowadays, the food demand is augmenting, and everyone needs quality food. Agriculture is the custom of farming or cultivating livestock and plants. "Information and communication technology" (ICT) has aided in crafting remediation strategies for the husbandry sector's ecological devastation. The "Internet of Things" (IoT) qualifies sensors and apparatuses to interact and connect to interchange data in wireless or wired networks via communication protocols. These sensors and devices measure all kinds of data remotely and provide real-time information (e.g., soil moisture, chemical application levels, livestock health, dam level, etc.) to farmers in agribusiness. Albeit, they also unlocked the doors for attackers' outbreaks on IoT communication systems to get sensitive information. This chapter introduces IoT communication protocols, security, and privacy concerns. This chapter discusses attacks on communication

* Corresponding author: khudabukhsh1992@gmail.com

protocols. Also, it provides an IoT Blockchain-enabled secure communication architecture for agriculture evolving growing threats in industry 4.0.

1. Introduction

The 'Agriculture' word originated from the Latin 'Agri' and culture. Agriculture is cultivating, growing food plants, and planting on a piece of land. Taking care of animals for meat, eggs, and milk is also a category of agriculture. In Agriculture 1.0, from 1784 to 1870, the farmers relied on indigenous tools for cultivating. Such farming required manual labor till the end of the 19th-century. Agriculture 1.0's food production level was deficient [1, 47]. At the commencement of the 20th century, food production increased, and manual labor was reduced when machinery was introduced for sowing, weeding, harvesting, and seedbed preparation. Advancement in embedded systems, communication technologies, and software engineering. Agriculture 3.0 explored information technology for yield monitoring, precision agriculture, guidance farming system, and variable rate application [2, 48]. By 2050, a 31% rise in world population and climate alterations is expected. The demand for food production and natural resource will also increase by 71% due to the populational surge and type of weather change. One must move from traditional farming to smart farming (SF) technologies. Agriculture 4.0 is referred to the adoption of communication and information technologies from industry 4.0, viz., Machine Learning (ML), Artificial Intelligent (AI), Internet of Things (IoT), Blockchain, and Big Data (BD) to use these emerging technologies to deal with agricultural scenario changes and meet the food demand due to increase in population [3, 4, 49].

The IoT concept was propositioned by Kevin Ashton in 1999. The word Internet and Things mean worldwide or interconnected networks based on communication and sensor, information processes, and networking, which is the new era of communication and information technology [5, 50, 51]. IoT empowers sensors and devices to interact and connect to exchange data via communication protocols via wireless or wired networks [6]. These sensors and devices measure all kinds of data remotely and provide farmers with real-time information (chemical application, soil moisture, livestock health, dam level, etc.). IoT Applications use different communication protocols. Some examples are the Zigbee, "radio frequency identification" (RFID), Z-wave, "near field communication" (NFC), SigFox, "LORA (short for long range) wireless area network" (LoRaWAN), "message queuing telemetry transport" (MQTT), Bluetooth, Wi-Fi, and cellular networks established on the need. Notwithstanding, it also opened the doors for attackers' threats on IoT communication systems to get sensitive information.

Section 2 overviews interrelated work. Section 3 revises the most significant IoT communication protocols. Section 4 performs a comparison between communication protocols. IoT challenges are itemized in Section 5. Section 6 sketches a possible Blockchain-enabled IoT communication protocol architecture.

The proposed Architectural-Based Implementation Challenges and Limitations are debated in Section 7. Section 8 completes this chapter.

2. Related Work

LoRa is a wireless audio frequency technological advance in a spectrum of unlicensed radio frequencies. LoRa employs a physical layer protocol that exploits spread spectrum modulation and facilitates long-range communication at the expense of bandwidth. Smart agriculture (SA) exploiting IoT multi-sensors or watering management occurs in [7], whose solution utilizes a LoRa transmission module for an affordable and effective SA system in Vietnam, designed and realized to regulate environmental factors in agronomy. It arranges a solution for accurate real-time water management surveillance that ameliorates power consumption for SA with cost-savings, continuously observes measured parameter values, and broadcasts them to handlers with the assistance of networks and mobile applications to gauge the dehydration, humidity, Ph, and temperature values.

A drone and IoT architecture aids in detecting rice disease in its early stages with a position mapping system [8], employing a system that finds the rice disease and catches it with location mapping through IoT combined with drones. Drones' exploitation helps survey the field in a large-scale area, takes less time, and supports real-time data. This system can also classify what types of diseases appear on the leaf by extracting color and texture pattern features [52].

The work in [9] focuses on introducing a crop management smart drone where real-time "unmanned aerial vehicle" (UAV) records blend with IoT and Cloud Computing (CC) technologies to help build sustainable smart agriculture. The IoT and CC involvement in the agricultural sector would result in better production of crops by controlling the cost, monitoring performance, and maintenance [53].

This chapter takes an approach to monitoring environmental data on a farm and collecting this data using LoraWAN and a drone [10]. It monitors forestry with sensor nodes and a single gateway. The sensor nodes are located at each area in a farmstead collecting environmental data. Each node uses temperature and humidity sensors, a light sensor, a real-time clock module, and a micro-SD unit. The data are sent to a gateway connected to a drone and then transferred into a database server [54].

This chapter contains the studies and reviews about incorporating wireless networks and sensor/actuator technologies for IoT within SA systems [11], hinging on a "Remote Monitoring System" (RMS). It has a combined wireless communication approach with the Internet. This system affords real-time data to the farmer from the agriculture production environment, e.g., alerts through SMS, and suggests advice on crops, weather patterns, etc. [55].

3. IoT Communication Protocols

3.1 Zigbee

Some complex systems may employ this wireless communication protocol. It offers high security and consumes low power via the IEEE 802.15.4 standard, and partakes a 2.4-Ghz frequency with a 10–100 m approximate range and 250 kbps Data Rates [12, 13, 56]. Its robust security features include device management, key transport, cryptographic key establishment, and frame protection [14]. Zigbee relies on three types of keys: link, master, and network. All keys rely on 128-bit AES encryption. All network nodes utilize a network key. It inhibits unauthorized gadgets from joining the network and protects network frames. The device's master key uses for generating link keys. Each device partakes in a master key to create a link key. It provides a secret session between Two Communicating devices. The MAC layer includes security services support: (1) encryption, (2) access control, (3) frame integrity, and (4) sequential freshness. These services used in three modes: secure, insecure, and access control list [13]. Furthermore, Zigbee has some types, i.e., (a) XBee S1, (b) XBee S2, (c) XBee Pro S1, and (d) XBee Pro S. XBee uses star, mesh, and tree, while the S2 type uses mesh and tree depending on the communication type between Zigbee [15].

3.2 Radio Frequency Identification (RFID)

The RFID authentication protocol can authenticate/identify a being or object through radio waves. RFID is paramount in IoT to identify objects in various fields. RFID utilizes specific tags to authenticate/identify any object. Every RFID tag possesses unique identification and it is encrypted for the RFID reader. RFID tags' categories are active, passive, and semi-passive depending on the power supply method [16, 17].

3.3 Near Field Communication (NFC)

It is a pervasive wireless communication short-range-high frequency ISO/ IEC 18000-3 standard that works in 13.56 MHz (ISM) frequencies. NFC-enabled devices can converse with each other over data rates in the interval of 100–420 kbs and inside the 10 cm range (approximately) [18, 57]. Three data exchange processes/modes exist: (1) read/write, (2) card emulation, and (3) peer-to-peer. In read/write mode, it reads information from NFC tags that accumulate a small data amount in one's microchips for transforming to other NFC tag-enabled apparatuses, like smart devices. Card emulation mode enables these devices to act as contactless smart cards, and peer-to-peer mode allows two NFC-based devices to communicate and exchange information [19].

3.4 LoRaWAN

The "LORA wireless area network" (LoRaWAN), also called a "low-power, wide-area network" (LPWAN), remains the most potent communication protocol currently for Machine-to-Machine (M2M) communication. It is specially planned to provide inexpensive, secure IoT devices' bidirectional communication [2]. It also delivers a low rate and low power to transfer information developed to overcome IoT application challenges. LoRaWAN employs a star network topology, decreasing network complexity and saving battery life. The start network topology nodes only take their data [6, 58].

3.5 SigFox

It is a low-cost, wide-area network coverage communication protocol with SigFox Standard. It operates in an unlicensed (Industrial, Scientific, Medical) 900 Mhz frequency spectrum band intending to provide Machine-to-Machine (M2M) communication with a wide-area coverage range [20] 30–50 km rural environment and 3–10 km urban environments with a data range of 10–1000 bps. It is a better selection for remote equipment that runs on battery power for extended periods without changing the battery changes or maintenance. Its bidirectional communication functionality can also be used in IoT infrastructure (i.e., home and end-user goods, energy-related communication, retail, shelf, PoS, and transportation). The hardware and software of the point-of-sale (POS) system allow one to sell, collect payments, and check out clients. POS information is acquired during a sale transaction with a client. A POS system administers every in-store, online, and mobile purchase a consumer makes. It can arrange product transactions', inventory, customers' and employees' information.

3.6 Z-Wave

Silicon Labs owns the Z-Wave trademark and technology. It is a wireless communication protocol that is best for home automation projects. It works on 900 MHz frequencies within a 30 m range, has inexpensive power intake, and utilizes the mesh network topology.

3.7 Message Queuing Telemetry Transport (MQTT)

The "International Organization for Standardization" (ISO)/IEC 20922 standard works on the Application Layer of TCP/IP standard [21]. The MQTT is a lightweight message transmission protocol that transmits data up to 256 Mb. It is designed for restricted equipment with low bandwidth, undependable networks, and high latency. MQTT protocol uses three key elements: publisher, subscriber, and broker, which suits M2M, Wireless Sensors Networks, and IoT applications [22]. MQTT is good, reliable, and efficient for unreliable networks to transmit data [23]. Facebook is using MQTT in one's Facebook messenger application for a chat.

3.8 Bluetooth

Bluetooth is best for short-range digital data transmission [24] and suitable communication for sending small data chunks [25]. Bluetooth uses a 4.2 core specification standard with 2.4 Ghz (ISM) frequency to spread data up to 50–150 m with a 1-Mbps data rate.

3.9 Wi-Fi

Wi-Fi is excellent and the most popular IoT Wireless Communication protocol. Mainly used in homes and businesses, established on the IEEE 802.11n standard that exploits the frequency bands of 2.4 GHz and 5 GHz [26]. Wi-Fi is best for the local area network environment, offering a maximum data transfer rate of 600 Mbps. Data transmission also hangs on the channel frequency and the amount of antennas utilized. It commonly transfers data between 150 and 200 Mbps rates.

4. Comparison between Communication Protocols

This text segment compares different IoT communication protocols to help the researchers easily select the proper protocols for their research. The comparison criteria include standard, frequency, network, power, data rate, security, and spreading (Table 1).

Table 1. Comparison of criteria

Features	Zigbee	RFID	Bluetooth	NFC	SigFox	Z-Wave	Cellular
Standard	EEE802. 15.4	RFID	IEEE 802.15.1	ISO/IEC 14443 A&B, JIS X-6319-4	SigFox	Z-Wave	3GPP and GSMA, GSM/GPRS/EDGE (2G), UMTS/HSPA (3G), LTE (4G)
Frequency Band	2.4 GHz	125 kHz, 13.56 MHz, 902–928 MHz	2.4 Ghz	125 kHz 13.56 MHz 860 MHz	868 MHz (EU) 902 MHz (USA)	868 MHz–908 MHz	Common Cellular bands
Network	WPAN	Proximity	WPAN	P2P Network	LPWAN	WPAN	WNAN
Power	30 mA Low power	Ultra-low power	30 mA Low Power	50 mA low power Very Low	10 mW–100 mW	2.5 mA Low power consumption	High power consumption
Data Rate	250 kbps	1 Mbps	4 Mbps	106 212 or 424 kbps	100 bps (UL), 600 bps (DL)	40 kbps	NA
Security	AES	RC4	E0 Stream AES-128	RSA AES	Partially addressed	AES-128	RC4
Spreading	DSSS	DSSS	FHSS	GSMA	DSSS	No	DSSS

5. IoT Challenges

IoT is one of the most intensively growing technologies in the world. Many physical sensors and smart devices are connected to microcontrollers and share data over the Internet. It is predicted that more than 50 billion devices will be in IoT by 2025. As the number of gadgets rises, it opens the gates for unauthorized access by cybercriminals to these devices. There is massive competition between IoT device manufacturers who want to hand the latest devices first to the consumer. Manufacturers are considering the privacy and security caveats associated with IoT devices' data management and communication. There are the most prominent security and privacy challenges with IoT connected-devices example: Data security and privacy, insecure interfaces, IoT security Risks, sinkhole attacks, DoS threats, man-in-the-middle outbreaks [27, 59], poor device management, default password, remote vehicle access, lack of IoT awareness, botnet attack, etc. in this chapter's part will discuss on Security Challenges in IoT Network/ Communication Layer [60]. There are some necessary security services for IoT, as follows.

Confidentiality: Resources pass from one node to another and could easily be tampered with or intercepted by attackers. This justification means secure in-between nodes' communication is required in IoT. The simple solution is to keep end-to-end encryption or decryption.

Integrity: The node A sending data to node B in a network data should not be altered. The same data must be received on node B's side. Data integrity means information is checked from node A to node B during transmission.

Availability: IoT devices' continued services and data access must always be available in working mode. So, it is vital to impede and detect intrusion to guarantee IoT gadgets' availability of continuous services and data access. Reliability and data authentication in the network layer are the principal security challenges regarding the communication layer or IoT network. There are some security challenges in IoT Communication/Network Layer.

Denial of Service (DoS) Attack: A DoS outbreak is a form of cyber threat in which cyber criminals/attackers aims to interrupt a server or device's normal functioning and make the device inaccessible to the user [28, 29]. Malicious actors send multiple requests to the target machine until the normal traffic flow cannot access the server or device. It results in a DoS attack on the intended user. A simultaneous attack from a different source is a Distributed DoS (DDoS) occurrence.

Sinkhole Attack: Sinkhole threats create fake information and send this information to the neighbor node or cancel all the packets instead of delivering these packets to the destination [29]. It is a routing attack in the IoT environment that unlocks the gates for other threats like spoofing, discriminatory forwarding, and drop routing attacks [30].

Man-in-Middle Attack (MiMA): It assaults communication between two communicating nodes or devices [31]. It is a dangerous attack by an attacker to interrupt the communication between two devices where the attacker poses as an original sender. They trick the receiver of the message that they are receiving a message from the intended receiver.

Gateway Attack: This routing or DoS attack in which information is false to transfer from sensors/actuators/devices to the Internet. It attacks the communication between the Internet infrastructure and the sensor/actuator network. This attack cuts off the connection between sensors and the Internet [29].

6. Advised Blockchain-Enabled IoT Communication Protocol Architecture

The proposed blockchain system is an IoT-reliant structure with sensors that detect important properties like DHT11, soil moisture, Ph, light, and Nitrogen-Phosphorus-Potassium (NPK) nutrients. It will perform tasks taking real-time data from the soil like atmosphere temperature, atmosphere humidity, ground temperature, earth humidity, soil electrical conductivity, surface temperature, Ph levels, and NPK nutrients and sending these real-time data to a microcontroller (ESP32). Then, the information is transferred to fog devices using LoRaWan, which sends data securely and efficiently through LoRawan technology, covers a wide area from 60 km in rural and 1 km in urban sensor data transfer from an IoT gadget through LoRaWan to fog the data will be in urban sensor data transfer from IoT device through LoRaWan to fog then the data will be stored in a distributed ledger named a blockchain (Fig. 1).

6.1 IoT Device

Next, some fundamental agrarian modules and sensors are itemized.

6.1.1 ESP32

The ESP32 board is a chip-based microcontroller with low power and low cost with 2.4 Ghz dual Bluetooth and Wi-Fi. All the sensors relate to this microcontroller board [32]. The ESP32 program can perform project tasks employing an on-chip memory available on the board. The source code will be sore in his memory. It has a power and low-noise amplifier, filters, integrated antenna, and power controlling module.

6.1.2 LoRaWan

A LoRa "wireless area network" (LoRaWAN), also termed "low power wide-area network" (LPWAN), is the most potent communication protocol currently for Machine-to-Machine (M2M) communication (Fig. 2). It is predominantly

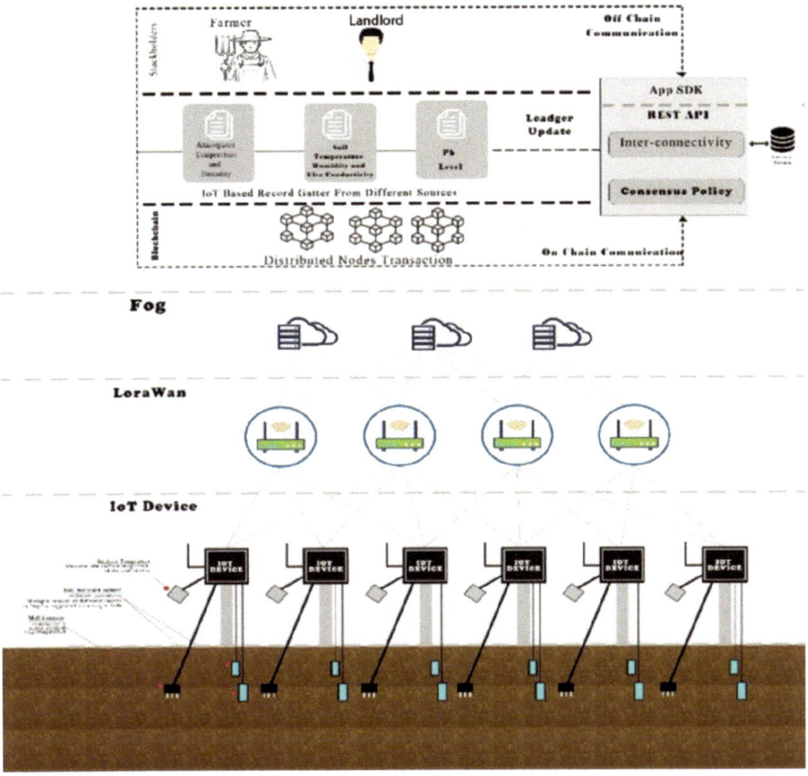

Fig. 1. Suggested architecture of Blockchain and IoT-enabled secure communication protocols architecture for smart-agriculture.

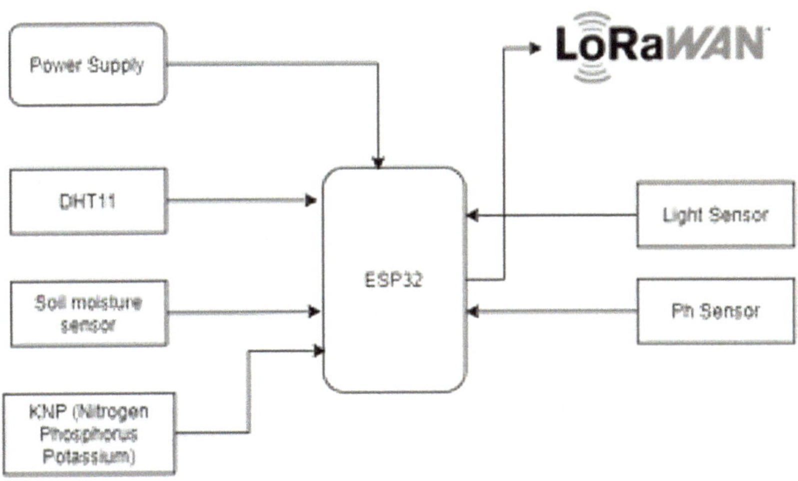

Fig. 2. Proposed IoT device circuit diagram.

designed to provide low-cost, secure bidirectional communication to IoT devices [2]. It also affords low rates and low power consumption to transfer information to overcome IoT application challenges. LoRaWAN is centered on star network topology, decreasing network complexity and saving battery life. The nodes in the start network topology only receive its data.

6.1.3 Light Sensor

The light sensor, ordinarily identified as a "photo sensor" or "photoelectric device", converts light energy into an electric signal. There are three common light sensor kinds: phototransistors, photoresistors, and photodiodes. A photoresistor is the most popular among the utilized in light sensor circuits, and it senses whether there is brightness or not. The photodiode is a different kind of sensor used in light sensing, which is unlike a photoresistor. It quickly changes light into an electric current. A photoresistor is an alternative to the photodiodes combined with photodiode + amplification. This type of sensor has its PN-junction, collector-base and reverse-biased, exposing it to the radiance source. The light sensor can determine whether crops get plenty or too much light. Illumination is amongst the critical parameters for crops to grow.

6.1.4 NPK Sensors

The NPK elements are the most important for growing plants. If any one of these nutrients is not enough, plants will fail. NPK is a soil fertility sensor suitable for detecting the contents of NPK and determining fertility [33]. Nitrogen helps leaves on the plants to synthetize phosphorus, which is responsible for fruit development, and root growth development of plants. Potassium assists the overall functioning of plants to grow. One can pick the plant variety by knowing the NPK values (proportions and concentrations) of fertilizer for a given land's soil, which will help you choose the type of plant to grow on one's land.

6.1.5 Moisture Measurement

A soil moisture sensor evaluates the soil's volumetric water level content. It can estimate stored water in the soil horizon by measuring some other properties of the earth related to water content [33]. Soil moisture sensor manufacturer uses different technologies for measuring soil moisture. To get high accuracy of soil water content, one must surround the soil moisture sensor with the soil to get accurate results or water level.

6.1.6 Ph Sensor

A Ph sensor gauges acidity and alkalinity. This tool is essential tool for water measurement in soil. The range of the Ph algorithm scale is from 0 to 14. The best range for water quality is from 6 to 8.5. A measurement below 7 signposts an acidic solution and above 7, an alkaline solution [34].

6.2 Fog Computing

The term fog, fog computing (FC), or fog network means decentralized computing infrastructure that provides storage, networking, and computing services between edge devices and cloud computing data centers [35, 36]. Where several heterogeneous smart devices communicate and cooperate in providing storage and processing tasks without the interference of any third party. It can be deployed on a vehicle, railway track, factory floor, power pole top, oil rig smart grids, smart traffic lights, and software-defined network computing system [37]. The devices that provide connective storage, networking, and computing services can be fog nodes, including switches, industrial controllers, video surveillance cameras, embedded servers, and routers. Locating fog nodes closer to the IoT-ready, smart devices, rather than storing them on the cloud and analyzing IoT data, conserves network bandwidth, minimizes latency, enhances security, reduces operating cost, boosts business agility, improves reliability, and expands insights without losing privacy. The common thing between fog applications and IoT is monitoring and scrutinizing real-time data from network-connected devices and acting according to analyzing results. The action can be human-to-machine or machine-to-machine interaction. Examples include locking the door, sending an alert to security concerns, generating reports, creating bar chat, or zooming a camera.

6.3 Blockchain

A decentralized digital ledger (DDL) termed Blockchain tackles transaction records, called blocks, related to each other using cryptographic techniques. It is also described as an immutable, fully decentralized, and irreversible peer-to-peer data storage. When a new transaction is entered into the block, it spreads over the network of participants, referred to as. Each block interconnects with its prior block through the preceding block's hash [38], containing transaction data and timestamp forming a chain. Therefore, Blockchains are resistant to data modification of stored data. The user or group cannot alter the data without altering all blocks. Blockchain applications entail cryptocurrencies, smart contracts, financial services, video games, energy trading, and healthcare [39]. The three blockchain pillars are consensus, most trustworthy, and dissemination.

6.4 Peer-to-Peer

Peer-to-Peer (P2P) is a distributed network defined as an individual or group of nodes connected without any centralized node to communicate and share transactions or resources. Every P2P network node has equal rights/power and executes the same tasks. P2P architecture is secure and protected from cyber-attack because of no centralized node or server [40]. The P2P networks can be threefold: structured, unstructured, and hybrid. In an unstructured peer-to-peer network, the communication between nodes is random because the nodes are not defined strictly. That is why unstructured P2P schemes are suitable for a lot of activities. A structured network has nodes organized in a structured arrangement, and the nodes have a way to relate. A hybrid P2P network blends both structured and unstructured P2P architectural features, solving a central server possible need for P2P. P2P distributed network provides high scalability, security [41].

6.5 Consensus Algorithms (CAs)

One knows that blockchain is a distributed ledger that provides safety, privacy, immutability, and transparency without a third party to validate and verify transactions in distributed decentralized blockchain network [20], yet it is considered that every transaction is thoroughly verified and secured in blockchain. This possibility follows the Blockchain backbone: the consensus algorithm (CA), which is a process to agree all the nodes in a distributed network to reach an agreement in the current DDL state. The CA protocol achieves trust, agreement, and safety across the distributed, decentralized network.

6.6 Farmer

A farmer is a person who fulfills the need for food that is essential for everyone's life and engages in agriculture to cultivate crops or land [42]. The entire planet population depends on agriculturalists, as one of the main backbones of every country. Thus, farmers are essential since they can afford food and other living assets. Different types of farmers worldwide grow crops like rice, wheat, etc. A second group of farmers cultivates fruits. A third one handles animals (livestock). Farmers can quickly check the atmosphere temperature, atmosphere humidity, soil temperature, soil humidity, soil electrical conductivity, surface temperature, Ph level, NPK concentrations, wind values, and rain prediction using mobile applications.

6.7 Landlord

Landlord or landowner is a person who has a large amount of land and leases pieces of this land for husbandry to the farmer with some rules and conditions. The landlord or landowner can quickly check the atmosphere temperature, atmosphere humidity, soil temperature, soil humidity, soil electrical conductivity, surface temperature, Ph level NPK concentrations, and wind and rain prediction through mobile applications.

6.8 Smart Contracts

Smart contracts (SCs) were proposed for electronic commerce applications in the early 1990s by Szabo [43], but now a day smart contracts have been used in distributed ledger and blockchain technology. The SC is a line of code between all participants in a distributed blockchain network that runs when some specific conditions are met. Without any time lose and third party, the SC automatically accomplishes the agreement with the aim of all participants in the distributed blockchain network can be confident about the outcome [44, 61].

7. The Proposed Architectural-Based Implementation Challenges and Limitations

There are some implementation challenges and limitations for this proposed architecture.

7.1 Regulatory Compliance

The denotation of compliance is to follow the order or rule. Regulatory compliance is an organization obeying guidelines, laws, and specifications for its business process. A business that violates regulatory compliance receives legal punishment. Examples of laws/regulations include the "Federal Information Security Management" (FISM) act, "Sarbanes-Oxley Act" (SOA), and "California Consumer Privacy Act" (CCPA), to name a few.

7.2 Platform Interoperability

Platform interoperability aims to help machines and people access, analyze, discover, and integrate data. It inspirits the community to adopt the standard file format, identifiers, vocabularies, and metadata to work internationally to achieve its goal.

7.3 Scope of Data Privacy

Data privacy is a security branch that shields and prevents data from unauthorized access [45, 46]. This concern with the management of sensitive data, including personal information example: name, contact information, address, location, etc., and other sensitive material like intellectual property and financial records to meet regulatory requirements, while protecting the immutability and confidentiality of data.

8. Conclusion

Agribusiness comprises planting on land for oneself or growing food plants to trade, raising livestock, producing biomass/energy, and industrializing products within the agri-environment. Raising animals for milk, meat, manure or to balance an ecosystem is also a category of agriculture and The Food demand is augmenting, everyone asking for quality food. By 2050, 31% rise is expected in world population and change in climate. The demand of food production and natural resource will also increase 71% owing to the population upsurge and climate change. We must move from traditional farming to smart farming technologies. Eventually, people must move from traditional farming to smart farming technologies. The agriculture 4.0 is referred to the adoption of communication and information technologies such as IoT, AI, Big Data, and Blockchain. In this described framework, IoT devices collect and process evidence in real-time utilizing soil and ecological measurements like atmosphere temperature, atmosphere humidity, earth temperature, ground humidity, earth electrical conductivity, surface temperature, Ph level, concentrations of NPK, and sending these real-time data to microcontroller (ESP32). The data are then transformed by a microcontroller into fog devices, and stored in a distributed blockchain network. This chapter offers a secure communication architecture system for Agriculture 4.0 and discusses proposed architecture-based implementation challenges and limitations.

References

1. Rapela, M. (2019). *Fostering Innovation for Agriculture 4.0*. Springer Link.
2. Yandun Narvaez, F., Reina, G., Torres-Torriti, M., Kantor, G. and Cheein, F.A. (2017). A survey of ranging and imaging techniques for precision agriculture phenotyping. *IEEE/ASME Transactions on Mechatronics*, 22(6), 2428–2439, doi: 10.1109/TMECH.2017.2760866.
3. From Industry 4.0 to Agriculture 4.0: Current Status,... - Google Scholar. https://scholar.google.com/scholar?hl=en&as_sdt=0%2C5&q=From+Industry+4.0+to+Agriculture+4.0%3A+Current+Status%2C+Enabling+Technologies%2C+and+Research+Challenges&btnG= (accessed Sep. 09, 2021).
4. Lasso, E. and Corrales, J.C. (2018). Towards an alert system for coffee diseases and pests in a smart farming approach based on semi-supervised learning and graph similarity. *Advances in Intelligent Systems and Computing*, 687, 111–123. doi: 10.1007/978-3-319-70187-5_9.
5. Li, S., Li, D.X. and Zhao, S. (2015). The Internet of Things: A survey. *Information Systems Frontiers*, 17(2), 243–259. Springer., Accessed: Sep. 10, 2021. [Online]. Available: https://idp.springer.com/authorize/casa?redirect_uri=https://link.springer.com/content/pdf/10.1007/s10796-014-9492-7.pdf&casa_token=XuLj8nOOl1AAAAAA:i9sYFhqpUMZYdBboBRsTe2o_-SPKffqAsApw3ddyAWbRABSIobaR2-i4j7ka07lzmbZOejRoGh72agOKUQ.
6. Laghari, A.A., Wu, K., Laghari, R.A., Ali, M. and Khan, A.A. (2021, July). A review and state of art of Internet of Things (IoT). *Arch. Comput. Methods Eng.* doi: 10.1007/S11831-021-09622-6.
7. Khoa, T.A., Man, M.M., Nguyen, T.-Y, Nguyen, V. and Nguyen Hoang Nam (2019). Smart agriculture using IoT multi-sensors: A novel watering management system. mdpi.com. doi: 10.3390/jsan8030045.
8. Kitpo, N. and Inoue, M. (2018). Early rice disease detection and position mapping system using drone and IoT architecture. *2018 12th S.E.A Technical University Consortium (SEATUC)*. ieeexplore.ieee.org. Accessed: Sep. 19, 2021. [Online]. Available: https://ieeexplore.ieee.org/abstract/document/8788863/.

9. Namani, S. and Bilai, G. (2020). Smart agriculture based on IoT and cloud computing, ieeexplore.ieee.org, Accessed: Sep. 19, 2021. [Online]. Available: https://ieeexplore.ieee.org/abstract/document/9092136/.
10. Park, S., Yun, S., Kim, H., Kwon, R., J.G.-P. et al. (2018). Forestry monitoring system using lora and drone. dl.acm.org. Accessed: Sep. 19, 2021. [Online]. Available: https://dl.acm.org/doi/abs/10.1145/3227609.3227677?casa_token=mlC-tPDmpQEAAAAA:rBVI-z8BjlN1BbNcr4pZASGzykS3f1LsdGgb0r_pHXNrWiJd7QXa54EBOrGnT98-W6LH_GhzzzHhRU7-.
11. Patil, K. and Kale, N.R. (2016). A model for smart agriculture using IoT. *2016 International Conference on Global Trends in Signal Processing, Information Computing and Communication.* ieeexplore.ieee.org. Accessed: Sep. 25, 2021. [Online]. Available: https://ieeexplore.ieee.org/abstract/document/7955360/.
12. Kung, Y.F., Liou, S.W., Qiu, G.Z., Zu, B.C., Wang, Z.H. and Jong, G.J. (2018). Home monitoring system based Internet of Things. *Proceedings of 4th IEEE International Conference on Applied System Innovation* (*ICASI*) 325–327. doi: 10.1109/ICASI.2018.8394599.
13. Grabovica, M., Popic, S., Pezer, D. and Knezevic, V. (2016). Provided security measures of enabling technologies in Internet of Things (IoT): A survey. *2016 Zooming Innovation in Consumer Electronics International Conference* (*ZINC*), Jun., 28–31. doi: 10.1109/ZINC.2016.7513647.
14. NIST (National Institute of Standards and Technology),... - Google Scholar. https://scholar.google.com/scholar?hl=en&as_sdt=0%2C5&q=1%29%09National+Institute+of+Standards+and+Technology%2C+Common+Vulnerability+Scoring+System+Version+3.0+Calculator.+%5BOnline%5D.+Available%3A+https%3A%2F%2Fnvd.nist.gov%2Fvuln-metrics%2Fcvss%2Fv3. Accessed Jul. 27, 2021.
15. Adi, P.D.P. and Kitagawa, A. (2019). ZigBee Radio Frequency (RF) performance on Raspberry Pi 3 for Internet of Things (IoT) based blood pressure sensors monitoring. *Int. J. Adv. Comput. Sci. Appl.*, 10(5), 18–27. doi: 10.14569/ijacsa.2019.0100504.
16. Chopra, K., Gupta, K.A. and Lambora, A. (2019). Future internet: The internet of Things: A literature review. *2019 International Conference on Machine Learning, Big Data, Cloud and ParallelComputing (COMITCon)*, Faridabad, India, 135–139, doi: 10.1109/COMITCon.2019.8862269.
17. Wei L., Songyou, X., Jing, L., Kuan-Ching, L., Dafang Z., Keqin L. et al. (2019). A double PUF-based RFID identity authentication protocol in service-centric Internet of Things environments. Elsevier, 503, 129–147. doi: 10.1016/j.ins.2019.06.047.
18. Smith, N., Cahill, C., Scott Janus Kenneth, T. and Layton Michael, A. (2017). Continuous multi-factor authentication. US Patent 9 705,869. Google Patents. Accessed Aug. 05, 2021. [Online]. Available: https://patents.google.com/patent/US9705869B2/en.
19. Sutjiredjeki, E., Basrajuddin, N.C., Fajrin D. and Noor F. (2020). Development of NFC and IoT-enabled measurement devices for improving health care delivery of Indonesian children. iopscience.iop.org., 1450, 12072. doi: 10.1088/1742-6596/1450/1/012072.
20. Resilient, D.B. (2019). An in-depth Look of BFT Consensus in Blockchain: Challenges and Opportunities. jhellings.nl. 6–10, Dec. doi: 10.1145/3366625.3369437.
21. Xiong, F., Li, A., Wang, H. and Tang, L. (2019). An SDN-MQTT based communication system for battlefield UAV swarms. ieeexplore.ieee.org. Accessed: Aug. 13, 2021. [Online]. Available: https://ieeexplore.ieee.org/abstract/document/8808160/.

22. Uy, N.Q. and Nam, V.H. (2019). A comparison of AMQP and MQTT protocols for Internet of Things. *2019 6th Conference on Information and Computer Science (NICS)*. ieeexplore.ieee.org. Accessed Aug. 13, 2021. [Online]. Available: https://ieeexplore.ieee.org/abstract/document/9023812/.
23. Liu, X., Zhang, T., Hu, N. and Zhang, P. (2020). The method of Internet of Things access and network communication based on MQTT. *Computer Communications*, 153(8). Elsevier. Accessed: Aug. 13, 2021. [Online]. Available: https://www.sciencedirect.com/science/article/pii/S0140366419320171.
24. Fürst, J., Chen, K., Kim, H.-S. and Bonnet, P. (2018). Evaluating Bluetooth low energy for IoT. *2018 IEEE Workshop on Benchmarking Cyber-Physical Networks and Systems (CPSBench)*, Porto, Portugal, 1–6, doi: 10.1109/CPSBench.2018.00007.
25. Lonzetta, A., Cope, P., Campbell, J., Joseph C., Bassam, J. Mohd and Thaier, H. (2018). Security vulnerabilities in Bluetooth technology as used in IoT. mdpi.com. Accessed: Sep. 30, 2022. [Online]. Available: https://www.mdpi.com/317722.
26. Sadowski, S. and Sparchos, P. (2020). Wireless technologies for smart agricultural monitoring using Internet of Things devices with energy harvesting capabilities. *Computers and Electronics in Agriculture*, 172, 105338. Elsevier. Accessed Sep. 30, 2022. [Online]. Available: https://www.sciencedirect.com/science/article/pii/S0168169919318381.
27. Virat, M.S, Bindu, S.M., Aishwarya, B., Dhanush, B.N. and Kounte, M.R. (2018). Security and privacy challenges in Internet of Things. *2018 2nd International Conference on Trends in Electronics and Informatics (ICOEI)*, Tirunelveli, India, 454–460, doi: 10.1109/ICOEI.2018.8553919.
28. Sonar, K. and Hardik U. (2014). A survey: DDOS attack on Internet of Things. academia.edu, 10(11), 58–63. Accessed Sep. 26, 2022. [Online]. Available: https://www.academia.edu/download/36275768/I10115863.pdf.
29. Rao, T. and Haq, E. (2018). Security challenges facing IoT layers and its protective measures. *International Journal of Computer Applications*, 179(27), March. researchgate.net. Accessed Sep. 26, 2022. [Online]. Available: https://www.researchgate.net/profile/Tariq-Rao/publication/323892938_Security_Challenges_Facing_IoT_Layers_and_its_Protective_Measures/links/5ab1ccc9a6fdcc1bc0bff2d0/Security-Challenges-Facing-IoT-Layers-and-its-Protective-Measures.pdf.
30. An Enhanced Technique to Detect Sinkhole Attack in... - Google Scholar. https://scholar.google.com/scholar?hl=en&as_sdt=0%2C5&q=An+Enhanced+Technique+to+Detect+Sinkhole+Attack+in+Internet+of+Things&btnG=. Accessed Sep. 26, 2021.
31. Jayapandian Natarajan (2020). Cyber secure man-in-the-middle attack intrusion detection using machine learning algorithms. igi-global.com. Accessed: Sep. 30, 2021. [Online]. Available: https://www.igi-global.com/chapter/cyber-secure-man-in-the-middle-attack-intrusion-detection-using-machine-R learning-algorithms/236343.
32. Pravalika,V. and Prasad, R. (2019). Internet of Things based home monitoring and device control using Esp32. *Int. J. of Rec. Tech. & Engg.* (*IJRTE*), researchgate.net, 8, 2277–3878. Accessed Sep. 29, 2021. [Online]. Available: https://www.researchgate.net/profile/Ch-Rajendra-Prasad-2/publication/334226986_Internet_of_Things_Based_Home_Monitoring_and_Device_Control_Using_Esp32/links/5d1de7fe458515c11c0ff074/Internet-of-Things-Based-Home-Monitoring-and-Device-Control-Using-Esp32.pdf.

33. Madhumathi, R., Arumuganathan, T. and Shruthi, R. (2020). Soil NPK and moisture analysis using wireless sensor networks. *2020 11th International Conference on Computing, Communication and Networking Technologies (ICCCNT)*, Kharagpur, India, 1–6, doi: 10.1109/ICCCNT49239.2020.922554.
34. Daigavane, V. and Gaikwad, M.A. (2017). Water quality monitoring system based on IoT. ripublication.com., 10(5), 1107–1116. Accessed Sep. 25, 2021. [Online]. Available: http://www.ripublication.com/awmc17/awmcv10n5_24.pdf.
35. Bonomi, F., Milito, R., Zhu, J. and Addepalli, S. (2012). Fog computing and its role in the Internet of Things. *MCC'12 - Proc. 1st ACM Mob. Cloud Comput. Work.*, 13–15. doi: 10.1145/2342509.2342513.
36. Yi, S., Li, C. and Qun, L. (2015). A survey of fog computing: Concepts, applications and issues, dl.acm.org, 37–42, Jun. doi: 10.1145/2757384.2757397.
37. Gollaprolu, H., Nagaraju, Basavoju H. and Mazeeda, S. (2015). Fog computing and its real time applications, academia.edu., 9001(6), 266, 2008, Accessed Sep. 18, 2021. [Online]. Available: https://www.academia.edu/download/43419591/IJETAE_fog.pdf.
38. Jumani, A.K., Laghari, A.A. and Khan, A.A. (2021). Blockchain and Big Data: Supportive Aid for Daily Life. *Secur. Issues Priv. Concerns Ind. 4.0 Appl.*, 141–178, Jun. doi: 10.1002/9781119776529.CH8.
39. Tasatanattakool, P. and Techapanupreeda, C. (2018). Blockchain: Challenges and applications. *2018 International Conference on Information Networking (ICOIN)*, Chiang Mai, Thailand, 473–475, doi: 10.1109/ICOIN.2018.8343163.
40. Tapscott, A. and Tapscott, D. (2017). How blockchain is changing finance. *Howard Business Review*. capital.report. Accessed: Sep. 29, 2021. [Online]. Available: https://capital.report/Resources/Whitepapers/40fc8a6a-cdbd-47e6-83f6-74e2a9d36ccc_finance_topic2_source2.pdf.
41. Li, Z., Barenji, A. and Huang, G. (2018). Toward a blockchain cloud manufacturing system as a peer to peer distributed network platform. *Robotics and Computer-integrated Manufacting*, Elsevier. Accessed: Sep. 30, 2021. [Online]. Available: https://www.sciencedirect.com/science/article/pii/S073658451830022X.
42. Anwarudin, O. and Haryanto, Y. (2018). The role of farmer-to-farmer extension as a motivator for the agriculture young generation. ijsser.org. Accessed: Sep. 30, 2021. [Online]. Available: https://ijsser.org/more2018.php?id=29.
43. Governatori, G., Idelberger, F., Milosevic, Z., Riveret, R., Sartor, G. and Xu, X. (2018). On legal contracts, imperative and declarative smart contracts, and blockchain systems. *Artif. Intell. Law*, 26(4), 377–409. doi: 10.1007/S10506-018-9223-3.
44. Cong, L. and He, Z. (2019). Blockchain disruption and smart contracts. *Rev. of Financial Studies*, 32(5), 1754–1797. academic.oup.com., 2018, Accessed Sep. 28, 2021. [Online]. Available: https://academic.oup.com/rfs/article-abstract/32/5/1754/5427778.
45. Khan, A.A., Laghari, A.A., Awan, S. and Jumani, A.K. (2021). Fourth industrial revolution application: Network forensics cloud security issues. *Secur. Issues Priv. Concerns Ind. 4.0 Appl.*, 15–33, doi: 10.1002/9781119776529.CH2.
46. Khan, A.A., Laghari, A.A., De-Sheng Liu, Shaikh, Aftab, A., Dan-An Ma, Chao-Yang Wang, Wagan, Asif, A. (2021). EPS-Ledger: Blockchain hyper ledger sawtooth-enabled distributed power systems chain of operation and control node privacy and security. *Electronics*, 10(19), 2395. https://doi.org/10.3390/electronics10192395.

47. Kuo, T. and Ohno-Nachado, L. (2018). Model chain: Decentralized privacy-preserving healthcare predictive modeling framework on private blockchain networks. arXiv. arXiv:1802.01746.
48. Dagher, G.G., Mohler, J., Milojkovic, M. and Marella, P.B. (2018). Ancile: Privacy-preserving framework for access control and interoperability of electronic health records using blockchain technology. *Sustain. Cities Soc.*, 39, 283–297.
49. Zhang, A. and Lin, X. (2018). Towards secure and privacy-preserving data sharing in e-health systems via consortium blockchain. *J. Med. Syst.*, 42, 140.
50. Monteiro, A.C.B., França, R.P., Estrela, V.V., Razmjooy, N., Iano, Y. and Negrete, P.D.M. (2021). Metaheuristics applied to blood image analysis. *In*: Razmjooy N. et al. (Eds.), *Metaheuristics and Optimization in Computer and Electrical Engineering*. Lecture Notes in Electrical Engineering, 696. Cham: Springer.
51. Deshpande, A., Razmjooy, N. and Estrela, V.V. (2021). Introduction to computational intelligence and super-resolution. *In*: Deshpande, A. et al. (Eds.), *Computational Intelligence Methods for Super-Resolution in Image Processing Applications*. Cham, Switzerland: Springer.
52. Chaabane, L., Khelassi, A., Terziev, A., Andreopoulos, N., Jesus, M.A. and Estrela V.V. (2021). Particle swarm optimization with tabu search algorithm (PSO-TS) applied to multiple sequence alignment problem. *In*: Khelassi, A., Estrela, V.V. (Eds.), *Advances in Multidisciplinary Medical Technologies: Engineering, Modeling and Findings*. Cham, Switzerland: Springer.
53. Kalantari, S., Ramezani, M., Madadi, A. and Estrela, V.V. (2021). Reduction AWGN from digital images using a new local optimal low-rank approximation method. *In*: Iano Y. et al. (Eds.), *Proc. 5th Brazilian Technology Symposium* (*BTSym 2019*). *Smart Innovation, Systems, and Technologies*, 201. Cham.: Springer.
54. Khan, A.A., Shaikh, A.A., Cheikhrouhou, O., Laghari, A.A., Rashid, M., Shafiq, M. and Hamam, H. (2021). *IMG-forensics: Multimedia-enabled Information Hiding Investigation Using Convolutional Neural Network*. IET Image Processing.
55. Razmjooy, N. and Estrela, V.V. (2019). *Applications of Image Processing and Soft Computing Systems in Agriculture*, 1–300, Hershey, PA, USA: IGI Global.
56. Hemanth, J. and Estrela, V.V. (2017). *Deep Learning for Image Processing Applications:* 31 (*Advances in Parallel Computing*). Amsterdam, Netherlands: IOS Press.
57. Loschi, H.J., Estrela, V.V., Hemanth, D.J., Fernandes, S.R., Iano, Y., Laghari, A.A. et al. (2020). Communications requirements, video streaming, communications links, and networked UAVs. *In*: Estrela V.V. et al. (Eds.), *Imaging and Sensing for Unmanned Aircraft Systems*, 2(6), 113–132. London, UK: IET.
58. Khan, A.A., Uddin, M., Shaikh, A., Laghari, A.A. and Rajput, A. (2021). MF-Ledger: Blockchain Hyperledger Sawtooth-enabled Novel and Secure Multimedia Chain of Custody Forensic Investigation Architecture. *IEEE Access*.
59. Padilha, R., Iano, Y., Monteiro, A.C., Arthur, R. and Estrela, V.V. (2019). Betterment proposal to multipath fading channels potential to MIMO systems. *In*: Iano Y. et al. (Eds.), *Proc. 4th BTSym.* Cham, Switzerland: Springer.
60. Khan, A.A., Laghari, A.A. and Awan, S.A. (2021). Machine learning in computer vision: A review. *FAI Endorsed Transactions on Scalable Information Systems*, 8(32).
61. Cachin, C., Schubert, S. and Vukolic, M. (2016). Non-determinism in byzantine fault-tolerant replication. *Proc. Int. Conf. Princ. Distrib. Syst.* (OPODIS 2016), Madrid, Spain, 13–16 December, pp. 1–20.

CHAPTER

14

Applied Composting in Education – From Solid Waste to Organic Fertilizer and Beyond in Agriculture 4.0

Ingrid de Souza Siqueira[1][0009-0009-4636-4604], **Alef da Silva Sousa**[2][0009-0007-7492-1506], **Gabriela Petroceli Mota**[1][0009-0001-9744-1173], **Glacielen Ribeiro de Souza**[1][0009-0003-0773-1092], **Analyze Villanueva Gaete**[1][0000-0003-1471-7443], **Gabriel Quintanilha-Peixoto**[3][0000-0002-7476-5845], **Carlos Eduardo de Souza Teodoro**[2][0000-0001-8591-0502] **and Aline Chaves Intorne**[1,4*][0000-0001-8015-6926]

[1] Laboratory of Physiology and Biochemistry of Microorganisms, Universidade Estadual do Norte Fluminense Darcy Ribeiro (UENF), Campos dos Goytacazes, RJ, 28013-602, Brazil

[2] Biotechnology Laboratory, Universidade Federal Fluminense (UFF), Volta Redonda, RJ, 2629-5000

[3] Laboratory of Molecular and Computational Biology of Fungi, Federal University of Minas Gerais (UFMG), Belo Horizonte, MG, 31270-901, Brazil

[4] Laboratory of Chemistry and Biology, Federal Institute of Education, Science and Technology of Rio de Janeiro (IFRJ), Volta Redonda, RJ, 27213-100, Brazil

Agriculture 4.0 meets environmental education (EE) in this work via organic waste (OW) compost and as per developing countries' existing state of affairs, albeit emphasizing "Mexico, Central and South America" (MCASA). Biotechnology and digital tools (ICT) were a university extension project's inspiration deemed Science Today (Ciencia Hoje) as a form to learn from a finalized agro-industrial products' perspective, i.e., a social technology. Professors and students investigated locally collected data and green nomenclature repercussions, such as "composting employing social media" (i.e., YouTube and Instagram). These ICT instruments afford data from feeds/views that can engender STEAM tools and craft social, cultural, economic, statistical, and educational knowledge. The core experimental compost fallouts in teaching practice were remarkable.

* Corresponding author: aline.intorn@ifrj.edu.br

1. Introduction

The uptake of agronomical "information and communication technologies" (ICT) entails new skills and knowledge for agrarians, as they are indispensable to raising awareness and organizing global, regional, and local training, especially to reach small and medium-sized farmsteads. Some agribusinesses not recurrently profiting from innovations can be stimulated to embrace ICT [1]. "Agricultural knowledge and innovation" (AKI) frameworks [2] can propel and encourage reciprocated learning to produce, share, and dispose of AKI information related to agronomy digitization. The widespread, pervasive digital education [3] simplifies knowledge acquisition, opens new learning opportunities, engenders other critical lifelong instruction aspects, and acknowledges a translational understanding that new job openings often require [4, 5]. All training programs should be adaptive and advantageous enough to employ these novel learning schemes besides fostering translational expertise. Infusing the syllabuses supported by digital learning and AKI since the simplest formalized schooling stages throughout higher education can also handle the digital divide [6–8]. Although old-style training is vital in acquiring new skills, the complementary ICT self-learning role is also becoming increasingly important. ICT single-handedly can engender viable didactic transformations like employing online, immersive classes and games [9, 10]. Mention should include personal learning clouds, educational videos, substantial open online courses, and interactive problem sets with built-in automatic feedback and classification. ICT can be inserted into rural and urban farmers' lives in several forms: (a) basic/high school teaching and technical courses where workforces can attain specific competencies at any time in their lives, and (b) professional training without unescapably enrolling in formal courses and by exchanging experiences.

The Agriculture 4.0 concept this chapter addresses revolves around applications targeting education from a recycling viewpoint as it dramatically affects people's bounds and raises awareness. Composting mixes ingredients to turn them into plant fertilizers that improve the ground's biological, physical, and chemical traits. It commonly stems from decomposing "organic matter" (OM) (e.g., manure). The final mixture has plenty of plant nutrients and helpful organisms, for instance, bacteria, fungi, protozoa, and nematodes. Compost improves fertility in horticulture, gardens, landscaping, metropolitan (aka urban) husbandry, and organic farming, dropping dependency on commercial, unnatural chemical fertilizers. Composted-induced farming has benefits comprising crops' fertilization, producing energy through, e.g., greenhouse gas (GHG) emissions, conditioning soil, and augmenting the terrain's humus, besides fostering valuable microbes to restrain pathogens and lessen soil-yielded diseases/plagues [2, 3, 11–14].

At the most superficial level, composting transforms a combination of green waste (i.e., nitrogen-rich materials, viz., leaves, grass, and food leftovers) and carbon-rich brown waste (e.g., wood materials, paper, and tree branches). The OM may take months to become humus. Composting methodologies can be multistep, i.e., closely monitored processes whose inputs may comprise measured water, air, and carbon- and nitrogen-rich materials. Decomposition is aided by shredding OM, adding water, and permitting proper aeration by repeatedly turning the mixture through open piles (windrows). Earthworms, fungi, plus other detritivores promote the OM breakup. Aerobic bacteria and fungi transform the green and brown inputs into carbon dioxide (CO_2), heat, and ammonium ions. Composting practices can immensely advance "waste management" (WM) since compostable materials constitute approximately 20% of the litter in landfills. As a result of anaerobic prerequisites, these constituents take longer to biodegrade. Composting outdoes OM for landfills ecologically because composting diminishes greenhouse gas (GHG) emissions (CO_2, methane, etc. from anaerobic processes and arranges for positive financial and eco-friendly effects. Composted outcomes can also help land/watercourse recuperation, wetland ecological construction, and serve as landfill cover [15–25].

Composting handling requires expertise, time, know-how, equipment, and effort. The composting method and product drawbacks must be pondered against the benefits. Composting is ideal for unstable, nonuniform, hard-to-store organic byproducts. Biosolids, manures, and food-related residuals must be warehoused periodically. After composting, manure becomes drier, more homogeneous, and biologically stable, making it useful beyond land application. Farmers and feedlot owners value composted manures more. Biosolids dry out as composts and are easier to land-apply. Pathogenic derivatives are nontoxic after high-temperature composting. Compost products are slow-release fertilizers with an elevated ratio of carbon-to-nitrogen (C/N) [26–36].

Composting byproducts have limitations: equipment costs and site preparation, a long treatment period, targeting the final compost product, and ecological difficulties like odors and dust. Depending on the procedure, composting can take weeks to stabilize. A good compost market is essential to justify the extra work in generating compost. Biological composting can cause odors if not appropriately controlled. Slow-release composted nutrients demand more expressive application rates than ordinary byproducts to sustain plant responses. More elevated application rates claim more material and field trips than the by-product.

Composting benefits and drawbacks help byproduct generators estimate cost vs. return for initiating a composting process. The environment may regulate the final treatment and beneficial byproduct use. The suitable byproduct to co-compost or utilize with composts can yield value-added goods to be sold. The aerobic OM breakdown of co-composting mixtures produces compost for fertilizer or soil enhancement.

A bibliometric analysis was carried out to identify works in the English language in the *Scopus Elsevier* database that worked on the topic to understand the panorama of composting in the educational context. Then, we searched using the keywords 'composting' AND 'education' in titles, abstracts, and keywords. Three hundred and fifteen works were found, first published in 1969, with a maximum of 23 publications in 2020. It is possible to observe in Fig. 1 that publications have large fluctuations over time. As for countries, the USA stands out with 78 publications, and among the top 15, Brazil is in seventh place, being the only "Mexico, Central (CA) and South America (SA)" (MCASA) country to stand out. Among the other MCASA countries, Chile and Colombia presented 2 works each.

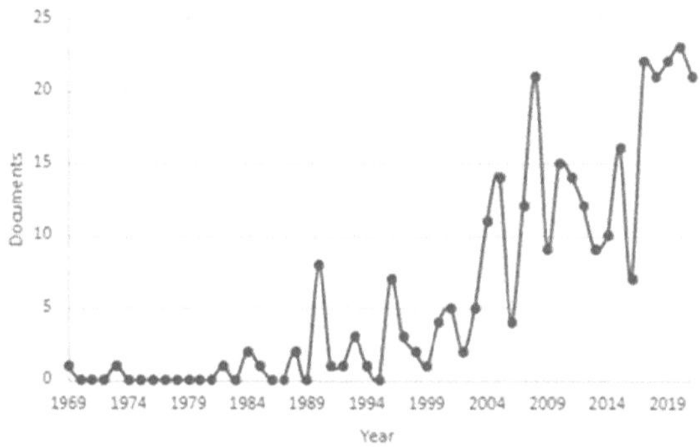

Fig. 1. Documents published between 1969 and 2021 encompassing the keywords 'composting' AND 'education' in the title, abstract, and keywords. (*Source*: *Scopus Elsevier*, accessed on December 16, 2021)

After briefly discussing Agriculture 4.0, this article scrutinizes a variety of composting agendas. Section 3 features cutting-edge compost characteristics in MCASA to impulse underprivileged nations to embrace and adapt these ideas. Section 4 discusses disseminating knowledge and "social technology" (ST) rationales, especially in developing nations. Section 5 discusses composting within "environmental education" (EE) with a breakdown of elementary and high school biotechnological didactic spaces.

2. Agriculture 4.0

The Agriculture 4.0 rationale discards disproportionate water, fertilizers, and insecticide expenditure. Contemporary agriculturalists are already struggling to minimalize the uptakes of these essential requisites via advanced ICT (e.g., automation, cloud/fog computing, genetics, epigenetics, biotechnology, and unpolluted energy applications) to contend with farmsteads more selectively. The rapid ICT development has provided agriculturalists with ample crop cultivation choices and new farming approaches, such as 3D-printed foodstuff, artificial/natural meat production, safer genetic modifications, and seawater desalination. Akin to other trades, automated production contexts encompassing robots, environmental sensors, ground/aerial/spatial surveillance systems, and "Global Positioning Systems" (GPSs) are critical to boosting production and well-being, as well as crafting eco-friendly farms [37–45].

Cutting-edge, sustainable technologies concerns will support the pursuit and engendering of eco-friendly, more efficient farming methods that satisfy the value chain's requirements and societal demands, especially those of consumers. Agriculture 4.0 intends to go beyond applying trailblazing technologies to farmlands' operation and management to overcome forthcoming nutrition shortages by encouraging agriculturalists to fully exploit innovations and reengineering their production methods to maximize profits while reorienting their strategies en route for an optimized value chain.

Sophisticated technologies and processes encompassing precise sensors for supervising manifold parameters (like temperature, porosity, humidity, and so on), myriad actuators (for instance, agrarian robots, aka agribots), advanced equipment/devices, and other automated entities will fuel future farms operation and management. Agribots, myriad sensors/actuators for handling manifold parameters (like temperature, humidity, etc.), and aerial surveillance/observation within GPS-reliant devices will be used continuously. Nevertheless, operation and management frameworks with optimized efficiency will assure agriculturalists more profits while consenting them to run agribusinesses in an efficient, sustainable mode [40–45].

Traditional husbandry overutilizes water, fertilizers, and pesticides. Agriculture 4.0 partisans focus on either being minimalized or abandoned (i.e., exchanged by other safer means). Also, upcoming agribusiness expertise will assist planters in cultivating wastelands with eco-friendly energy sources and conserve abundant ecological reserves, including saltwater and other green supplies unavailable locally but obtainable by sustainable processes. Measuring GHG emissions is paramount to sustainable, vigorous agribusiness models and environment conservation.

3. Composting as a Development Instrument

In MCASA, 541,000 tons (t) of "urban solid waste" (USW) are discarded daily, 27% of which is sent to landfills. Considering that around 50% of these materials are organic, which, in contact with other waste, makes recycling impossible or difficult, which in the region is about 10%, and also contributes to the GHG release, it is urgent that public policies to reduce "organic waste" (OW) and adequately manage any excess. This also stimulates the growth of the circular economy, income generation by informal collectors of recyclable materials, the participation of society in decision-making, and investments in the economic sustainability of the sector [46]. In Brazil, the PNRS report establishes in section V of Article 36: "implement a composting system for organic solid waste" (SW) and coordinate with economic and social agents ways of using the compost produced; as a shared responsibility in the life cycle of the products (**LAW Nº 12,305**) [47]. Public OW treatment policies have become essential in the urban environment, and composting is a bet in many cities.

São Paulo municipality in Brazil (the main SA city), where 51% of waste is classified as biological, has an Integrated Solid WM Plan. In the city, waste collected from street markets is taken to so-called composting yards, which go through the composting and vermicomposting processes. After 120 days, it returns to the city's green areas as fertilizer. In total, around 10 t are collected daily from the markets. Food waste from food courts and pruning plants is also taken to composting and vermicomposting. With these initiatives, the city hall hopes to recognize waste's economic and social value, guaranteeing a new cycle and encouraging citizen participation in food/nutritional insecurity issues. The initiatives are also expected to bring scenic beauty to the city, accelerate the fight against disease vectors, and promote a sustainable city [48]. Therefore, the social technologies of agriculture, urban gardens, and community composting are a network to stimulate the solidarity economy and generate income.

Lima-Peru is the second LA city with the most inhabitants [49]. In 2014, the city generated more than 9 thousand tons of household waste daily, and 53% of USW is OM [50]. The campaign "Yo compost en casa, en la escuela y en la comunidad" encourages and values composting, where 160 families participated by installing compost bins in their homes, helping with the implementation of the project and the valorization of 4.2 t of OW in 2019. The city also has a 450 m^2 composting center that reused (in 2019) 101.38 t of waste from the Mercado de Abasto and the pruning and maintenance of green areas [51].

Table 1. Activities carried out by UENF-IFRJ between 2016 and 2021

Activity	Achievements
Monographies	1
Dissertations	1
In-person extension activities	141
Extension activities through social media	20
Total	163

Bogotá-Colombia, the third most densely inhabited SA city, with 7,181,569 [52] citizens, generates 7.5 t of USW daily, 51% equivalent to OM. A 2018 "National Administrative Department of Statistics" (DANE) survey disclosed that 62% of households reported practising selective collection [53]. In 2006, the Bogotá Comprehensive "Solid Waste Management" (SWM) Master Plan was ratified [54, 55]. Now, the city's OW must be treated in composting places. The fertilizer generated as agricultural, agro-industrial, and forestry inputs can also help recover soil from degraded areas. Thus, the waste is reinserted into the market, with added economic and green value.

For these and other initiatives related to composting to succeed, the population must understand how the inadequate disposal of USW impacts their lives at regional and global levels. Investing in organic SWM strategies in "environmental education" (EE) started with USA's initiatives in the 1960s and 1970s. This accomplishment aimed at bringing urban folks closer to citizenship practices to unravel difficulties in the natural environment as urban areas expanded [56–58].

"EE aims at producing a well-informed citizenry about the biophysical environment and accompanying eco problems, seeking forms to solve these complications, and motivated to work toward their elucidation." Work from 2016 to 2021 stimulated the Brazilian State of Rio de Janeiro (RJ) population toward composting. Since 2016, the "Environmental Microbiology Research Group" from UENF–IFRJ has utilized composting as a research object and brought it into face-to-face university extension methodologies and social media (Table 1).

4. Composting as Social Technology Transfer

The Institute of Social Technology – Brazil, understands the practice of social technology (ST) as follows:

> "Set of techniques, transformative methodologies, developed and/or applied in interaction with the population and appropriated by them, which represent solutions for social inclusion and improvement of living conditions" [58].

ST's major action dimensions are fourfold: (i) knowledge, science, and technology; (ii) participation, citizenship, and democracy; (iii) education; and (iv) social relevance. ST elevates the humane quality of life with an emphasis on societal growth through tactics that tackle problems in a community so that it participates in the system, considering sociocultural and economic factors. Composting can transform communities, being an easy-to-replicate alternative for people's daily lives [59–61].

The Brazilian Ministry of the Environment set domestic compost bin guidelines. They propose containers that alternate OM and soil layers and recommend this OW handling destination (Fig. 2). To close the system, one can use the container lid or some fabric that inhibits insects from entering. Compost

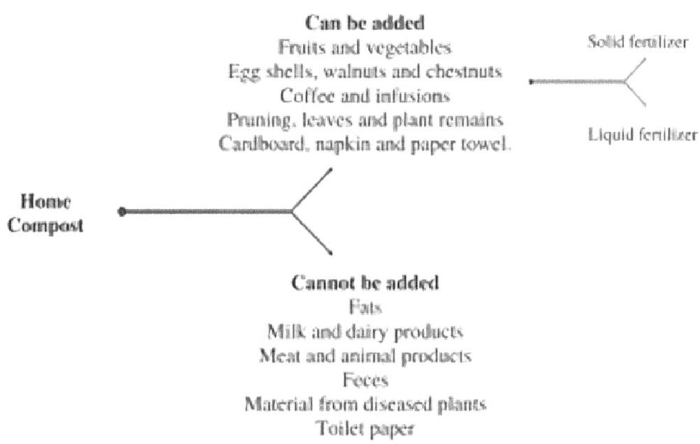

Fig. 2. Materials that may or may not be added to the home compost bin and the final products.

bins must be sheltered from rain, and their material must be turned over whenever more OW is added and at least twice a week to maintain the microbiological systemic health. High-volume systems must have a way to separate liquid from organic fertilizer, e.g., by installing a tap. Within 30 days, the resultant solid fertilizer is ready for consumption, contingent on waste kind and amount [62].

The group worked on composting as ST at different times, as in Fig. 3, totaling 141 in-person actions. The most diverse groups were affected, from early childhood education to postgraduate students, employees from partner companies, science and biology teachers, and the general community. The investigation in [63] showed how working with the public on composting impacts their view of waste, their relationships with nature, and the importance of creating sustainable changes for the world. Between the second half of 2018 and the first half of 2019 alone, the public reached 14,959 in 20 cities and 44 institutions.

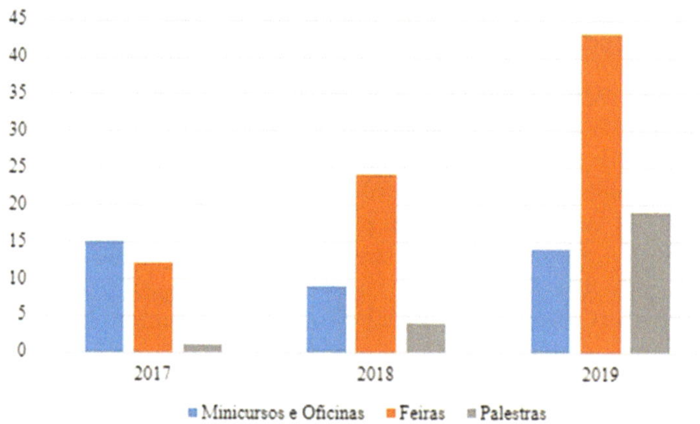

Fig. 3. Composting activities in 2017 (series 1), 2018 (series 2), and 2019 (series 3) and repetitions.

"Science fairs" (SFs) began in Mexico, Central and South America during the 1960s under North American education influence as an opportunity for primary school students to present their scientific productions to the school community [64]. The SF practiced in this project presents composting as an alternative for managing household OW, contextualizing environmental issues, the need for individual actions, and breaking paradigms regarding microorganisms. However, from a new perspective, the exhibitors are undergraduate, postgraduate, and Open University students from the group. Seventy-nine SFs happened for all basic education segments, undergraduate and postgraduate students, and the general community. Using the SF moment to expose scientific knowledge that correlates with the community is another way of bringing society and science closer together. The authors from [65] evaluated the knowledge high school students constructed after visiting a university's SF and identified that students built knowledge about discussed SF subjects. They indicate that students show that:

"learn through interaction with the S monitors, referring to them as a transmitter of knowledge in a language accessible to the public, in a way that did not exclude them, but rather guided them to the knowledge they did not possess" [65].

With teachers and undergraduate students, composting is treated as a pedagogical strategy for the interdisciplinary teaching of science and microbiology through short courses (Fig. 4). First, it indicates how to approach the subject in the classroom and which subjects and content it can relate to. Subsequently, it is shown how to set up a home compost bin using recyclable materials of different volumes, how to maintain it, how to encourage students to take care of it, and how the process is considered finished. Solid fertilizers for seed germination and liquid fertilizers like foliar fertilizers have been proposed and presented to this audience on 5 occasions. Moresco et al. [66] worked on the continuing training of basic education teachers, using microbiology as a focus in the workshops. The result was positive, as 60% of participants started to use the practical approaches learned during the meetings in their classes; Dani [67] showed several strategies for integrating the teacher in training into local social problems, including composting and applying scientific knowledge to improve the community – another social technology transfer strategy.

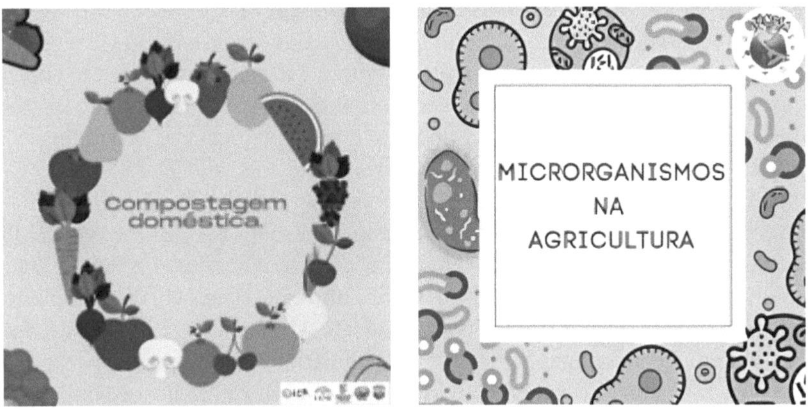

Fig. 4. Examples of content related to home composting published on the "Ciênciapra Gente" (Science for People) Instagram® feed. (*Source:* Instagram® Science for People)

Composting was also discussed in lectures offered to high school students, undergraduate and postgraduate students, and company employees; Kinchescki [68] investigated how EE in the corporate environment is linked to business leaders' ecological perspective and the company's need to adapt to the new socio-environmental directions that relationships outside the corporate environment demand. The work presented positive results when working on the waste topic focused on collecting electronic waste. It also showed how important partnerships are to encourage debates about sustainability in the business environment.

In total, 141 EE actions focused on domestic composting. The composting project is part of the research group's collaboration through the Science for People Extension Project UENF-IFRJ, where students from the Young Talents for Science Program-FAPERJ (JT) and project scholarship holders are also involved. University extension was established in Brazil in 1996 with the Education Guidelines and Bases Law [69]. In other MCASA countries, composting is commonly practised as an EE project.

In Colombia, the "A Commitment of All" project for creating a composting station in the municipality of Puente Nacional is an example of Participatory Action Research, which has made it possible to improve the quality of life of an entire community. Since 1998, constant work together with the municipal administration, the company 'Acuapuente', teachers, young students, and the urban and rural community have made pedagogy together with the strategies applied around the theme of collection, classification, and transformation of OW leads to the strengthening and success of the project. Mainly what has to do with the empowerment of women in sustainable, productive projects, such as organic home gardens, whose fertilizer used is the compost produced at the composting station and obtained free of charge for the area of influence or at a meager cost for the rest of the municipality, contributing to the reduction of poverty and improving the quality of life of the residents of this municipality [70].

In the Carhuaz district, Peru, a comprehensive environmental SWM plan was developed in a participatory manner with all districts and with the support of the national government and several organizations. In the district, garbage collection covers the Carhuaz city and some populated centers, averaging 4.77 tons daily. Agricultural activity predominates in the area, and 75.75% of USW is organic. The promotion of separation is carried out through home visits with the delivery of informative material, training, and workshops to different associations (educational institutions, milk glass mothers' club, neighborhood councils, health establishments, etc.), broadcasts in media and social networks, campaigns, implementation of legal instruments and inspection of commercial establishments. According to the municipality, 60% of the population served separate waste [71].

In El Salvador, in the municipality of Suchinoto [72], to respond to the growing generation of waste that the population increase and the advance of consumerism brought with them, in the mid-1990s, the City Council requested technical support from the organization CESTA-Friends of the Earth El Salvador for the design and implementation of an SWM program in the urban area that prioritized the recovery of the organic fraction. In 1997, the USW separation into organic, inorganic, plastic, and glass fractions began, and the system continued. According to the municipality, 80% of the urban population separates waste, and 52% of the waste collected is transformed into compost. The municipality highlights that to sustain participation, it is essential not to neglect the area of awareness and education. According to data from the municipality, 24 t of OW are collected per week in the urban area, which enters the composting plant. The remainder is processed for recycling or disposed of in a landfill [72].

Population-academia contact occurs through university extension, where scientific dissemination is practiced. Scientific dissemination "is the dissemination in simple terms of science as a process, of the principles established in it, of the methodologies it employs" [73]; it is the transformation of scientific information using language accessible to the public. Working on composting within the scope of scientific dissemination through university extension has brought the group awards and recognition in the area.

Due to the COVID-19 pandemic decreed by the World Health Organization (WHO) at the beginning of 2020, in-person actions were paused, and composting began to be promoted through social media. In MCASA, 430 million people use the internet, with Chile being the country with the highest percentage of the population using social media and Brazil being the country with the most media users. Media allows people worldwide to connect and have faster access to news [74]. Therefore, social media is a tool that scientists can use to publicize their research [75]. The Environmental Microbiology Group used it to continue promoting composting, approaching the topic differently.

For Instagram®, 7 feed publications were created, and two in stories dealing with waste separation, the process, and the benefits of composting. Engagement rates were analyzed, which is the sum of viewers' interactions with the content, reach (how many unique accounts were reached), and visualizations or views (how many times the content was viewed), collected by *Instagram Analytics*. On Youtube®, two lives were held on the topic of composting, where the average visualization rates (how many times viewers saw the video thumbnail), and views (legitimate views on the channel), the lives also generated 252 chat messages, collected by *Youtube Analytics*. Table 2 details these data as discussed by [76].

Table 2. Activities carried out by Environmental Microbiology Research Group UENF–IFRJ between 2020 and 2021 on Instagram® e Youtube® of the Science for People Extension Project [76]

Media	Engagement Rate	Range	Print	Print Rate	Views
Instagram®	>1,2%	1.334	1.663	N/D	N/D
Youtube®	N/D	N/D	1.593	8,3%	408

5. Experimental Results of Composting Usage as a School Practice Tool

Debating composting education is an alternative to working on education and environmental awareness [77–80]. Using simple, low-cost, and easy-to-obtain materials, one can set up a home composting system to use at school or in homes, reusing OM from kitchens, and the material can be used in vegetable gardens [80]. Composting is a microbial process. Therefore, it can be a pedagogical tool in microbiology classes, which face challenges in primary education [77–79].

The microscope, its handling by trained personnel, and the necessary materials make studying living beings invisible to the naked eye difficult for students [80–82]. One can also approach composting as an interdisciplinary strategy by debating the physical-chemical and mathematical issues of the process [77–79]. Composting can be a pedagogical procedure at all educational levels, bringing classroom content to students' practice and raising awareness about waste problems. Research from [83] produced a booklet for teachers on topics that can be discussed for each group of students about composting, how to set up a compost bin, and how to encourage students to participate in the program.

5.1 Composting in Elementary School

In 2016, a monograph on the Biological Sciences Degree course – CEDERJ/UENF used composting as a learning alternative for EE [84]. First, environmental perception work was carried out with students from the 6th to the 9th year of Elementary School II (i.e., current Brazilian fundamental schooling), totaling 63 students. The questionnaire aimed to understand students' relationships with household waste, its disposal, and composting. Around 63% of students were unaware of the technique, and 89% believed composting as a waste destination solves environmental problems related to OM disposal.

A compost bin was set up with the cooperation of students and school staff. Around 300 kg of waste from the school kitchen was reused during the operation, producing 80 kg of solid fertilizer. The students monitored the humidity, temperature, and aeration of the process. The compost was considered stable after around 3 months and used in the school garden and garden.

Using compost promoted the teaching-learning process and made learners aware of the waste problem as per [85]. Working on interdisciplinary EE and occupying school spaces makes the composting project a starting point for developing healthy and sustainable habits. Also, the compost bin was kept at the school for at least 1 year after the strategy that gave rise to this monograph, as per [85, 86] used composting as a teaching strategy for science and geography subjects in elementary school, an approach that involved teachers and students, raising environmental awareness. Workshops from [87] addressed the SW issue with elementary school students. The result was positive at the end of the approach, allowing students to learn more about USW management in their region.

5.2 Composting in High School

In 2019, [77–79] worked on composting with 25 second-year high school students, approaching the entire process as interdisciplinary (Table 3) and threefold.

First, the 25 students' perception of microorganisms and the practice of composting was investigated. At this stage, 126 questionnaires applied to students in the 9th year of elementary school, 1st, 2nd, and 3rd year of high school were also analyzed, totaling 151 questionnaires. It was observed that 77% of participants

Table 3. Curricular components and central axes related to composting in the 2nd year of high school [77–79]

Curricular component	Central axis
Biology	Basic sanitation Environmental impacts National Solid Waste Program Microbial decomposition Photosynthesis Production of energy Food chain Biogeochemical cycles
Physical	Temperature Moisture Evaporation Condensation
Mathematics	Absolute frequency Relative frequency Proportionality relationship
Chemical	pH variations Exothermic chemical reactions Redox reactions

attributed microorganisms to the harm that some of these can cause, and 80% were not familiar with composting. The 25 second-year high school students were separated into 5 groups, so each team had its own compost bin and the responsibilities involved. The students were also accompanied by a high school student linked to the JT Program.

The second moment comprised the pedagogical intervention developed between practical and theoretical classes, totaling 24 class periods dedicated to the project. The process utilized OM from the school kitchen, using the EMBRAPA protocol [88] and the Brazilian Ministry of the Environment [62]. Dry OM from pruning was also added to these compost bins, which has a structuring function and controls humidity. The composters were mounted on water bottles, with a system for removing liquid fertilizer called organic leachate. Weekly, throughout the entire process, students monitored and measured the physical-chemical processes of condensation, temperature, pH, aeration, humidity, and evaporation. The importance of these processes in the system's health was discussed.

In the ninth project week, the students participated in a school SF, where each group was responsible for contextualizing their compost bin with some theme on the subject: disposal of organic SW and environmental issues; OW used in the compost bin; how to build a home compost bin; decomposition process and evaluation of physical-chemical parameters; importance of organic fertilizer. The activity ended 3 months after implementation, using the fertilizer to set up a vegetable garden in the school environment. Students actively participated in each step of the process. According to [77–79], the students showed a lot of interest in participating. It was possible to observe a greater commitment in classes and an improvement in academic performance, a fact also noted by other teachers in the class.

The last stage occurred one month after the completion of the composting project, in which the questionnaire was administered again to the class. Only 4% of the students associated microorganisms only with harm, and 100% knew the composting process and responded that it was possible to use them in organic SWM. When comparing the results with the first questionnaire, it can be said that composting in an interdisciplinary proposal provides contextualized learning and promotes the scientific literacy necessary for citizenship.

The results showed that the proposed pedagogical intervention brought academic students' gains, improved social interaction in the school environment by actively contributing to their nutrition, and stimulated their curiosity and critical sense. For at least 1 year, the composting project remained active at the school, maintained by the author and student of a JT applied project.

In [88], there is a debate on how transdisciplinarity improves EE conversations. Investigations from [88–90] showed how children's concepts about microorganisms can be focused on pathogenicity. In [88–90], it was investigated how microbiology is debated in the classroom and how practical classes benefit students' understanding. In [77–79], the activity shows the potential of composting in discussing microbiology and EE in the classroom and in reaching different societal sectors through the transfer of social technology.

6. Discussion and Prospects

Organic and inorganic byproducts stem from OW abundantly. Composting is an ST that spreads globally. Emerging nations must transpose an ample distance to progress, embracing and calling for urgent actions, data, and standards. This picture exacerbates in low-GDP, post-pandemic markets.

6.1 Main Compost End Products

The most popular OW compost (OWC) gain is an ample waste decline. OWC can be rerouted from landfills and incinerators to the market value chain, shrinking waste disposal losses and upcycling those constituents into productive, environmentally advantageous products. Likewise, recycling, reprocessing, as well as landfills outcomes follow:

- **Soil Health Improvement:** Compost-resultant nutrients and biota addition to terrains can bring back their operational, geotechnical, biological, engineering-linked, and chemical well-being, which help lead to healthy fauna/flora with better social technologies, pave the way to cost-effective projects, and enhance human life quality.

- **Stormwater Management (SM):** Compost aids in controlling water streams/floods, behaving as a hands-on, sustainable SM tool.
- **Soil Erosion Prevention:** Binding soil together, augmenting infiltration, and decelerating surface water flow can remediate topsoil erosion.
- **Water Conservation:** Compost can safeguard hydric resources thanks to retaining while efficiently allocating/reallocating water, besides assisting in SM. Thus, it helps ecologically and makes a project more drought-resistant.
- **Climate Change Mitigation:** Compost cuts down on atmospheric GHGs. Thus, it robustifies ecosystems, turning them into more resilient to escalating temperatures.
- **Lower Maintenance Expenditures:** Compost can abbreviate designs' maintenance budgets by fortifying vegetal growth, dropping mortality, and consequent replacement expenses. Similarly, it stimulates drought resistance. Hence, non-natural irrigation expenditures are curbed.
- **Wetland Reclamation Instrument:** Compost-reliant road/transportation departments can guard, reclaim, or institute wetlands activities in a superior manner. Pollutants can be filtered by compost installed on roads and wetlands. What is more, it is a remarkable additive to corroborate wetland restoration ventures as eco-friendly units entail highly organic material.

6.2 Compost as a Social Technology

This STEAM educational study employed compost as a teaching aid and catalyst for pedagogical subject comprehension and assimilation, collaborative work, and the UN SDGs. Preliminary results point toward this strategy as quite auspicious in university extension settings encompassing fundamental and secondary schooling. Nevertheless, it has great prospects in curricular goings-on related to fundamental, secondary, undergraduate, and graduate activities. Agriculture 4.0 comes in as an agent to facilitate and boost the learning experience. This comes from incorporating ICT practices into the extension activities. The STs work as a soft power form of fostering collective engagement in a transdisciplinary way. Here, the focus was on biotechnology and elementary/high school instruction.

6.3 Novel Usages

Although the knowledge addresses the biotechnological realm, it can enhance engineering and computer science curricula. Composting can create other products besides fertilizers and by-products to improve terrain handling from the plain cultivation activity. There are gasification results, for instance, methane, hydrogen, and heat, for utilizing them as energy sources. Intelligent noses, relying on gas sensors, can help maximize and redirect these elements to the agribusiness control stations. Actuators can redirect by-products, taking commands from decision-making controllers. If necessary, they help change the local energy matrix configuration and even exchange energy with other agribusiness facilities.

Energy matrices encompass several energy source types to avoid deteriorating resources. A regular supply of materials changes crucial physicochemical features that can drive other agricultural needs, independent of size.

7. Conclusion

This chapter has depicted composting in the context of Agriculture 4.0 and social technology, touching up hands, arms, and hearts to foster highly productive supply chains. Agriculture 4.0 meets EE in this work via organic waste (OW) compost and as per developing countries' existing state of affairs, albeit emphasizing "Mexico, Central and South America" (MCASA). The whole project applied STEAM in environmental education (EE). Composting played the roles of a teaching aid and catalyst that can foster subject comprehension and assimilation, collaborative work, and the UN SDGs in two educational echelons: elementary and high school.

Biotechnology and digital tools (ICT) were a university extension project's inspiration deemed Science Today (Ciencia Hoje) as a form to learn from a finalized agroindustrial products' perspective, i.e., a social technology. Professors and students investigated locally collected data and green nomenclature repercussions, such as "composting employing social media" (i.e., YouTube and Instagram). Most local, in-site task leaders were UENF's undergraduate and graduate students. These ICT instruments afford data from feeds/views that can engender STEAM tools and craft social, cultural, economic, statistical, and educational knowledge. The core experimental compost fallouts in teaching practice were remarkable. Preliminary results point toward this strategy as quite promising in a university extension setting involving fundamental schooling. Nevertheless, it has boundless prospects in curricular activities related to fundamental, high school, undergraduate, and graduation activities.

References

1. Benos, L., Makaritis, N. and Kolorizos, V. (2022). From precision agriculture to agriculture 4.0: Integrating ICT in farming. *In*: Bochtis, D.D., Sørensen, C.G., Fountas, S., Moysiadis, V., Pardalos, P.M. (Eds.), *Information and Communication Technologies for Agriculture—Theme III: Decision. Springer Optimization and Its Applications*, 184. Cham.: Springer, doi: 10.1007/978-3-030-84152-2_4.
2. MacPherson, J., Voglhuber-Slavinsky, A., Olbrisch, M., Schöbel, P., Dönitz, E. et al. (2022). Future agricultural systems and the role of digitalization for achieving sustainability goals. A review. *Agronomy for Sustainable Development*, 42.
3. Pogorelskaia, I. and Várallyai, L. (2020). Agriculture 4.0 and the role of education. *J. Agricultural Informatics*, 11(1). doi: 10.17700/jai.2020.11.1.571.
4. Reynolds, M., Borrell, A., Braun, H., Edmeades, G., Flavell, R. et al. (2019). Translational research for climate resilient, higher yielding crops. *Crop Breeding,*

Genetics and Genomics (Hapres), 1, e190016. London, UK. doi: 10.20900/cbgg20190016.
5. Silva, F.T.d., Baierle, I.C., Correa, R.G.d.F., Sellitto, M.A., Peres, F.A.P. and Kipper, L.M. (2023). Open innovation in agribusiness: Barriers and challenges in the transition to Agriculture 4.0. *Sustainability*, 15(11), 8562. MDPI. doi: 10.3390/su15118562.
6. Soma, T. and Nuckchady, B. (2021). Communicating the benefits and risks of digital agriculture technologies: Perspectives on the future of digital agricultural education and training. *Front. Comm.*, 6, 762201. doi: 0.3389/fcomm.2021.762201.
7. Shepherd, M., Turner, J.A., Small, B. and Wheeler, D. (2020). Priorities for Science to overcome hurdles thwarting the full promise of the 'Digital Agriculture' revolution. *J. Sci. Food Agric.*, 100(14), 5083–5092. doi:10.1002/jsfa.9346e.
8. Klerkx, L., Jakku, E. and Labarthe, P. (2019). A review of social science on digital agriculture, smart farming, and agriculture 4.0: New contributions and a future research agenda. *NJAS: Wageningen J. Life Sciences*, 90–91(1), 1–16. Informa UK Limited. doi: 10.1016/j.njas.2019.100315.
9. de Jesus, M.A., Estrela, V.V., Aroma, R.J., Razmjooy, N., da Silva, S.E.B. et al. (2021). Building bridges and remediating illiteracy: How intergenerational cooperation foster better engineering professionals. *In*: Khelassi, A., Estrela, V.V. (Eds.), *Advances in Multidisciplinary Medical Technologies: Engineering, Modeling and Findings*. Cham.: Springer. doi: 10.1007/978-3-030-57552-6_3.
10. Wu, Y., Zhong, X. and Takács-György, K. (2023). Digital agricultural education in Hungary. *Gradus*, 10(1), Budapest, Hungary. doi: 10.47833/2023.1.AGR.007.
11. Neugebauer, M. and Sołowiej, P. (2017). The use of green waste to overcome the difficulty in small-scale composting of organic household waste. *J. Cleaner Production*, Elsevier. doi: 10.1016/j.jclepro.2017.04.095.
12. Monteiro, A.C.B., Franca, R.P., Estrela, V.V., Fernandes, S.R., Khelassi, A. et al. (2020). UAV-CPSs as a test bed for new technologies and a primer to Industry 5.0. *In*: Estrela, V.V., Hemanth, J., Saotome, O., Nikolakopoulos, G., Sabatini, R. (Eds.), *Imaging and Sensing for Unmanned Aircraft Systems*, 2(1), 1–22. London, UK: IET. doi: 10.1049/PBCE120G_ch1.
13. Burdziakowski, P., Razmjooy, N., Estrela, V.V. and Hemanth, J. (2020). Open source software (OSS) and hardware (OSH) in UAVs. *In*: Estrela, V.V., Hemanth, J., Saotome, O., Nikolakopoulos, G., Sabatini, R. (Eds.), *Imaging and Sensing for Unmanned Aircraft Systems*, 2(3), 49–66, London, UK: IET. doi: 10.1049/PBCE120G_ch3.
14. Arshaghi, A., Razmjooy, N., Estrela, V.V., Burdziakowski, P., Nascimento, D.A. et al. (2020). Image transmission in UAV MIMO UWB-OSTBC system over Rayleigh channel using multiple description coding (MDC). *In*: Estrela, V.V., Hemanth, J., Saotome, O., Nikolakopoulos, G., Sabatini, R. (Eds.), *Imaging and Sensing for Unmanned Aircraft Systems*, 2(4), 67–90, IET, London, UK: IET. doi: 10.1049/PBCE120G_ch4.
15. Zambrano-Monserrate, M.A., Ruano, M.A. and Ormeño-Candelario, V. (2021). Determinants of municipal solid waste: A global analysis by countries' income level. *Environ Sci. Pollut.*, 28, 62421–62430.
16. Kaza, S., Yao, L., Bhada-Tata, P. and Van Woerden, F. (2018). What a waste 2.0: A global snapshot of solid waste management to 2050. *Urban Development*, Washington, DC: World Bank. CC BY 3.0 IGO.
17. Devadoss, P.S.M., Pariatamby, A., Bhatti, M.S., Chenayah, S. and Shahul Hamid, F. (2021). Strategies for reducing greenhouse gas emissions from municipal solid

waste management in Pakistan. *Waste Management & Research*, Sage, 39, 914–927. London, UK. doi: 10.1177/0734242X209839.
18. Cerda, A., Artola, A., Font, X., Barrena, R., Gea, T. and Sánchez, A. (2018). Composting of food wastes: Status and challenges. *Bioresource Technology*, 248(Pt A), 57–67.
19. Hameed, M., Bhat, R.A., Pandit, B.A., Ramzan, S., Dijoo, Z.K. and Wani, M.A. (2021). Qualitative assessment of compost engendered from municipal solid waste and green waste by indexing method. *J. The Air & Waste Management Association*, 72, 210–219.
20. Iqbal, M.K., Khan, R.A., Nadeem, A. and Hussnain, A. (2012). Comparative study of different techniques of composting and their stability evaluation in municipal solid waste. *J. Chem. Soc. Pak.*, 34(2), 273–282.
21. Ayilara, M.S., Olanrewaju, O.S., Babalola, O.O. and Odeyemi, O. (2020). Waste management through composting: Challenges and potentials. *Sustainability*, 12(11), 4456. MDPI.
22. Daud, N.M., Khalid, S.A., Nawawi, W.N. and Ramli, N. (2016). Producing fertilizer from food waste recycling using Berkeley and Bokashi method. *Ponte*, 72(4). doi: 10.21506/j.ponte.2016.4.11.
23. Lew, P.S., Nik Ibrahim, N.N.L., Kamarudin, S., Thamrin, N.M. and Misnan, M.F. (2021). Optimization of Bokashi-composting process using effective microorganisms-1 in smart composting bin. *Sensors*, 21(8), 2847. MDPI.
24. Manyapu, V., Mandpe, A. and Kumar, S. (2018). Synergistic effect of fly ash in in-vessel composting of biomass and kitchen waste. *Bioresource Technology*, 251, 114–120.
25. Yogamoorthi, T., Estrela, V.V., Edoh, T.O., Razmjooy, N., Khelassi, A., da Hora, H.R.M. et al. (2021). Digital garbage bin monitoring system (DGBMS). *In*: Iano, Y., Saotome, O., Kemper, G., Mendes de Seixas, A.C., Gomes de Oliveira, G. (Eds.), *Proc. 6th Brazilian Technology Symposium (BTSym'20)*. BTSym 2020. Smart Innovation, Systems and Technologies, 233. Cham.: Springer. doi: 10.1007/978-3-030-75680-2_54.
26. Fierer, N. (2017). Embracing the unknown: Disentangling the complexities of the soil microbiome. *Nat. Rev. Microbiol.*, 15, 579–590. doi: 10.1038/nrmicro.2017.87.
27. Luo, S., Chen, L., Lu, Y., Daia, S., Sun, D. et al. (2021). The microbial community structure of rhizosphere soil was influenced by different sugarcane varieties with different ratooning abilities. *Sugar Tech.*, 23, 1306–1316. doi: 10.1007/s12355-021-01019-w.
28. Rao, D.L.N., Aparna, K. and Mohanty, S.R. (2019). Microbiology and biochemistry of soil organic matter, carbon sequestration, and soil health. *Indian J. Fertil.*, 15, 124–138.
29. Huang, Y., Dai, Z., Lin, J., Li, D., Ye, H. et al. (2021). Labile carbon facilitated phosphorus solubilization as regulated by bacterial and fungal communities in *Zea mays*. *Soil Biology and Biochemistry*, 163, 108465. Elsevier. doi: 10.1016/j.soilbio.2021.108465.
30. Ma, B., Stirling, E., Liu, Y., Zhao, K., Zhou, J. et al. (2021). Soil biogeochemical cycle couplings inferred from a function-taxon network. *Research*, 2021, Article ID: 7102769.
31. Bonfim, F.P.G., Honorio, I.C.G., Reis, I.L., Pereira, A.J. and Souza, D.B. (2011). Caderno dos Microrganismos Eficientes (EM). *Instruções práticas sobre uso ecológico e social do EM* (2nd Edn.), Brazil.

32. Chatterjee, D., Nayak, A.K., Mishra, A., Swain, C.K., Kumar, U. et al. (2021). Effect of long-term organic fertilization in flooded rice soil on phosphorus transformation and phosphate solubilizing microorganisms. *J. Soil Sci. Plant Nutr.*, 21, 1368–1381.
33. Liu, X.J., Hayer, M., Mau, R.L., Schwartz, E., Dijkstra, P. and Hungate, B.A. (2021). Substrate stoichiometric regulation of microbial respiration and community dynamics across four different ecosystems. *Soil Biology and Biochemistry*, 163, 108458. Elsevier.
34. Zhang, M., Liu, Y., Wei, Q. and Gou J. (2021). Effects of short-term application of Moutai lees biochar on nutrients and fungal community structure in yellow soil of Guizhou. *Env. Sci. Pollut. Res.*, 28, 67404–67413. Springer. doi: 10.1016/j.soilbio.2021.108458.
35. Sáez, J., Clemente, R., Bustamante, M.A., Yañez, D.R. and Bernal, M.P. (2017). Evaluation of the slurry management strategy and the integration of the composting technology in a pig farm: Agronomical and environmental implications. *J. Environmental Management*, 192, 57–67.
36. Forte, A., Fagnano, M. and Fierro, A. (2017). Potential role of compost and green manure amendment to mitigate soil GHGs emissions in Mediterranean drip irrigated maize production systems. *J. Environmental Management*, 192, 68–78.
37. Monostori, L. (2014). Cyber-physical production systems: Roots, expectations, and R&D challenges. *Procedia CIRP*, 17, 9–13. Elsevier. doi: 10.1016/j.procir.2014.03.115.
38. Estrela, V.V., Saotome, O., Loschi, H.J., Hemanth, D.J., Farfan, W.S., Aroma, R.J. et al. (2018). Emergency response cyber-physical framework for landslide avoidance with sustainable electronics. *Technologies*, 6(42). doi:10.3390/technologies6020042.
39. Zambon, I., Cecchini, M., Egidi, G., Saporito, M.G. and Colantoni, A. (2019). Revolution 4.0: Industry vs. Agriculture in a Future Development for SMEs. *Processes*, MDPI, 7(1), 36.
40. Aroma, R.J., Raimond, K., Estrela, V.V. and de Jesus, M.A. (2023). A coastal band spectral combination for water body extraction using Landsat 8 images. *Int. J. Environ. Sci. Technol.* doi: 10.1007/s13762-023-05027-z.
41. Khan, A.A., Laghari, A., Gadekallu, T.R., Shaikh, Z.A., Javed, A.R., Rashid, M. et al. (2022). A drone-based data management and optimization using metaheuristic algorithms and blockchain smart contracts in a secure fog environment. *Computers and Electrical Engineering*, 102, 108234.
42. Deshpande, A., Estrela, V.V. and Razmjooy, N. (2021). *Computational Intelligence Methods for Super-Resolution in Image Processing Applications*. Zurich, Switzerland: Springer Nature. doi: 10.1007/978-3-030-67921-7.
43. Estrela, V.V., Deshpande, A., Sroufer, R., Lopes, R.T., Grata, E.G.H. et al. (2023). Digital twin framework for intelligent healthcare facilities through ISO/IEEE 11073. *In*: Estrela, V.V. (Ed.), *Intelligent Healthcare Systems*, 14, 279–297. London, UK: CRC Press.
44. Clercq, M., De Vats, A. and Biel, A. (2018). *Agriculture 4.0: The Future of Farming Technology*. The World Government Summit, Dubai. https://www.worldgovernmentsummit.org. Accessed on October 24, 2023.
45. Laghari, A.A., Zhang, X., Shaikh, Z.A., Khan, A., Estrela, V.V. and Izadi, S. (2023). A review on quality of experience (QoE) in cloud computing. *J. Reliable Intell. Environ.* doi: 10.1007/s40860-023-00210-y.
46. Savino, A., Solórzano, G., Quispe, C. and Correal, M.C. (2018). *Waste Management Outlook for Latin America and the Caribbean*. UN Environment Programme, Latin America and the Caribbean Office; Panama City, Panama: 2018.

47. https://legislacao.presidencia.gov.br/atos/?tipo=LEI&numero=12305&ano=2010&ato=e3dgXUq1keVpWT0f1. Accessed on October 24, 2023.
48. https://www.prefeitura.sp.gov.br/cidade/secretarias/subprefeituras/amlurb/noticias/?p=169047. Accessed on October 15, 2023.
49. https://www.mimp.gob.pe/adultomayor/regiones/Lima_Metro2.html#:~:text=Lima%20(Metropolitana)%20cuenta%20con%20una,poblaci%C3%B3n%20adulta%20mayor%20(963%2C353). Accessed on Oct 12, 2023.
50. PLANRES, 2017 Plan Nacional de Gestión Integral de Residuos Sólidos 2016-2024 (PLANRES) (2017). Ministerio del Ambiente – MINAM, Peru. https://sinia.minam.gob.pe/documentos/plan-nacional-gestion-integral-residuos-solidos-2016-2024. Accessed on Oct 24, 2023.
51. PIGARS, s/a https://smia.munlima.gob.pe/uploads/documento/927d34ddcb1597e8.pdf. Accessed on October 12, 2023.
52. National Administrative Department of Statistics (DANE) (2018). Colombia. https://www.dane.gov.co/index.php/estadisticas-por-tema/demografia-y-poblacion/proyecciones-de-poblacion. Accessed on Oct 24, 2023.
53. https://bogota.gov.co/yo-participo/blogs/basura-en-bogota-una-responsabilidad-de-todos-los-ciudadanos. Accessed on Oct 24, 2023.
54. https://www.alcaldiabogota.gov.co/sisjur/normas/Norma1.jsp?i=21059. Accessed on October 24, 2023.
55. https://www.alcaldiabogota.gov.co/sisjur/normas/Norma1.jsp?i=28150. Accessed on October 24, 2023.
56. Stapp, W.B. (1969). The concept of environmental education. *J. Environmental Education*, 1(1), 30–31.
57. Estrela, V.V., Andreopoulos, N., Sroufer, R., Jesus, M.A., Mamani, W.D. and Peixoto, A.B. (2021). Transmedia ecosystems, quality of experience and quality of service in fog computing for comfortable learning. *Proc. 2021 IEEE Global Eng. Education Conf.* (*EDUCON*), 1003–1009.
58. Schwab, D. and Freitas, C.C.G. (2016). Tecnologia social: Implicações e desafios da implantação. *Revista Tecnologia e Sociedade*, 12(26), Universidade Tecnologica Federal do Parana (UTFPR), Parana, Brazil. doi: 10.3895/rts.v12n26.3794.
59. Albuquerque, L.C. (2009). Tecnologias sociais ou tecnologias apropriadas? O Resgate de um termo. *In*: Otterloo, A. (Ed.). *Tecnologias sociais: Caminhos para a sustentabilidade*, RTS, Brasília, Brazil.
60. Rotz, S., Gravely, E., Mosby, I., Duncan, E.W., Finnis, E., Horgan, M. et al. (2019). Automated pastures and the digital divide: How agricultural technologies are shaping labour and rural communities. *J. Rural Studies*, 68, 112–122. Elsevier. doi:10.1016/j.jrurstud.2019.01.023.
61. Mothé, G.P.B., Santo, P.S. do E., Bandeira, R.F.S., Souza, G.R., Siqueira, I.S. et al. (2021). Compostagem como ferramenta na educação ambiental e inclusiva. *In:* Meio ambiente: Questões éticas x progresso tecnológico. *Atena Editora, Brazil*, 14–25. doi: 10.22533/at.ed.4942117062
62. Ministério do Meio Ambiente – MMA (2017). *Compostagem doméstica, comunitária e institucional de resíduos orgânicos: Manual de orientação*. Centro de Estudos e Promoção da Agricultura de Grupo, Serviço Social do Comércio, Brasília, DF, Brazil.
63. Christie, B. and Waller, V. (2018). Community learnings through residential composting in apartment buildings. *The J. Env. Education*, 50(2), 97–112. Informa UK. doi: 0.1080/00958964.2018.1509289.
64. Ministry of Education and Technology (MEC). (2006). Secretaria de Educação

Básica. Programa Nacional de Apoio às Feiras de Ciências da Educação Básica: FENACEB. Brasília, DF, Brazil, MEC/SEB.
65. Francisco, W. and de Castro, M.C. (2017). Relações com o saber constituídas por estudantes durante visitação a uma feira de Ciências. *Chemical Education in Point of View*. ISSN: 2527-0915, Rede Latino-Americana de Pesquisa em Educação Química – ReLAPEQ. Brazil, 01(01). doi: 10.30705/eqpv.v1i1.900 .
66. Moresco, T.R., Carvalho, M.S., Klein, V., Lima, A.D., Barbosa, N.V. et al. (2017). Ensino de microbiologia experimental para Educação Básica no contexto da formação continuada. *Revista Electrónica de Enseñanza de las Ciencias*, 16(3), 435–457. Universidad de Vigo, Spain.
67. Dani, D. (2019). A community and place-based approach to middle childhood science teacher education. *Middle School J.*, 50(2), 45–52, Informa, UK. doi: 10.1080/00940771.2019.1576581.
68. Kinchescki, R. (2020). Educação corporativa ambiental: O estudo de caso do primeiro tabelionato de notas e protesto de palhoça na campanha de coleta de resíduos eletroeletrônicos. *Revista de Extensão de Iniciação Científica da UNISOCIESC*, 5(2), 1–19. https://reis.unisociesc.com.br/index.php/reis/article/view/70. Accessed on October 24, 2023.
69. MEC. (1996). Lei de Diretrizes e Bases da Educação Nacional, LEI Nº 9.394. https://legislacao.presidencia.gov.br/atos/?tipo=LEI&numero=9394&ano=1996&ato=3f5o3Y61UMJpWT25a. Accessed on Oct. 24, 2023.
70. Avila, K.S. (2020). Planta de Compostaje de Puente Nacional "Un Compromiso de Todos". Universidad Nacional Abierta y a Distancia, Escuela de Ciencias Sociales, Artes y Humanidades, Programa de Sociología, Puente Nacional, Santander, Colombia. Technical Report. Repositorio Institucional UNAD. https://repository.unad.edu.co/handle/10596/39008. Accessed on Oct 24, 2023.
71. Ruiz, J.B. (2005). Educação ambiental e os temas transversais. *Revista de Ciências Humanas da UNIPAR, Akrópolis, Umuarama, Parana, Brazil*, 13, (1), jan./mar. Accessed on October 24, 2023.
72. Moreno, S.C. and Aedo, M.V.A. (2021). Análisis de las políticas públicas e iniciativas privadas que apoyan el uso de las tecnologías digitales en las mipymes agrícolas y agroindustriales en El Salvador, Documentos de Proyectos (LC/TS.2021/64). Santiago, Comisión Económica para América Latina y el Caribe (CEPAL), UN. Accessed on October 24, 2023.
73. Massarani, L.M. and Alves, J.P. (2019). A visão de divulgação científica de José Reis. *Ciência e Cultura*, 71(1), 56–59. FapUNIFESP (SciELO), Sao Paulo, SP, Brazil, doi: 10.21800/2317-66602019000100015.
74. https://www.statista.com/topics/6394/social-media-usage-in-latin-america/#dossierKeyfigures. Accessed on December 21, 2022
75. Habibi, S.A. and Salim, L. (2021). Static vs. dynamic methods of delivery for science communication: A critical analysis of user engagement with science on social media. *PLOS ONE*, 16(3), e0248507. Public Library of Science (PLoS), San Francisco, CA, USA. doi: 10.1371/journal.pone.0248507.
76. Sousa, Á.S., Siqueira, I.D.S., Souza, G.R.D. and Intorne, A.C. (2021). Compostagem: Uma Tecnologia Social ao Alcance de Todos. *In*: Anais do II Congresso On-line Internacional de Sustentabilidade. II Congresso On-line Internacional de Sustentabilidade. Revista Multidisciplinar de Educação e Meio Ambiente, Brazil. doi: 10.51189/rema/2099.
77. da Silva, G.D.P. and Hesselberg, T. (2019). A review of the use of black soldier

fly larvae, Hermetiaillucens (Diptera: Stratiomyidae), to compost organic waste in tropical regions. *Neotrop Entomol.*, 2020, 49(2), 151–162. doi: 10.1007/s13744-019-00719-z. Epub 2019 November 8. PMID: 31701478.

78. Pinto da Silva, C., Almeida, B.P.M., de, Xavier and de Campos, S. (2019). Monitoring and characterization of compost obtained of sludge of ultra-processed food industry by conventional and spectroscopic analyses. *Semina: Ciências Exatas E Tecnológicas*, 40(1), 39–46. Ponta Grossa, Parana, Brazil. doi: 10.5433/1679-0375.2019v40n1p39.
79. Silva, A.C., Rocha, P., Antelo, J., Bento, F., Bettencourt, P. et al. (2019). Characterization of organic matter at different stages of a composting process. *Proc. WASTES: Solutions, Treatments and Opportunities.* 5th International Conference September 4th–6th, 2019.
80. CONAMA, Resolução nº 481/17, 2017. http://conama.mma.gov.br/?option=com_sisconama&task=arquivo.download&id=702. Accessed on October 24, 2023.
81. de Jesus, M.A., Estrela, V.V., Huacasi, W.D., Razmjooy, N., Plaza, P. and Peixoto, A.B.M. (2020). Using transmedia approaches in STEM. *Proc. 2020 IEEE Global Engineering Education Conference (EDUCON)*, 1013–1016. doi: 10.1109/EDUCON45650.2020.9125239.
82. Barbosa, F.H.F. and Barbosa, L.P.J.L. (2010). Alternativas metodológicas em microbiologia: Viabilizando atividades práticas. *Revista de Biologia e Ciências da Terra*, 10(2), 134–143, Brazil. Accessed on October 24, 2023.
83. Schwarz, M. and Bonhotal, J.F. (2017). School Composting – Let's Get Growing – A guide for student leaders and teachers. Report from the Cornell Waste Management Institute, Cornell University, Cornell, USA. https://ecommons.cornell.edu/items/2c52fd06-902a-4e2f-8366-abbe494b4744. Accessed on October 24, 2023.
84. León, C.C., Ríos, M.J., Alamo, N.F., Berriel, C.H. and Robles, O.J. (2016). Biosorption of lead (II) and chromium with *Zea mays* rachis in aqueous solutions: Characterization studies. *Afinidad*, 73(574), 140–147. Barcelona, Spain.
85. Granato, T.M., Berriel, O.P. and Intorne, A.C. (2017). Composting at school: A case study at a rural area of the city of São Fidélis/RJ. *Revista de Extensao da UENF, Campos dos Goytacazes*, RJ, Brazil. https://uenf.br/publicacoes/revista-de-extensao/wp-content/uploads/sites/4/2017/08/dezembro-15.compressed.pdf. Accessed on Oct 24, 2023.
86. da Costa, A.P. and Silva, W.C.M. (2011). A compostagem como recurso metodológico para o ensino de ciências naturais e geografia no ensino fundamental. *Enciclopédia Biosfera*, 7(12), 1–12. Centro Científico Conhecer - Goiânia, Goias, Brazil.
87. Phan Hoang, T.T. and Kato, T. (2016). Measuring the effect of environmental education for sustainable development at elementary schools: A case study in Da Nang city, Vietnam. *Sustainable Environment Research*, 26(6), 274–286, Springer. doi: 10.1016/j.serj.2016.08.005.
88. EMBRAPA. (2009). Compostagem de Resíduos Para Produção de Adubo Orgânico na Pequena Propriedade – Circular Técnica (Technical Report) 59. Aracaju, SE, Brazil.
89. Annerstedt, M. (2010). Transdisciplinarity as an inference technique to achieve a better understanding in the health and environmental sciences. *Int'l J. Env. Research and Public Health*, 7(6), 2692–2707. MDPI AG. doi: 10.3390/ijerph7062692.
90. Redfern, J., Burdass, D. and Verran, J. (2013). Practical microbiology in schools: A survey of UK teachers. *Trends in Microbiology*, 21(11), 557–559. Elsevier. doi: 10.1016/j.tim.2013.09.002.

Conclusion

In the closing chapters of *Intelligent Designs, Innovations, and Sustainability in Agriculture 4.0*, we reflect upon a transformative journey through the crossroads of technology, agriculture, and sustainability. As the collective insights of contributors from diverse backgrounds converge, a mosaic of possibilities emerges—a vision of agriculture that transcends traditional boundaries and embraces the intelligent, the innovative, and the sustainable.

The narrative woven throughout this book has illuminated the paradigm shift occurring within the agricultural landscape. Agriculture 4.0 is not just a technological evolution; it is a holistic approach that harmonizes human ingenuity with the rhythms of the natural world. It is about leveraging intelligent designs to optimize efficiency, innovations to propel productivity, and sustainability to ensure the resilience of our food systems.

The threads of this discourse are intertwined with the acknowledgment that the challenges facing agriculture are multifaceted. From the demands of a growing global population to the imperative of mitigating climate change, the solutions lie in the intersection of cutting-edge technology and sustainable practices. The chapters within this book have explored the potential of Artificial Intelligence, the Internet of Things, robotics, and data analytics to revolutionize farming practices, all while promoting ecological balance.

Sustainability emerges as the linchpin in this journey. Beyond the buzzword, it encapsulates a commitment to stewarding our planet responsibly. Precision agriculture, smart irrigation systems, and regenerative practices are not just tools in the agricultural toolbox; they are pathways toward a future where food production is both bountiful and respectful of the Earth's finite resources.

As we conclude this exploration, we are reminded that the future of agriculture is not predetermined; it is shaped by the decisions we make today. The collective knowledge and innovations encapsulated in these pages serve as a compass, guiding us toward a future where intelligent designs and sustainable practices converge to create resilient, equitable, and thriving agricultural systems. We extend our deepest gratitude to the contributors who have shared their expertise and perspectives. Their work enriches the dialogue surrounding Agriculture 4.0 and lays the groundwork for continued exploration and innovation. As we close this chapter, may the insights within these pages inspire ongoing conversations, collaborations, and actions that propel us toward a more sustainable and intelligent future for agriculture.

Index